AF880414

SpringerWienNewYork

SpringerWienNewYork

CISM COURSES AND LECTURES

Series Editors:

The Rectors
Giulio Maier - Milan
Franz G. Rammerstorfer - Wien
Jean Salençon - Palaiseau

The Secretary General
Bernhard Schrefler - Padua

Executive Editor
Paolo Serafini - Udine

The series presents lecture notes, monographs, edited works and proceedings in the field of Mechanics, Engineering, Computer Science and Applied Mathematics.
Purpose of the series is to make known in the international scientific and technical community results obtained in some of the activities organized by CISM, the International Centre for Mechanical Sciences.

INTERNATIONAL CENTRE FOR MECHANICAL SCIENCES

COURSES AND LECTURES - No. 526

IMPACT ENGINEERING OF COMPOSITE STRUCTURES

EDITED BY

SERGE ABRATE
SOUTHERN ILLINOIS UNIVERSITY,
CARBONDALE, IL, USA

SpringerWienNewYork

This volume contains 237 illustrations

This work is subject to copyright.
All rights are reserved,
whether the whole or part of the material is concerned
specifically those of translation, reprinting, re-use of illustrations,
broadcasting, reproduction by photocopying machine
or similar means, and storage in data banks.
© 2011 by CISM, Udine
Printed in Italy
SPIN 80030352

All contributions have been typeset by the authors.

ISBN 978-3-7091-0522-1 SpringerWienNewYork

PREFACE

This book is a much improved version of a set of notes developed for a one-week course taught at the International Center for Mechanical Sciences (CISM) in Udine, Italy, July 21-25, 2005. The support and encouragement of the CISM staff and its Rector, Professor Giulio Maier are gratefully acknowledged for providing the opportunity to bring together different aspects of the problem of impact on composite structures. The opportunity to work with experts in different areas from different countries and to teach a very diverse group of students was also very much appreciated.

As the course coordinator, I want to express my thanks to the colleagues who participated in this long term venture: Professor Giovanni Belingardi from the Politecnico di Torino, Italy; Professor Wesley Cantwell from the University of Liverpool, England; Professor Uday Vaidya, University of Alabama-Birmingham, USA, and Professor Ramon Zaera from the University Carlos III, Madrid, Spain. It is also a pleasure to acknowledge the contributions of Professor Jorge Ambrosio from the University of Lisbon, Portugal who taught a portion of the course in Udine.

The book is intended for beginning graduate students and practitioners in industry who need an introduction with a strong technical background to this subject, one that enables them to pursue their own research or design activities and to pursue further studies on their own. We have attempted to present a broad range of topics. The two common threads throughout the book are that it deals with structures made of composite materials and that those structures are subjected to impacts. The structures can be part of aircrafts, motor vehicles, or armored military vehicles, for example. The impacts can be tool drops, ballistic projectiles or vehicle crashes. The book examines ways to model the impact event, to determine the size and severity of the damage and discusses general trends observed during experiments.

Serge Abrate

CONTENTS

Introduction to the Mechanics of Composite Materials

Serge Abrate

Department of Mechanical Engineering and Energy Processes, Southern Illinois University, Carbondale, IL, USA

This chapter recalls the basic notions of stress and strain, the equations of motion of linear elasticity, and the constitutive equations for linear isotropic and orthotropic materials. Then, we introduce some of the most commonly used criteria for predicting failure inside a lamina and the delamination of the interface between adjacent layers. Finally, we discuss two types of approaches used to predict the propagation of delaminations.

1- Stress

This section introduces the concept of stress, the stress tensor as a quantity defining the intensity of the loading at a point, and the equations of motion of a solid in terms of these stress components. To define the intensity of the loading at a given point P in a solid, one can make an imaginary cut through that point. A small area ΔA surrounding P, will be subjected to a force $\overline{F}$ that has a component F_n in the normal direction and a component F_s in the tangential direction. Dividing a force by the area it acts on gives a measure of the intensity of the loading. Expecting this load to vary from point to point, we define the normal stress σ_n as the limit of the ratio $F_n / \Delta A$ as $\Delta A \to 0$. The shear stress σ_s is defined as the limit of the ratio $F_s / \Delta A$ as $\Delta A \to 0$. Since the shear stress can have an arbitrary orientation on the surface, it is usually split into two orthogonal components to account for its orientation. Therefore, on a surface passing through a point P there are three stress components: a normal stress and two shear stresses. The stresses have units of pressure and vary from point to point and also with the orientation of the surface. In the next section we discuss how to fully describe the state of stress at one point.

1.1- Stress Components Acting on a Small Element

Consider a small parallepiped with dimensions $\Delta x.\Delta y.\Delta z$. A face is called an x-face if the outside normal to that surface is oriented in the x direction. Similarly, for a y-face, the normal is oriented in the y direction and for a z-face the normal is in the z direction. A face of that element is called a positive face if the outside normal to that surface is pointing in the positive axis direction.

On a given face, three are 3 stress components: one normal stress and two shear stresses. On a positive x-face (Figure 1), the sign convention is that the stresses are positive in the positive x, y, and z direction. The stress components are designated by two indices. The first index denotes the surface that component is acting on. The second index indicates the direction in which that stress component is acting. For example, Figure 1 shows the three stress components acting on a positive x face. The first index is x for all three components because they are all acting on the x-face and then, σ_{xx}, σ_{xy} and σ_{xz} are acting in the positive x, y and z direction respectively. The same convention is shown to apply to the three stress components acting on a positive z-face in Figure 2 and Figure 3 shows the sign convention applied to a negative y-face.

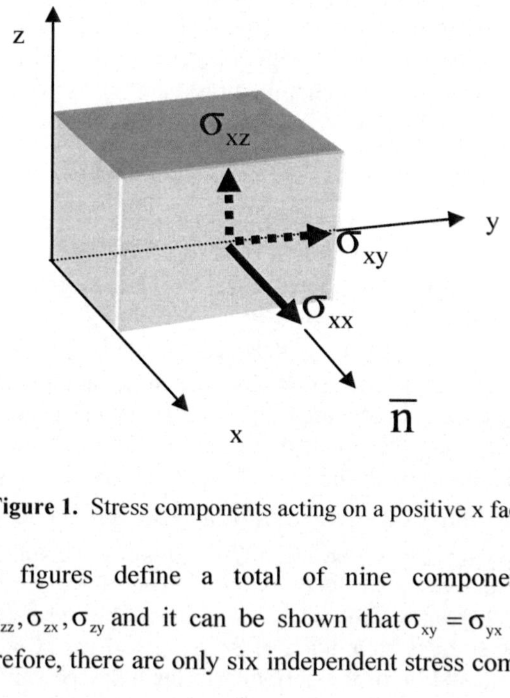

Figure 1. Stress components acting on a positive x face

These three figures define a total of nine components $\sigma_{xx}, \sigma_{xy}, \sigma_{xz}$, σ_{yy}, σ_{yx}, $\sigma_{yz}, \sigma_{zz}, \sigma_{zx}, \sigma_{zy}$ and it can be shown that $\sigma_{xy} = \sigma_{yx}$, $\sigma_{xz} = \sigma_{zx}$, and $\sigma_{yz} = \sigma_{zy}$. Therefore, there are only six independent stress components that can be written as

$$[\sigma]=\begin{bmatrix}\sigma_{xx} & \sigma_{xy} & \sigma_{xz}\\ \sigma_{xy} & \sigma_{yy} & \sigma_{yz}\\ \sigma_{xz} & \sigma_{yz} & \sigma_{zz}\end{bmatrix} \tag{1}$$

$[\sigma]$ is called the stress tensor. A tensor is a quantity that follows transformation laws. One way to look at it is to think of a vector in space as a quantity with three components. If this vector joins two points in space or represents a force, it is a physical quantity with a given magnitude and orientation. If its components are known in one coordinate system, one can calculate the components of the vector in another coordinate system. Tensors are a generalization of vectors and follow some transformation laws when changing coordinate systems.

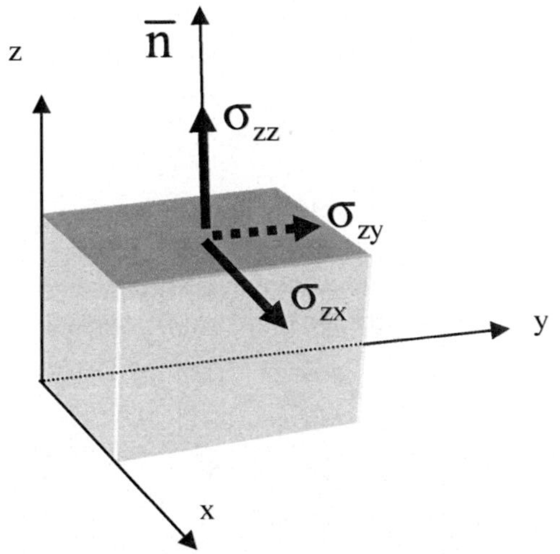

Figure 2. Stress components acting on a positive z face

1.2- Surface Tractions on an Arbitrary Surface

From the stress tensor (Equation 1), one can determine the loads acting on any surface passing through that point. Consider a wedge with three faces oriented in the x-, y- and z-directions and a fourth face oriented in the direction of an arbitrary vector $\bar{n}$ (Figure 4). The stresses acting on the first three faces are components of $[\sigma]$, the stress tensor in the xyz coordinate system. On the face oriented in the $\bar{n}$ direction, we define the surface tractions t_x, t_y and t_z which are forces per unit area acting in the x, y, and z directions. Stresses and surfaces tractions have the same dimension but stresses have components that are normal or in the place of the surface they are acting on while surface tractions are acting in the x, y, z directions.

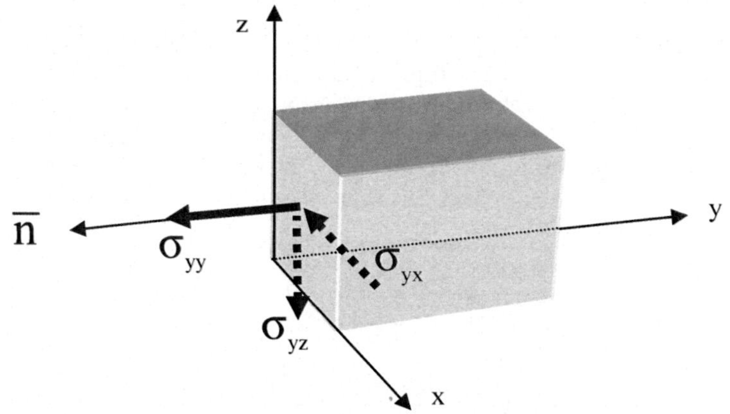

Figure 3. Stress components acting on a negative y face

To establish the relationship between the stress tensor $[\sigma]$ and the surface tensions $\{t\}$ we consider the equilibrium of the wedge (Figure 4) and find that

$$\begin{Bmatrix} t_x \\ t_y \\ t_z \end{Bmatrix} = \begin{bmatrix} \sigma_{xx} & \sigma_{xy} & \sigma_{xz} \\ \sigma_{xy} & \sigma_{yy} & \sigma_{yz} \\ \sigma_{xz} & \sigma_{yz} & \sigma_{zz} \end{bmatrix} \begin{Bmatrix} n_x \\ n_y \\ n_z \end{Bmatrix} \quad \text{or} \quad \{t\} = [\sigma]\{n\} \tag{2}$$

where i can be either x, y or z. Equation 2 shows that the stress tensor $[\sigma]$ completely defines the state of stress at that point because, given $[\sigma]$, one can calculate the surface tractions t_x, t_y and t_z acting on an arbitrary surface defined by its normal $\bar{n}$. The normal stress on an arbitrary surface is the dot product of the surface traction vector and the normal to the surface. That is

$$\sigma_N = \begin{bmatrix} t_x & t_y & t_z \end{bmatrix} \begin{Bmatrix} n_x \\ n_y \\ n_z \end{Bmatrix} = t_x n_x + t_y n_y + t_z n_z \tag{3}$$

The shear stress acting on that surface is

$$\{\tau\} = \begin{Bmatrix} t_x \\ t_y \\ t_z \end{Bmatrix} - \sigma_N \begin{Bmatrix} n_x \\ n_y \\ n_z \end{Bmatrix} \tag{4}$$

Equation 4 gives the components of that shear stress in the xyz coordinate system. Eqs. (3, 4) show that, given $[\sigma]$ in one coordinate system, the stresses acting on any surface passing through that point can be determined. The stress tensor $[\sigma]$ completely defines the state of stress at that point.

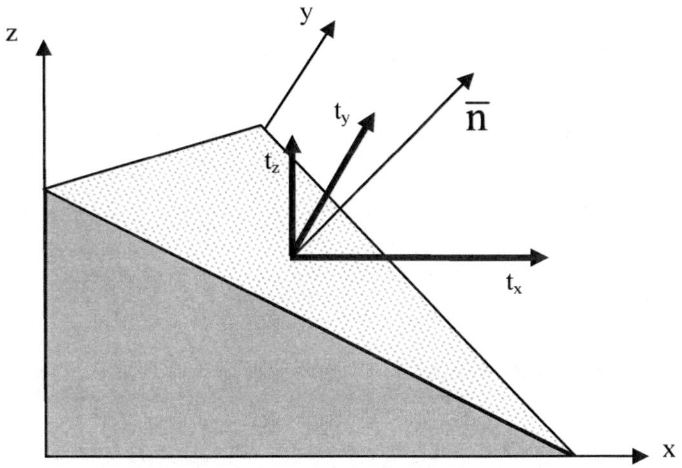

Figure 4. Surface tractions acting on an arbitrary surface

1.3- Equations of Motion

Consider a portion of a body with a volume V and a surface S. The surface is subjected to surface tractions $\bar{t}$, the body is also subjected to body forces $\overline{B}$, and the position of an arbitrary point inside the body is defined by the vector $\bar{r}$. Using Newton's law, the sum of the forces acting on the body gives

$$\int_S \bar{t}\, dS + \int_V \bar{B}\, dV = \int_V \rho \ddot{\bar{r}}\, dV \quad (5)$$

where ρ is the density of the material. In this expression we note that:

- $\int_S \bar{t}\, dS$ is the some of the forces acting on the external surface S
- $\int_V \bar{B}\, dV$ is the resultant of the body forces acting on the volume V
- $-\int_V \rho \ddot{\bar{r}}\, dV$ is the inertia force

Using Equation (2) and applying the divergence theorem, the first term becomes

$$\int_S \bar{t}\, dS = \int_S \bar{\bar{\sigma}}.\bar{n}\, dS = \int_V \bar{\nabla}.\bar{\bar{\sigma}}\, dV \quad (6)$$

Then, for this volume V, the motion is governed by

$$\int_V \left[\bar{\nabla}.\bar{\bar{\sigma}} + \bar{B} - \rho \ddot{\bar{r}} \right] dV = 0 \quad (7)$$

This equation holds for any volume inside the body so we must have

$$\bar{\nabla}.\bar{\bar{\sigma}} + \bar{B} - \rho \ddot{\bar{r}} = 0 \quad (8)$$

This vector equation (Equation 8) can be written as three scalar equations

$$\frac{\partial \sigma_{xx}}{\partial x} + \frac{\partial \sigma_{yx}}{\partial y} + \frac{\partial \sigma_{zx}}{\partial z} + X = \rho \ddot{u}$$

$$\frac{\partial \sigma_{xy}}{\partial x} + \frac{\partial \sigma_{yy}}{\partial y} + \frac{\partial \sigma_{zy}}{\partial z} + Y = \rho \ddot{v} \quad (9)$$

and

$$\frac{\partial \sigma_{xz}}{\partial x} + \frac{\partial \sigma_{yz}}{\partial y} + \frac{\partial \sigma_{zz}}{\partial z} + Z = \rho \ddot{w}$$

X, Y, and Z are the components of the body force vector $\bar{B}$. Equations 9 are the equations of motion of the linear theory of elasticity in terms of stresses and displacements. In the following, we will see how the stresses are related to the deformation of the body and how the three equations of motion can be expressed in terms of the three displacements u, v, and w.

2-Strain

Since stresses define the loading at a point, we also need to describe the deformation at that point and then relate those two measures knowing the behavior of the material. To describe the deformation at one point we consider three orthogonal line segments. After deformation, the length of each segment will be different and the angle between any two of those segments will be different from 90 degrees. We define three linear strains from the changes in length of these three line segments and three shear strains from the changes in the angles between them. These six strains will then describe the deformation at that point.

2.1- Linear Strains

Consider a small line segment of length L oriented in the x direction before deformation (Figure 5). After deformation, its length becomes $L+\Delta L$ and the linear strain in the x-direction is defined as $\varepsilon_{xx} = \Delta L / L$. In terms of u, the displacement in the x-direction, $\Delta L = \frac{\partial u}{\partial x}.L$ so the linear strain in the x-direction is $\varepsilon_{xx} = \frac{\partial u}{\partial x}$.

Similarly, considering line segments oriented in the y and z directions we can define the other two linear strain components: $\varepsilon_{yy} = \frac{\partial v}{\partial y}$ and $\varepsilon_{xx} = \frac{\partial w}{\partial z}$.

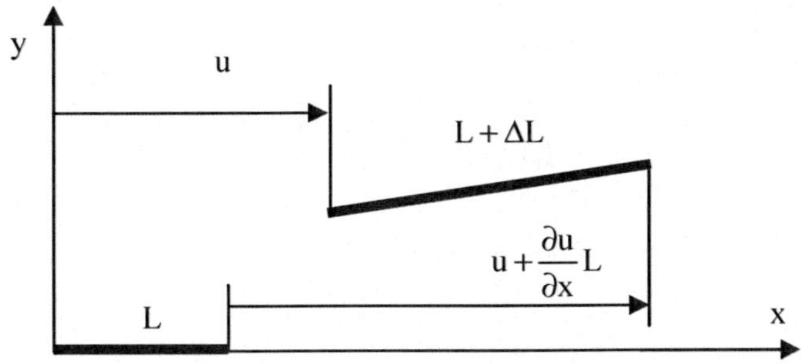

Figure 5. Deformation of a small line element in the x direction

2.2- Shear Strains

Shear strains provide another measure of the deformation at a point. Consider three small line segments initially oriented along the x, y and z directions respectively. These three segments are perpendicular to each other initially but, after deformation, the angle between them is no longer $\pi/2$. For example, the angle between the segments oriented in the x and y directions will be $\frac{\pi}{2}-\varepsilon_{xy}$ and ε_{xy} is defined as the shear strain in the xy plane. Figure 7 shows that $\varepsilon_{xy}=\frac{\partial u}{\partial y}+\frac{\partial v}{\partial x}$. Similarly, $\varepsilon_{xz}=\frac{\partial u}{\partial z}+\frac{\partial w}{\partial x}$ and $\varepsilon_{yz}=\frac{\partial v}{\partial z}+\frac{\partial w}{\partial y}$.

The six strain components defined here can be shown to from a tensor

$$[\varepsilon]=\begin{bmatrix} \varepsilon_{xx} & \varepsilon_{xy} & \varepsilon_{xz} \\ \varepsilon_{xy} & \varepsilon_{yy} & \varepsilon_{yz} \\ \varepsilon_{xz} & \varepsilon_{yz} & \varepsilon_{zz} \end{bmatrix} \tag{10}$$

called the strain tensor. It can be shown that the strain tensor defines the deformation at that point. That is, knowing $[\varepsilon]$, one can calculate the change in length of any short line segment passing through that point or the change of angle between any two line segments at that point.

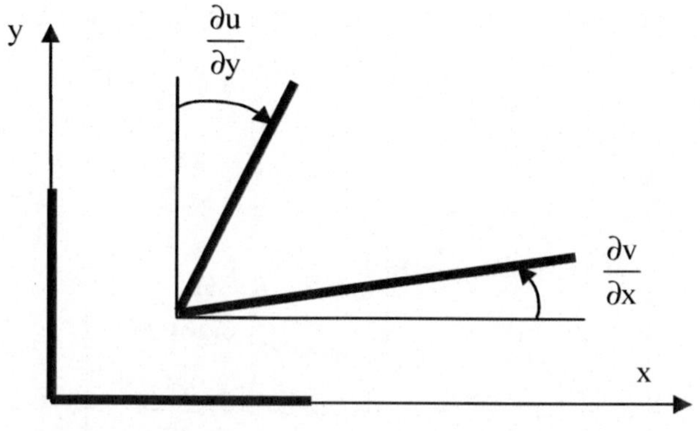

Figure 6. Deformations in the xy plane

3- Stress-Strain Behavior

In describing the mechanical behavior of materials, several definitions must be introduced. A material is isotropic if the behavior is the same in all directions. A material is homogeneous if its properties are the same as we move from point to point inside the material. In general, steel and cast iron are isotropic and are considered to be homogeneous on a macroscopic scale even if under a microscope one can observe different grains with different shapes and properties. Sometimes metals are not isotropic after rolling operations for example. Composite materials often consist of strong fibers embedded in a soft matrix. On a microscopic scale those materials are not homogeneous but on a macroscopic scale (many times the diameter of a fiber) these materials can be considered to be homogeneous. Composite materials are not isotropic either since they are usually stiffer and stronger in the fiber direction than in the transverse direction.

Another important concept in discussing the behavior of materials is whether or not the material is elastic. A material is elastic if it recovers it original length after being unloaded. Plotting the applied force versus the elongation or stress versus strain, a certain path is followed during the loading process and the same path is followed in reverse during the unloading. If for a material the unloading follows a different path, the material is said to be inelastic this can be due to strain rate effects (viscoelasticity), the introduction of permanent deformations (plasticity) or both. A material is said to be linear if the stress-strain curves are straight or, in other words, stress is proportional to strain. It should be pointed out that a material can be elastic without being linear which is the case for rubber.

In the following, we will first discuss the stress-strain behavior for an isotropic material and introduce different concepts such as Hooke's law and material properties such as the modulus of elasticity, Poisson's ratio and the shear modulus. The more general case of an orthotropic material is considered next.

3.1- Hooke's Law

For isotropic materials, most of the information needed to characterize the mechanical behavior of the material can be obtained from a tensile test. For such a uniaxial loading, a normal stress produces an elongation in the direction of the applied stress and a contraction in the transverse direction but no shear deformation. The normal stress σ is directly proportional to ε, the normal strain in that direction

$$\sigma = E\varepsilon \tag{11}$$

for what are called linear elastic solids. Eqn. (11) is called Hooke's law and the proportionality constant E is the modulus of elasticity or Young's modulus of the material.

3.2- Poisson's Ratio

Under a uniaxial normal stress in the x direction for example, the material contracts in the transverse direction (y or z direction) and the ratio between the transverse and normal strains

$$\nu = -\varepsilon_{yy} / \varepsilon_{xx} \qquad (12)$$

is a constant and a property of the material. This ratio is called Poisson's ratio. Note that when $\varepsilon_{xx} > 0$ in a tensile test, $\varepsilon_{yy} < 0$ and the negative sign in Equation 12 is introduced so that $\nu > 0$. Since the material is isotropic, $\nu = -\varepsilon_{zz} / \varepsilon_{xx}$ also.

3.3- Stress-Strain Relation in Shear

When the material is subjected to shear the shear stress is directly proportional to the shear strain so, for example,

$$\sigma_{xy} = G\varepsilon_{xy} \qquad (13)$$

where G is the shear modulus. It can be shown that, for isotropic materials, the shear modulus is related to the modulus of elasticity and Poisson's ratio by

$$G = \frac{E}{2(1+\nu)} \qquad (14)$$

Therefore, only two independent constants are needed to characterize the elastic behavior of the material. Some authors give E and G while others provide E and ν.

3.4- Stress-Strain Relations for an Orthotropic Solid

Composite materials are heterogeneous on a microscopic scale which can be defined by the diameter of a fiber: typically $10\ \mu m$. However, the analysis of a structure takes place on a much larger scale where dimensions are measured in millimeters. On such a macroscopic scale it is possible to consider the heterogeneous mixture of fibers and matrix materials as a homogeneous material with different properties in different directions. Considering a rectangular block of fiber reinforced material with fibers oriented in the x_1 direction and subjected to a normal stress σ_{11} (Figure 7). The elongation of a line segment in the x_1 direction such as AB will define the strain ε_{11} and according to Hooke's law,

$$\sigma_{11} = E_1 \varepsilon_{11} \tag{15}$$

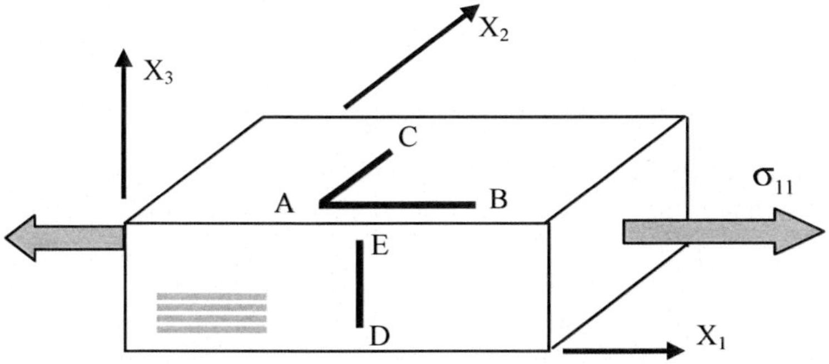

Figure 7. Orthotropic material subjected to tension in the fiber direction

The strains in the x_2 and x_3 directions determine the changes in length of the line segments AC and DE respectively. We define two Poisson's ratios

$$\nu_{12} = -\frac{\varepsilon_{22}}{\varepsilon_{11}} \qquad \nu_{13} = -\frac{\varepsilon_{33}}{\varepsilon_{11}} \tag{16}$$

so that

$$\varepsilon_{22} = -\nu_{12}\frac{\sigma_{11}}{E_1} \quad \text{and} \quad \varepsilon_{33} = -\nu_{13}\frac{\sigma_{11}}{E_1} \tag{17}$$

Note that ν_{12} was defined for the case when the load was applied in direction 1 and the transverse direction was direction 2. Similarly, for ν_{13} the load was applied in direction 1 and the transverse direction that was considered was direction 3. The first index is the direction in which the load is applied and the second index designates the transverse direction.

For the same orthotropic block, applying σ_{22} alone will produce strains

$$\varepsilon_{22} = \frac{\sigma_{22}}{E_2}, \quad \varepsilon_{11} = -\nu_{21}\frac{\sigma_{22}}{E_2}, \quad \varepsilon_{33} = -\nu_{23}\frac{\sigma_{22}}{E_2} \tag{18}$$

Similarly, applying σ_{33} alone will produce

$$\sigma_{33} = E_3\varepsilon_{33}, \quad \varepsilon_{11} = -\nu_{31}\frac{\sigma_{33}}{E_3}, \quad \varepsilon_{22} = -\nu_{32}\frac{\sigma_{33}}{E_3} \tag{19}$$

Combining the effects of these three normal stress components we have

$$\begin{Bmatrix} \varepsilon_{11} \\ \varepsilon_{22} \\ \varepsilon_{33} \end{Bmatrix} = \begin{bmatrix} \frac{1}{E_1} & \frac{-\nu_{21}}{E_2} & \frac{-\nu_{31}}{E_3} \\ \frac{-\nu_{12}}{E_1} & \frac{1}{E_2} & \frac{-\nu_{32}}{E_3} \\ \frac{-\nu_{13}}{E_1} & \frac{-\nu_{23}}{E_2} & \frac{1}{E_3} \end{bmatrix} \begin{Bmatrix} \sigma_{11} \\ \sigma_{22} \\ \sigma_{33} \end{Bmatrix} \tag{20}$$

In those equations, six Poisson's ratios have been introduced. It can be shown that the compliance matrix in Equation (20) must be symmetric so that

$$\frac{\nu_{21}}{E_2} = \frac{\nu_{12}}{E_1}, \quad \frac{\nu_{23}}{E_2} = \frac{\nu_{32}}{E_3}, \quad \frac{\nu_{31}}{E_3} = \frac{\nu_{13}}{E_1} \tag{21}$$

and the number of independent Poisson's ratios is reduced to three. In shear, the stress-strain relations are

$$\sigma_{12} = G_{12}\varepsilon_{12}, \quad \sigma_{13} = G_{13}\varepsilon_{13}, \quad \sigma_{23} = G_{23}\varepsilon_{23} \tag{22}$$

where three independent shear moduli are introduced. The elastic behavior of orthotropic materials is characterized by 9 elastic constants: $E_1, E_2, E_3, \nu_{12}, \nu_{13}, \nu_{23}, G_{12}, G_{13}, G_{23}$. Combining Equations 20 and 22, the constitutive equations for an orthotropic material can be written as

$$\begin{Bmatrix} \varepsilon_{11} \\ \varepsilon_{22} \\ \varepsilon_{33} \\ \varepsilon_{23} \\ \varepsilon_{13} \\ \varepsilon_{12} \end{Bmatrix} = \begin{bmatrix} \frac{1}{E_1} & \frac{-\nu_{21}}{E_2} & \frac{-\nu_{31}}{E_3} & 0 & 0 & 0 \\ \frac{-\nu_{12}}{E_1} & \frac{1}{E_2} & \frac{-\nu_{32}}{E_3} & 0 & 0 & 0 \\ \frac{-\nu_{13}}{E_1} & \frac{-\nu_{23}}{E_2} & \frac{1}{E_3} & 0 & 0 & 0 \\ 0 & 0 & 0 & \frac{1}{G_{23}} & 0 & 0 \\ 0 & 0 & 0 & 0 & \frac{1}{G_{13}} & 0 \\ 0 & 0 & 0 & 0 & 0 & \frac{1}{G_{12}} \end{bmatrix} \begin{Bmatrix} \sigma_{11} \\ \sigma_{22} \\ \sigma_{33} \\ \sigma_{23} \\ \sigma_{13} \\ \sigma_{12} \end{Bmatrix} \tag{23}$$

and, for an isotropic material,

$$\begin{Bmatrix} \varepsilon_{xx} \\ \varepsilon_{yy} \\ \varepsilon_{zz} \\ \varepsilon_{yz} \\ \varepsilon_{zx} \\ \varepsilon_{xy} \end{Bmatrix} = \begin{bmatrix} \frac{1}{E} & \frac{-\nu}{E} & \frac{-\nu}{E} & 0 & 0 & 0 \\ \frac{-\nu}{E} & \frac{1}{E} & \frac{-\nu}{E} & 0 & 0 & 0 \\ \frac{-\nu}{E} & \frac{-\nu}{E} & \frac{1}{E} & 0 & 0 & 0 \\ 0 & 0 & 0 & \frac{1}{G} & 0 & 0 \\ 0 & 0 & 0 & 0 & \frac{1}{G} & 0 \\ 0 & 0 & 0 & 0 & 0 & \frac{1}{G} \end{bmatrix} \begin{Bmatrix} \sigma_{xx} \\ \sigma_{yy} \\ \sigma_{zz} \\ \sigma_{yz} \\ \sigma_{zx} \\ \sigma_{xy} \end{Bmatrix} \tag{24}$$

The usual convention for unidirectional composites is to take the x_1 axis in the fiber direction, the x_2 axis in the plane of the lamina, and x_3 to be perpendicular to that plane. This is called a material principal coordinate system and, in that system, normal stresses produce linear strains but no shear deformations. Similarly, shear stresses produce shear strains but no linear strains.

3.5- Stress-Strain Relations for a Lamina under Plane Stress

Consider a thin lamina of orthotropic material in material principal coordinates. If the top and bottom surfaces are stress free and the thickness is small, it may be assumed that the stress components $\sigma_{33}, \sigma_{31}, \sigma_{32}$ are negligible. The lamina is then in a state of plane stress and, for inplane loading, the constitutive equations are

$$\begin{Bmatrix} \varepsilon_{11} \\ \varepsilon_{22} \\ \varepsilon_{12} \end{Bmatrix} = \begin{bmatrix} \frac{1}{E_1} & \frac{-\nu_{21}}{E_2} & 0 \\ \frac{-\nu_{12}}{E_1} & \frac{1}{E_2} & 0 \\ 0 & 0 & \frac{1}{G_{12}} \end{bmatrix} \begin{Bmatrix} \sigma_{11} \\ \sigma_{22} \\ \sigma_{12} \end{Bmatrix} \tag{25}$$

Inverting this relationship gives

$$\begin{Bmatrix} \sigma_{11} \\ \sigma_{22} \\ \sigma_{12} \end{Bmatrix} = \begin{bmatrix} Q_{11} & Q_{12} & 0 \\ Q_{12} & Q_{22} & 0 \\ 0 & 0 & Q_{66} \end{bmatrix} \begin{Bmatrix} \varepsilon_{11} \\ \varepsilon_{22} \\ \varepsilon_{12} \end{Bmatrix} \tag{26}$$

where the reduced stiffness constants are

$$Q_{11} = \frac{E_1}{1-\nu_{12}\nu_{21}}, \quad Q_{22} = \frac{E_2}{1-\nu_{12}\nu_{21}}, \quad Q_{12} = \frac{\nu_{12}E_2}{1-\nu_{12}\nu_{21}}, \quad Q_{66} = G_{12}$$

In this coordinate system, extension and shear deformations are uncoupled as indicated by the zeroes in the matrix [Q]. Applying any combination of stresses σ_{11} and σ_{22} will produce linear strains ε_{11} and ε_{22} but no shear strain. Similarly, a shear stress σ_{12} will induce a shear strain but no linear strain. The behavior of such a layer is governed by four independent constants $(E_1, E_2, \nu_{12}, G_{12})$.

3.6- Coordinate Transformation

Composite structures often consist of a laminate with many layers oriented in different directions. For each layer, the constitutive equations can be written in a local coordinate system but then it is necessary to use one global coordinate system for all the layers. The stress components acting on a surface with any orientation can be found using a coordinate transformation law. If the three stress components in the xy coordinate system ($\sigma_{xx}.\sigma_{yy}$, and σ_{xy}) are known, the stresses in any coordinate system x'y' oriented at an angle θ (Figure 8) are given by

$$\begin{Bmatrix} \sigma_{x'x'} \\ \sigma_{y'y'} \\ \sigma_{x'y'} \end{Bmatrix} = \begin{bmatrix} c^2 & s^2 & 2cs \\ s^2 & c^2 & -2cs \\ -cs & cs & c^2 - s^2 \end{bmatrix} \begin{Bmatrix} \sigma_{xx} \\ \sigma_{yy} \\ \sigma_{xy} \end{Bmatrix} \tag{27}$$

where $c = \cos\theta$, $s = \sin\theta$.

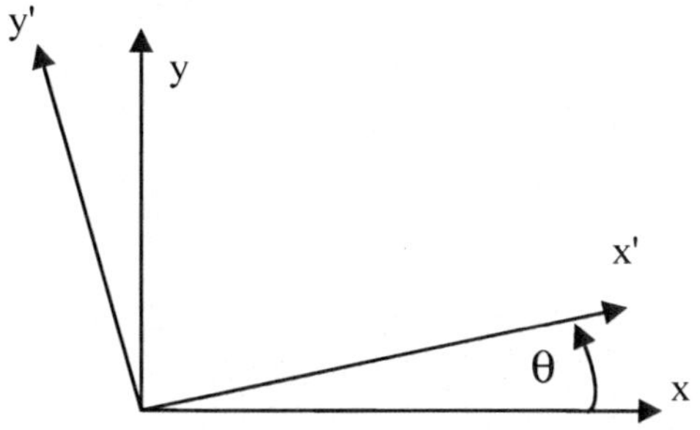

Figure 8. Coordinate transformation in the plane of a lamina

The stresses in the new coordinate system are related to the stresses in the old coordinate system by

$$\{\sigma'\} = [T]\{\sigma\} \tag{28}$$

where [T] is the coordinate transformation matrix. The displacements in the new coordinate system are related to those in the old system by

$$u' = u\ \cos\theta + v\ \sin\theta, \qquad v' = -u\ \sin\theta + v\cos\ \theta \tag{29}$$

The strains in the new coordinate system are

$$\varepsilon_{x'x'} = \frac{\partial u'}{\partial x'} = \frac{\partial u'}{\partial x}\frac{\partial x}{\partial x'} + \frac{\partial u'}{\partial y}\frac{\partial y}{\partial x'}$$

$$\varepsilon_{y'y'} = \frac{\partial v'}{\partial y'} = \frac{\partial v'}{\partial x}\frac{\partial x}{\partial y'} + \frac{\partial v'}{\partial y}\frac{\partial y}{\partial y'} \tag{30}$$

$$\varepsilon_{x'y'} = \frac{\partial u'}{\partial y'} + \frac{\partial v'}{\partial x'} = \frac{\partial u'}{\partial x}\frac{\partial x}{\partial y'} + \frac{\partial u'}{\partial y}\frac{\partial y}{\partial y'} + \frac{\partial v'}{\partial x}\frac{\partial x}{\partial x'} + \frac{\partial v'}{\partial y}\frac{\partial y}{\partial x'}$$

The coordinates in the new and old systems are related by

$$x = x' \cos\theta - y' \sin\theta\ , \quad y = x' \sin\theta + y' \cos\theta \tag{31}$$

Then, the relationships between the strain components in the new coordinate system and the strain components in the old system are

$$\varepsilon_{x'x'} = \varepsilon_{xx} \cdot \cos^2\theta + \varepsilon_{yy} \cdot \sin^2\theta + \sin\theta\ \cos\theta\ \varepsilon_{xy}$$

$$\varepsilon_{y'y'} = \varepsilon_{xx} \cdot \sin^2\theta + \varepsilon_{yy} \cdot \cos^2\theta + \sin\theta \cos\theta\ \varepsilon_{xy} \tag{32}$$

$$\varepsilon_{x'y'} = -\ \varepsilon_{xx} . 2\sin\theta\ \cos\theta + \varepsilon_{yy} . 2\ \cos\theta\ \sin\theta + \left(\cos^2\theta\ -\ \sin^2\ \theta\right)\ \varepsilon_{xy}$$

These strain transformation equations (Eqs. 32) can be written in matrix form as

$$\begin{Bmatrix} \varepsilon_{x'x'} \\ \varepsilon_{y'y'} \\ \varepsilon_{x'y'}/2 \end{Bmatrix} = \begin{bmatrix} c^2 & s^2 & 2sc \\ s^2 & c^2 & 2sc \\ -sc & sc & (c^2 - s^2) \end{bmatrix} \begin{Bmatrix} \varepsilon_{xx} \\ \varepsilon_{yy} \\ \varepsilon_{xy}/2 \end{Bmatrix} \tag{33}$$

or, in short hand notation,

$$\{\varepsilon'\} = [T]\,\{\varepsilon\} \tag{34}$$

where the transformation matrix T is the same matrix used for stress transformations (Equation 28). Eqns. (33) show that, in two-dimensions, three strain components are sufficient to define the state of deformation at one point. The strain in any other direction can be found using the transformation equations.

In a laminate, the material principal directions for a given layer are oriented at an angle θ from the global coordinate system xy (Figure 9). The stress-strain relations in the material principal coordinates (Equation 26) can be written as

$$\begin{bmatrix} c^2 & s^2 & 2cs \\ s^2 & c^2 & -2cs \\ -cs & cs & c^2-s^2 \end{bmatrix} \begin{Bmatrix} \sigma_{xx} \\ \sigma_{yy} \\ \sigma_{xy} \end{Bmatrix} = \begin{bmatrix} Q_{11} & Q_{12} & 0 \\ Q_{12} & Q_{22} & 0 \\ 0 & 0 & Q_{66} \end{bmatrix} \begin{bmatrix} c^2 & s^2 & 2cs \\ s^2 & c^2 & -2cs \\ -2cs & 2cs & c^2-s^2 \end{bmatrix} \begin{Bmatrix} \varepsilon_{xx} \\ \varepsilon_{yy} \\ \varepsilon_{xy} \end{Bmatrix}$$

or,

$$\begin{Bmatrix} \sigma_{xx} \\ \sigma_{yy} \\ \sigma_{xy} \end{Bmatrix} = \begin{bmatrix} c^2 & s^2 & -2cs \\ s^2 & c^2 & 2cs \\ cs & -cs & c^2-s^2 \end{bmatrix} \begin{bmatrix} Q_{11} & Q_{12} & 0 \\ Q_{12} & Q_{22} & 0 \\ 0 & 0 & Q_{66} \end{bmatrix} \begin{bmatrix} c^2 & s^2 & 2cs \\ s^2 & c^2 & -2cs \\ -2cs & 2cs & c^2-s^2 \end{bmatrix} \begin{Bmatrix} \varepsilon_{xx} \\ \varepsilon_{yy} \\ \varepsilon_{xy} \end{Bmatrix} \quad (35)$$

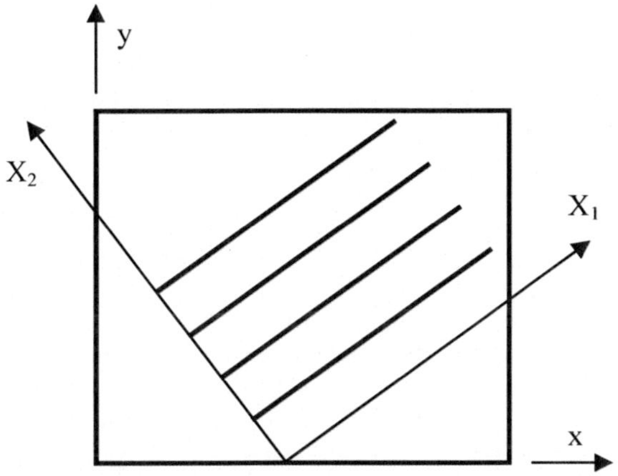

Figure 9. Material principal coordinates for an orthotropic lamina with fibers oriented at an angle θ from an arbitrary coordinate system xy.

Finally, the constitutive equations for a single orthotropic layer are oriented at an angle θ from the x-axis can be written as

$$\begin{Bmatrix} \sigma_{xx} \\ \sigma_{yy} \\ \sigma_{xy} \end{Bmatrix} = \begin{bmatrix} \overline{Q}_{11} & \overline{Q}_{12} & \overline{Q}_{16} \\ \overline{Q}_{12} & \overline{Q}_{22} & \overline{Q}_{26} \\ \overline{Q}_{16} & \overline{Q}_{26} & \overline{Q}_{66} \end{bmatrix} \begin{Bmatrix} \varepsilon_{xx} \\ \varepsilon_{yy} \\ \varepsilon_{xy} \end{Bmatrix} \quad (36)$$

where

$$\overline{Q}_{11} = Q_{11}c^4 + 2(Q_{12} + 2Q_{66})c^2s^2 + Q_{22}s^4,$$

$$\overline{Q}_{12} = (Q_{11} + Q_{22} - 4Q_{66})c^2s^2 + Q_{12}(c^4 + s^4)$$

$$\overline{Q}_{22} = Q_{22}c^4 + 2(Q_{12} + 2Q_{66})c^2s^2 + Q_{11}s^4,$$

$$\overline{Q}_{16} = Q_{11}c^3s - (Q_{12} + 2Q_{66})(c^2 - s^2)cs - Q_{22}s^3c$$

$$\overline{Q}_{26} = -Q_{22}c^3s + (Q_{12} + 2Q_{66})(c^2 - s^2)cs + Q_{11}s^3c,$$

$$\overline{Q}_{66} = (Q_{11} + Q_{22} - 4Q_{66})c^2s^2 + Q_{66}(c^2 - s^2)^2$$

with $c = \cos(\theta)$ and $s = \sin(\theta)$. When the coordinate system is not oriented in the material principal direction $(\theta \neq 0 \text{ or } 90^\circ)$, the $\overline{Q}_{16}$ and $\overline{Q}_{26}$ terms are not zero and there will be coupling between the extension and shear deformations. A normal stress will produce linear strains and a shear strain. Similarly a shear stress will produce both extensions and shear deformations.

4- Failure Criteria for Composite Materials

Under some given loads, the analysis of a composite structure determines stresses and strains that satisfy the equations of motion and the constitutive equations. It is also important to determine whether or not the structure will fail. In a composite structure, failure can occur inside a ply or at interfaces between plies. Many failure criteria have been proposed for predicting failure of composites and this section will describe some of the most commonly used criteria. A common feature to all the criteria described here is that there are based on the composite stresses (or strains). That is on the stresses calculated assuming that each ply is a homogeneous orthotropic solid as opposed to detailed stress distribution inside a fiber and the surrounding matrix.

First we examine criteria for predicting failure of the fibers, failure of the matrix, or delamination failure. It is often thought that it is easier to develop criteria for individual failure modes and that, when successful, they bring insight into the behavior of the composite. Then, we discuss criteria capable of predicting failure under any combination of stresses. The use of a single criterion for predicting failure of a lamina is easier to implement but the drawback is that, usually, they cannot predict the mode of failure. In each case we introduce the various failure criteria in order of increasing complexity.

4.1- Criteria for Fiber Failure

The simplest approach is to assume that fiber failure depends only on the normal stress in the fiber direction and that failure occurs when the magnitude of that stress exceeds a critical value. Following Hashin (1980), tensile fiber failure is predicted using a maximum stress criterion

$$\sigma_{11}/X_T = 1 \qquad \text{when} \qquad \sigma_{11} > 0 \tag{37.a}$$

This criterion is used by Green et al (2000), Luo et al (1999, 2001), Foo et al (2008) and Cesari et al (2007). Foo et al (2008) assumed that fiber compressive failure occurs when

$$|\sigma_{11}|/X_C = 1 \qquad \text{when} \qquad \sigma_{11} < 0 \tag{37.b}$$

X_T is the strength of the composite in the fiber direction and X_C is the strength in compression and in general $X_C < X_T$. Other researchers have proposed ways to account for the effects of other stress components. Lee (1982) and Hatami-Marbini and Pietruszczak (2007) use Equations 37 to predict tensile and compressive fiber failure. In addition, fiber failure due to shear is predicted using

$$\left(\frac{\sigma_{12}}{S_1}\right)^2 + \left(\frac{\sigma_{13}}{S_1}\right)^2 = 1 \tag{38}$$

where S_1 is the shear strength in the plane perpendicular to the material principal axis 1. This implies that the material is transversely isotropic and that the effect of normal and shear stresses are independent.
Yamada and Sun (1978) and Sun and Yamada (1978) considered that the shear stress σ_{12} also has an effect on fiber failure and used the quadratic criterion

$$\left(\frac{\sigma_1}{X}\right)^2 + \left(\frac{\sigma_{12}}{S_{12}}\right)^2 = 1 \tag{39}$$

where $X = X_t$ for $\sigma_{11} > 0$ and $X = X_c$ for $\sigma_{11} < 0$. S_{12} is the shear strength in the plane of the lamina. In the $\sigma_{12} - \sigma_1$ plane, the failure curve consists of two half ellipses: one for $\sigma_{11} > 0$ and the other for $\sigma_{11} < 0$. The strength of a ply is not the same when it is tested by itself compared with that of a ply inside a laminate. Sun and Yamada (1978) noted that the in-situ shear strength in a laminate is often two or three time higher than that measured on a single layer. Hou et al

(2000) and Zhang (2002) used Equation 39 for $\sigma_{22} > 0$. Hou et al (2000) also included the effect of the transverse shear stress σ_{13} on the tensile fiber failure

$$\left(\frac{\sigma_1}{X_T}\right)^2 + \left(\frac{\sigma_{12}^2 + \sigma_{13}^2}{S_f^2}\right) \geq 1 \tag{40}$$

Gomez del Rio et al (2005) used a slightly different version of Equation 40

$$\left(\frac{\sigma_1}{X_T}\right)^2 + \left(\frac{\sigma_{12} + \sigma_{13}}{S_f}\right)^2 \geq 1 \tag{41}$$

Hashin's fiber failure criteria where

$$\left(\frac{\sigma_{11}}{X_T}\right)^2 + \left(\frac{\sigma_{12}}{S_{12}}\right)^2 + \left(\frac{\sigma_{13}}{S_{13}}\right)^2 = 1 \quad \text{when } \sigma_{11} > 0 \tag{42}$$

and

$$|\sigma_{11}| = X_C \text{ when } \sigma_{11} < 0 \tag{43}$$

was used by Zhang et al (2002) and Li et al (2002). Zhang et al (2002) also introduced a modified Hashin criterion by replacing stresses by strains in Equations (42, 43).

4.2- Criteria for Matrix Failure

Like when the material is loaded in the fiber direction, failure under normal stress in the transverse direction can be thought to depend only on the magnitude of that stress σ_{22}. To predict the onset of matrix cracking, Li et al (2006) used the maximum stress criterion

$$\sigma_{22} / Y_T = 1 \text{ when } \sigma_{22} > 0 \text{ or } |\sigma_{22}| / Y_c = 1 \text{ when } \sigma_{22} < 0 \tag{44}$$

which accounts for the different strength in tension and compression when loaded in the transverse direction but does not account for the effect of shear stresses. In addition to Equations 44, Lee (1982) and Hatami-Marbini and Pietruszczak (2007) predicted matrix failure due to shear using

$$\left(\frac{\sigma_{12}}{S_2}\right)^2 + \left(\frac{\sigma_{13}}{S_2}\right)^2 = 1 \tag{45}$$

where S_2 is the shear strength in the plane perpendicular to the material principal axis 2. Equation 45 predicts failure due to shear stresses independently of the normal component σ_{22}.

In a laminate, stresses inside a ply vary through the thickness. The criterion proposed by Choi and Chang (1992)

$$\left(\frac{{}^n\overline{\sigma}_{22}}{{}^nY}\right)^2 + \left(\frac{{}^n\overline{\sigma}_{23}}{{}^nS_i}\right)^2 = e_M^2 \tag{46}$$

where ${}^nY = {}^nY_t$ if $\sigma_{yy} \geq 0$ and ${}^nY = {}^nY_c$ if $\sigma_{22} < 0$ was used in several publications (Her and Linag (2004), Krishnamurthy et al (2001), Mahanta et al (2004), Pradhan and Kumar (2000), Rahul et al (2005, 2006), Zheng et al (2006)). Overbars in Equation 46 indicate that the stresses are averaged through the thickness of layer n. When $\overline{\sigma}_{23}$ is plotted versus $\overline{\sigma}_{22}$ this failure criterion is represented by half of an ellipse when $\overline{\sigma}_{22} > 0$ and another half ellipse when $\overline{\sigma}_{22} < 0$. This criterion accounts for interactions between the normal and shear stress components in the yz plane.

Green et al [2] and Luo et al [3,4] assumed that matrix failure depends on the three stress components σ_{22}, σ_{12}, and σ_{23}

$$\left(\frac{\sigma_{22}}{Y_T}\right)^2 + \left(\frac{\sigma_{12}}{S_{12}}\right)^2 + \left(\frac{\sigma_{23}}{S_{23}}\right)^2 = 1 \quad \text{when} \quad \sigma_{22} \geq 0 \tag{47}$$

Green et al (2002) and Luo et al (2001) further assumed that $S_{12}=S_{23}$. Cesari et al (2007) used this criterion to predict failure in tension. They also used it for compression but with different strengths.

Green et al (2002), Luo et al (1999, 2001), Hou et al (2000), Zhang (2002), Li et al (2002) and Gómez-del Río et al (2005) assume that matrix cracking occurs when

$$\left(\frac{\sigma_{22}}{Y_T}\right)^2 + \left(\frac{\sigma_{12}}{S_{12}}\right)^2 = 1 \qquad \text{when} \quad \sigma_{22} \geq 0 \tag{48}$$

or

$$\frac{1}{4}\left(\frac{\sigma_{22}}{S_{12}}\right)^2+\frac{Y_c\sigma_{22}}{4S_{12}^2}-\frac{\sigma_{22}}{Y_c}+\left(\frac{\sigma_{12}}{S_{12}}\right)^2\geq 1 \quad \text{when} \quad \sigma_{22}<0 \tag{49}$$

Sun et al (1996) proposed a criterion that accounts for the reduction in shear strength when $\sigma_{22} < 0$

$$\left(\frac{\sigma_{22}}{Y}\right)^2+\left(\frac{\sigma_{12}}{S-\mu\sigma_{22}}\right)^2=1 \text{ with } \quad \mu=\begin{cases}\mu_o & \text{when} \quad \sigma_{22}<0\\ 0 & \text{when} \quad \sigma_{22}>0\end{cases} \tag{50}$$

where Y=Y_t for $\sigma_{22} > 0$ and Y=Y_c for $\sigma_{22} < 0$.

Zhang et al (2002) and Kim et al (1997, 2007), Foo et al (2008) used Hashin's criteria for predicting matrix failure. Matrix tensile failure occurs when $(\sigma_{22}+\sigma_{33})>0$ and

$$\left(\frac{\sigma_{22}+\sigma_{33}}{Y_T}\right)^2+\frac{\sigma_{23}^2-\sigma_{22}\sigma_{33}}{S_{23}^2}+\frac{\sigma_{12}^2}{S_{12}^2}+\frac{\sigma_{13}^2}{S_{13}^2}=1 \tag{51}$$

Similarly, compressive matrix failure is expected to occur when $(\sigma_{22}+\sigma_{33})<0$ and

$$\left(\frac{\sigma_{22}+\sigma_{33}}{2\sigma_{12}}\right)^2+\frac{(\sigma_{22}+\sigma_{33})}{Y_C}\left[\left(\frac{Y_C}{2S_T}\right)^2-1\right]+\frac{\sigma_{23}^2-\sigma_{22}\sigma_{33}}{S_{23}^2}+\frac{\sigma_{12}^2}{S_{12}^2}+\frac{\sigma_{13}^2}{S_{13}^2}\geq 1 \tag{52}$$

Kim et al (1997, 2007) averaged the stress components though the thickness of the layer in order to predict matrix failure (Equations 51, 52). They also assumed that $S_{12} = S_{23}$.

4.3- Delamination Failure Criteria

Debonding between adjacent layers depends on the stresses acting on that interface: the normal component σ_{33} and the two shear stresses σ_{13} and σ_{23}. Lee (1982), Zhang (2002), and Hatami-Marbini and Pietruszczak (2007) predicted delamination using a maximum stress criterion for the normal stress and a quadratic criterion for the two shear components

$$\sigma_{33} / Y_T = 1 \quad \text{or} \quad \left(\frac{\sigma_{23}}{S_3}\right)^2 + \left(\frac{\sigma_{13}}{S_3}\right)^2 = 1 \tag{53}$$

where S_3 is the shear strength in the plane perpendicular to the material principal axis 3. It is assumed that no failure occurs when $\sigma_{33} < 0$. The criterion proposed by Christensen and DeTeresa (2004) and mentioned by Cesari et al (2007)

$$\left(\frac{\sigma_{13}}{S_{13}}\right)^2 + \left(\frac{\sigma_{23}}{S_{23}}\right)^2 \geq 1 \tag{54}$$

allows for different strengths for σ_{13} and σ_{23} but does not account for interactions between the normal and shear stresses acting at the interface.

Several criteria accounting for the interaction between the three stress components acting at the interface have been introduced. Green et al (2000), Luo et al (1999, 2001), Cesari et al (2007), Zhao and Cho (2004, 2007), Wagner et al (2001), and Hou et al (2000, 2001), postulated that the onset of delamination is governed by

$$\left(\frac{\sigma_{33}}{Z_T}\right)^2 + \left(\frac{\sigma_{13}}{S_{13}}\right)^2 + \left(\frac{\sigma_{23}}{S_{23}}\right)^2 \geq 1 \quad \text{when } \sigma_{33} > 0 \tag{55}$$

Again, Green et al (2000) and Luo et al (2001), Wagner et al (2001), Hashin (1982) assumed that $S_{12} = S_{23}$. Cesari et al (2007) used the same criterion when $\sigma_{33} < 0$ but with Z_c instead of Z_T in that case. The ellipsoid defined by Equation 55 accounts for the interaction between the three stress components acting at the interface.

Her and Liang (2004), Kim et al [27, 28], Zhang et al (2002), Li (2002), Gómez-del Río et al (2005), Huang and Lee (2003), Lee and Huang (2003), Fuoss et al (1998) use the delamination criterion proposed by Choi and Chang (1992)

$$D_a \left[\left(\frac{{}^{n}\bar{\sigma}_{yz}}{{}^{n}S_i}\right)^2 + \left(\frac{{}^{n+1}\bar{\sigma}_{xz}}{{}^{n+1}S_i}\right)^2 + \left(\frac{{}^{n+1}\bar{\sigma}_{yy}}{{}^{n+1}Y}\right)^2 \right] = e_D^2 \tag{56}$$

where ${}^{n+1}Y = {}^{n+1}Y_t$ if $\sigma_{yy} \geq 0$ and ${}^{n+1}Y = {}^{n+1}Y_c$ if $\sigma_{yy} < 0$. D_a is a scaling parameter and failure occurs when $e_D \geq 1$. Note the similarity between Equations

55 and 56. In the latter, stresses are averaged over the thickness of the layer above the interface (layer n+1) and the strengths can be those of either the layer above or the layer below (n).

The quadratic delamination criterion of Brewer and Lagace (1988) is similar to Equations (55, 56) and can be written as

$$\left(\frac{\sigma_{33}}{Z}\right)^2+\left(\frac{\sigma_{13}}{S_{13}}\right)^2+\left(\frac{\sigma_{23}}{S_{23}}\right)^2=1 \tag{57}$$

where $Z=Z_t$ when $\sigma_{33}>0$ and $Z=Z_c$ when $\sigma_{33}<0$. Naik et al (2000, 2001) used Equation 57 and averaged the values of the stresses through the thickness of the ply. Li et al (2008) used the Brewer-Lagace criterion as defined when $\sigma_{33}>0$ and omitted the effect of the transverse normal stress when $\sigma_{33}<0$. In that case, we recover the criterion proposed by Yeh and Kim (2004) which predicts that tensile delaminations occurs when $\sigma_{33}>0$ and

$$\left(\frac{\sigma_{33}}{Z_T}\right)^2+\left(\frac{\sigma_{13}}{S_{13}}\right)^2+\left(\frac{\sigma_{23}}{S_{23}}\right)^2\geq 1 \tag{58}$$

and shear delaminations occurs when $\sigma_{33}<0$ and

$$\left(\frac{\sigma_{13}}{S_{13}}\right)^2+\left(\frac{\sigma_{23}}{S_{23}}\right)^2\geq 1 \tag{59}$$

Equation 59 is identical to Equation 54, the criterion proposed by Christensen and DeTeresa (2004). Huang and Lee (2003), Liu and Wang (2007) used Yeh's criterion (Equations 58, 59). Zhao and Cho (2007) used only the tensile part of the criterion (Equation 58). The modified Yeh criteria (2004) was used by Huang and Lee (2003) and Lee and Wang (2003). This criterion predicts a tensile delamination mode when

$$\left(\frac{\varepsilon_{33}}{Z_T^{\varepsilon}}\right)^2+\left(\frac{\varepsilon_{13}}{S_{13}^{\varepsilon}}\right)^2+\left(\frac{\varepsilon_{23}}{S_{23}^{\varepsilon}}\right)^2\geq 1 \qquad (\varepsilon_{33}>0) \tag{60}$$

and a shear delamination mode when

$$\left(\frac{\varepsilon_{13}}{S_{13}^{\varepsilon}}\right)^2+\left(\frac{\varepsilon_{23}}{S_{23}^{\varepsilon}}\right)^2\geq 1 \qquad (\varepsilon_{33}<0) \tag{61}$$

Chen (2004) included the effect of the in-plane transverse stress σ_{22}

$$\left(\frac{{}^{n}\sigma_{33}}{Z_C}\right)^2+D_1\left(\frac{{}^{n}\sigma_{23}^2+{}^{n+1}\sigma_{13}^2}{S^2}\right)+D_2\left(\frac{{}^{n+1}\sigma_{22}}{Y}\right)^2=e_D \tag{62}$$

Hou et al (2000, 2001) assumed that delamination occurs when

$$\left(\frac{\sigma_{33}}{Z_T}\right)^2+\frac{\sigma_{23}^2+\sigma_{13}^2}{S_{13}^2\left(d_{ms}\,d_{fs}+\delta\right)}=1 \qquad \text{when} \quad \sigma_{33}\geq 0 \tag{63}$$

or

$$\frac{\sigma_{23}^2+\sigma_{13}^2-8\sigma_{33}^2}{S_{13}^2\left(d_{ms}d_{fs}+\delta\right)}=1 \ \text{ when } \ -\sqrt{\left(\sigma_{23}^2+\sigma_{13}^2\right)/8}\leq\sigma_{33}<0 \tag{64}$$

They also assumed that no delamination occurs when

$$\sigma_{33}<-\sqrt{\left(\sigma_{23}^2+\sigma_{13}^2\right)/8} \tag{65}$$

In Eqs. (63, 64), d_{ms} is a damage coefficient of matrix cracking and d_{fs} is a damage coefficient of fibre failure and δ is the ratio between interlaminar stresses before and after matrix or fiber failure.

Zou et al (2002-a,b) proposed a single criterion that accounts for different strength for tension and compression in the transverse direction and for the effect of the two transverse shear stresses

$$\frac{\sigma_{33}^2}{Z_tZ_c}+\frac{\sigma_{13}^2+\sigma_{23}^2}{S^2}+\left(\frac{1}{Z_t}-\frac{1}{Z_c}\right)\sigma_{33}=1 \tag{66}$$

Fenske and Vizzini (2001) extended the Brewer-Lagace criterion by including the effects of inplane stresses.

These various criteria attempt to predict the onset of delamination at the interface between two adjacent plies in terms of the stresses acting at that interface. It is important to remember that the behavior is very different depending on the sign of the transverse normal stress.

4.4- Stress Invariants

Many yield or failure criteria are expressed in terms of invariants of the stress tensor, invariants of the deviatoric stress tensor, or in terms of mean stress and equivalent stress. These quantities are recalled here for future reference. The state of stress at one point is defined be the stress tensor $[\sigma]$ and its six components. The surface tractions acting on a surface oriented by a vector $\{n\}$ are given by $\{t\}=[\sigma]\{n\}$. The principal stress directions are such that only normal stresses are acting on the surfaces oriented in those directions. In other words, the vector $\{t\}$ acts in the normal direction or

$$\{t\}=\lambda\{n\} \qquad (67)$$

where the scalar quantity λ is called the principal stress. Equation 67 can be written as

$$[\sigma]\{n\}=\lambda\{n\} \quad \text{or} \quad ([\sigma]-\lambda[I])\{n\}=0 \qquad (68)$$

where [I] is the 3x3 identity matrix. Solving this eigenvalue problem (Equation 68) means finding values of the principal stresses λ and the corresponding vectors $\{n\}$ that satisfy Equation 68. Non-trivial solutions occur when $\left| [\sigma_{ij}]-\lambda[I] \right|=0$ which leads to the cubic algebraic equation

$$-\lambda^3+I_1\lambda^2+I_2\lambda+I_3=0 \qquad (69)$$

where

$$I_1=\sigma_{xx}+\sigma_{yy}+\sigma_{zz},$$

$$I_2=\sigma_{xy}^2+\sigma_{xz}^2+\sigma_{yz}^2-\sigma_{xx}\sigma_{yy}-\sigma_{yy}\sigma_{zz}-\sigma_{yy}\sigma_{zz} \qquad (70)$$

and

$$I_3=\sigma_{xx}\sigma_{yy}\sigma_{zz}+2\sigma_{xy}\sigma_{yz}\sigma_{zx}-\sigma_{zz}\sigma_{xy}^2-\sigma_{yy}\sigma_{xz}^2-\sigma_{xx}\sigma_{yz}^2$$

The state of stress at a point being independent of the coordinate system used to describe it, the principal stresses must be independent of the coordinate system. That means that the coefficients I_1-I_3 in Equation 69 must remain constant under coordinate transformation. For this reason, I_1-I_3 are called the invariants of the stress tensor. Solving Equation 69 gives the values of the three principal stresses and substituting into Equation 68, one can find the principal directions. These principal directions and principal stresses are important because they give the lowest and the highest normal stresses acting at that point and they are often used to define failure criteria. The mean stress or hydrostatic pressure is usually defined as

$$\sigma_m = I_1/3 = (\sigma_{xx} + \sigma_{yy} + \sigma_{zz})/3 \tag{71}$$

Both I_1 and σ_m are used in the development of many yield or failure criteria. The deviatoric stress tensor is defined as

$$[s_{ij}] = \begin{bmatrix} s_{11} & s_{12} & s_{13} \\ s_{21} & s_{22} & s_{23} \\ s_{31} & s_{32} & s_{33} \end{bmatrix} = \begin{bmatrix} (\sigma_{11} - \sigma_m) & \sigma_{12} & \sigma_{13} \\ \sigma_{21} & (\sigma_{22} - \sigma_m) & \sigma_{23} \\ \sigma_{31} & \sigma_{32} & (\sigma_{33} - \sigma_m) \end{bmatrix} \tag{72}$$

The eigenvalue problem $\left| [s_{ij}] - \lambda[I] \right| = 0$ leads to

$$-\lambda^3 + J_1\lambda^2 + J_2\lambda + J_3 = 0 \tag{73}$$

where

$$J_1 = s_{xx} + s_{yy} + s_{zz},$$

$$J_2 = s_{xy}^2 + s_{xz}^2 + s_{yz}^2 - s_{xx}s_{yy} - s_{yy}s_{zz} - s_{yy}s_{zz} \tag{74}$$

and

$$J_3 = s_{xx}s_{yy}s_{zz} + 2s_{xy}s_{yz}s_{zx} - s_{zz}s_{xy}^2 - s_{yy}s_{xz}^2 - s_{xx}s_{yz}^2$$

Using the definitions of s_{11}, s_{22} and s_{33} (Equation 72), it is easy to show that the first invariant of the deviatoric stress tensor is zero (J_1=0). J_2 can be rewritten as

$$J_2 = s_{xx}^2 + s_{yy}^2 + s_{zz}^2 + s_{xx}s_{yy} + s_{yy}s_{zz} + s_{yy}s_{zz} + s_{xy}^2 + s_{xz}^2 + s_{yz}^2 \tag{75}$$

or in tensor notation, $J_2 = s_{ij}s_{ij}$. In terms of stress components, the second invariant is written as

$$J_2 = \frac{1}{3}\left\{\sigma_{xx}^2 + \sigma_{yy}^2 + \sigma_{zz}^2 - \sigma_{xx}\sigma_{yy} - \sigma_{yy}\sigma_{zz} - \sigma_{zz}\sigma_{xx}\right\} + \sigma_{xy}^2 + \sigma_{yz}^2 + \sigma_{zx}^2 \tag{76}$$

Alternatively, the second invariant J_2 can be expressed as

$$J_2 = \frac{1}{6}\left\{(\sigma_{xx}-\sigma_{yy})^2+(\sigma_{yy}-\sigma_{zz})^2+(\sigma_{zz}-\sigma_{xx})^2\right\}+\sigma_{xy}^2+\sigma_{yz}^2+\sigma_{zx}^2 \tag{77}$$

or, in terms of principal stresses,

$$J_2 = \frac{1}{6}\left\{(\sigma_1-\sigma_2)^2+(\sigma_2-\sigma_3)^2+(\sigma_3-\sigma_1)^2\right\} \tag{78}$$

The equivalent stress, or von Mises stress, is defined as

$$\sigma_e = \sqrt{3J_2} = \sqrt{\frac{1}{2}\left[(\sigma_1-\sigma_2)^2+(\sigma_2-\sigma_3)^2+(\sigma_3-\sigma_1)^2\right]} \tag{79}$$

and the octahedral shear stress is defined as

$$\tau_{oct} = \sqrt{\frac{2}{3}J_2} = \frac{1}{3}\sqrt{(\sigma_1-\sigma_2)^2+(\sigma_2-\sigma_3)^2+(\sigma_3-\sigma_1)^2} \tag{80}$$

The second invariant of the deviatoric stress tensor J_2, the equivalent stress σ_e and the octahedral stress τ_{oct} are used in the development of many yield or failure criteria.

4.5- Strain Energy Density for Isotropic Solids

This section recalls that, for isotropic solids, the strain energy density depends exclusively on I_1, the first invariant of the stress tensor and J_2, the second invariant of the deviatoric stress tensor. For isotropic solids, the stress-strain relations (Equations 24) can be used to write the strain energy density $U = \frac{1}{2}\sigma_{ij}e_{ij}$ as

$$U = \frac{J_2}{G} + \frac{1-2\nu}{6E}I_i^2 \qquad \text{or} \qquad U = \frac{1}{2E}\left[\sigma_e^2 + \beta^2\sigma_m^2\right] \tag{81}$$

where $\overline{E} = \frac{3E}{2(1+\nu)}$ and $\beta^2 = \frac{9(1-2\nu)}{2(1+\nu)}$. The first term in Equation (81) represents the distortional strain energy and the second term is the strain energy corresponding to the hydrostatic loading. This result explains why, as we shall see, many yield criteria for isotropic materials are written in terms the two invariants I_1 and I_2 or, equivalently, in terms of the equivalent stress and the mean stress σ_e and the mean stress σ_m.

4.6- Von Mises Criterion

The fact that the strain energy density depends on I_1 and J_2 lead to the development of a number of failure or yield criteria based on stress invariants. Experiments conducted by Bridgman (1947) showed that, for metals, yielding is independent of the hydrostatic pressure. Therefore, a number of criteria written in terms of stress invariants are independent of I_1. The von Mises yield criterion in terms of the second invariant of the deviatoric stress tensor is

$$J_2 = \frac{1}{3}\sigma_{YT}^2 = k^2 \tag{82}$$

where σ_{YT} is the yield strength in tension and k is the yield in pure shear. Therefore, yielding occurs when J_2 reaches a critical value. Recalling Equation 79, the definition of the equivalent stress σ_e, the Von Mises criterion states that yielding occurs when σ_e reaches the value of the yield strength in tension. With this criterion the strength in tension and compression are equal. For the biaxial loading case, in terms of the principal stresses,

$$\sigma_1^2 - \sigma_1\sigma_2 + \sigma_2^2 = \sigma_{YT}^2 \tag{83}$$

The yield curve in the $\sigma_1 - \sigma_2$ plane is an ellipse centered at the origin with its major axis orientted at 45° from the σ_2 axis and the lengths of the semi-axes are $a = \sigma_{YT}\sqrt{2}$ and $b = \sigma_{YT}\sqrt{2/3}$. In the $\sigma_m - \sigma_e$ plane, the Von Mises criterion represented by the horizontal lines $\sigma_e = \pm\sigma_{YT}/3$. The Von Mises criterion can also be written in terms of principal stresses as

$$\left\{\left[(\sigma_1 - \sigma_2)^2 + (\sigma_2 - \sigma_3)^2 + (\sigma_3 - \sigma_1)^2\right]/2\right\}^{1/2} = \sigma_Y \tag{84}$$

or, in terms of the regular stress components, as

$$\frac{1}{6}\left\{ \left(\sigma_{xx}-\sigma_{yy}\right)^2+\left(\sigma_{yy}-\sigma_{zz}\right)^2+\left(\sigma_{zz}-\sigma_{xx}\right)^2 \right\}+\sigma_{xy}^2+\sigma_{yz}^2+\sigma_{zx}^2=\frac{\sigma_{YT}^2}{3} \quad (85)$$

These forms of von Mises' criterion (Eqs. 84, 85) are given for comparison with criteria used for anisotropic materials.

4.7- Hill's anisotropic criterion and extensions

In 1948, Hill proposed a criterion capable of accounting for anisotropic behavior that can be seen as a generalization of the von Mises criterion to account for the effect of orthotropy. This criterion was developed for metals such as aluminum that were subjected to rolling operations and became orthotropic. We will show that it is inadequate for modeling foams and other materials for which the effects of hydrostatic pressure are significant and that have different behavior in tension and compression. An extension to this criterion that addresses these deficiencies is discussed next. The well-known Hill criterion (1948, 1950) is written as

$$H(\sigma_{11}-\sigma_{22})^2+F(\sigma_{22}-\sigma_{33})^2+G(\sigma_{33}-\sigma_{22})^2+2L\sigma_{23}^2+2M\sigma_{31}^2+2N\sigma_{12}^2=1 \quad (86)$$

The similarity with von Mises' criterion is obvious when comparing with Equation 85. The coefficients H-N are introduced to account for the orthotropy of the material. If σ_{1T}, σ_{2T} and σ_{3T} are the yield strengths in the principal directions of anisotropy, the strength parameters in Equation 86 are

$$F=\frac{1}{2}\left\{\frac{1}{\sigma_{2T}^2}+\frac{1}{\sigma_{3T}^2}-\frac{1}{\sigma_{1T}^2}\right\},\quad G=\frac{1}{2}\left\{\frac{1}{\sigma_{3T}^2}+\frac{1}{\sigma_{1T}^2}-\frac{1}{\sigma_{2T}^2}\right\} \quad (87)$$

$$H=\frac{1}{2}\left\{\frac{1}{\sigma_{1T}^2}+\frac{1}{\sigma_{2T}^2}-\frac{1}{\sigma_{3T}^2}\right\},\quad L=\frac{1}{2\left(\tau_{23}^S\right)^2},\quad M=\frac{1}{2\left(\tau_{31}^S\right)^2},\quad N=\frac{1}{2\left(\tau_{12}^S\right)^2}$$

where τ_{12}^S, τ_{23}^S, and τ_{31}^S are the yield stresses with respect to the axes of anisotropy. Hill's anisotropic criterion (Equation 86) does not include the effect of hydrostatic pressure. This can easily be seen by adding a pressure p to σ_{11}, σ_{22} and σ_{33} and substituting into Equation (86) where the term p will then drop out thus proving that point. The strength parameters F, G and H are based on

tensile strengths only and Equation 87 predicts equal strengths in tension and compression. Therefore, while capable of accounting for orthotropy, Hill's criterion (Equation 86) is inadequate for modeling foams because it does not address the other two complicating factors discussed here.
The von Mises criterion is inspired by the form of the strain energy density of isotropic solids. Similarly, Hill's criterion bears a similarity with the strain energy density in orthotropic solids. With stress strain relations given by Equations 24, the strain energy density can be written as

$$2U = \frac{\sigma_{11}^2}{E_1} + \frac{\sigma_{22}^2}{E_2} + \frac{\sigma_{33}^2}{E_3} - 2\frac{\nu_{12}}{E_1}\sigma_{11}\sigma_{22} - 2\frac{\nu_{13}}{E_1}\sigma_{11}\sigma_{33} - 2\frac{\nu_{23}}{E_2}\sigma_{22}\sigma_{33} + \frac{2}{G_{23}}\sigma_{23}^2 + \frac{2}{G_{31}}\sigma_{31}^2 + \frac{2}{G_{12}}\sigma_{12}^2 \quad (88)$$

The similarity between the strain energy density (Equation 88) and Hill's criterion (Equation 86) is obvious. As we shall see later, a number of authors use this analogy between the strain energy density, written as $U = \frac{1}{2}\{\sigma\}^T[S]\{\sigma\}$, to define an equivalent stress of the form $\sigma_e = \{\sigma\}^T[P]\{\sigma\}$. [P] being a matrix of the same form as the compliance matrix [S].

Caddell, Raghava and Atkins (1973) proposed a modified Hill criterion for polymers where

$$H(\sigma_{11} - \sigma_{22})^2 + F(\sigma_{22} - \sigma_{33})^2 + G(\sigma_{33} - \sigma_{11})^2 + 2L\sigma_{23}^2 + 2M\sigma_{31}^2 + 2N\sigma_{12}^2 + K_1\sigma_{11} + K_2\sigma_{22} + K_3\sigma_{33} = 1 \quad (89)$$

The last three terms on the left hand side of Equation (89) have been added in order to account for differences in yield strengths in tension and compression. Both Raghava and Caddell (1974) and Caddell and Kim (1981) used Equation 89 to model the yield strength of anisotropic polycarbonate.

4.8- Development of Quadratic Criteria

To better understand the similarities between the quadratic failure criteria that have been proposed by various authors and that are widely used, the following presents a step by step development of such a criterion. In a case where the loading consists of two stress components σ_1 and σ_2, for example, we start with the simple expression

$$A\sigma_1^2 + B\sigma_2^2 = 1 \tag{90}$$

which is the equation for an ellipse in the $\sigma_1 - \sigma_2$ plane with major axes in the σ_1 and σ_2 directions. With this expression the strength in tension and compression are the same and $A = 1/X^2$ and $B = 1/Y^2$ where X and Y are the strengths in directions 1 and 2. Often the failure locus is an ellipse with major axes at an angle from the σ_1 and σ_2 directions. Adding an additional term to Equation 90, the criterion

$$A\sigma_1^2 + B\sigma_2^2 + C\sigma_1\sigma_2 = 1 \tag{91}$$

represents an ellipse with a major axis at an angle θ from the σ_1 axis. That angle is given by

$$\tan 2\theta = \frac{C}{A - B} \tag{92}$$

When A=B, $\theta = 45°$ and the ratio C/A controls the ratio of the major and minor axes. For uniaxial tests in direction 1, Equation 91 predicts the same strength X in tension and compression. Similarly, in direction 2, the strength Y is the same in tension and compression. Then, $A=1/X^2$ and $B=1/Y^2$. With composite materials, tensile and compressive strength are very different. Therefore, the criterion has to allow for cases where X_t, the strength in tension, is different from X_c, the strength in compression. Adding linear terms to the previous equation gives

$$A\sigma_1^2 + B\sigma_2^2 + C\sigma_1\sigma_2 + D\sigma_1 + E\sigma_2 = 1 \tag{93}$$

Under uniaxial tension and compression in direction 1, this criterion gives

$$AX_t^2 + DX_t = 1 \quad \text{and} \quad AX_c^2 - DX_c = 1 \tag{94}$$

respectively. Solving these two equations, we find the constants A and D

$$A = \frac{1}{X_t X_c}, \qquad D = \frac{1}{X_t} - \frac{1}{X_c} \tag{95}$$

Similarly, from the uniaxial strengths in direction 2, we find the constants B and E

$$B = \frac{1}{Y_t Y_c}, \qquad E = \frac{1}{X_t} - \frac{1}{X_c} \tag{96}$$

The constant C must be determined from biaxial tests. When introducing a shear stress component σ_{12}, simply add a σ_{12}^2 term. No σ_{12} term is required since the shear strength S is the same for positive and negative shear stresses. Then, the criterion becomes

$$\frac{\sigma_1^2}{X_t X_c} + \frac{\sigma_2^2}{Y_t Y_c} + C\sigma_1\sigma_2 + \left(\frac{1}{X_t} - \frac{1}{X_c}\right)\sigma_1 + \left(\frac{1}{Y_t} - \frac{1}{Y_c}\right)\sigma_2 + \frac{\sigma_{12}^2}{S^2} = 1 \tag{97}$$

Extending this approach to three dimensions, we can add σ_3^2, $\sigma_1\sigma_3$, $\sigma_2\sigma_3$ and σ_3 terms to account for the third normal stress components and its interaction with the first two. σ_{13}^2 and σ_{23}^2 must also be added to account for the additional shear stress components. This leads to the three- dimensional Tsai-Wu criterion discussed below.

4.9- Three Dimensional Tsai-Wu Failure Criterion

The approach discussed in the previous section can be generalized to three dimensional loading of a unidirectional composite. In that case, we have six stress components: three normal stresses σ_{11}, σ_{22}, and σ_{33}, and three shear stresses σ_{23}, σ_{13}, and σ_{12}. For the normal stresses, the criterion should include three linear terms $(\sigma_{11}, \sigma_{22}, \text{and } \sigma_{33})$, three square terms $(\sigma_{11}^2, \sigma_{22}^2, \text{and } \sigma_{33}^2)$, and three interaction terms $(\sigma_{11}\sigma_{22}, \sigma_{11}\sigma_{33}, \text{and } \sigma_{22}\sigma_{33})$. For the shear stresses, only the square terms are needed. Then, we obtain what is called the Tsai-Wu failure criterion

$$\begin{aligned} &F_1\sigma_{11} + F_2\sigma_{22} + F_3\sigma_{33} + F_{11}\,\sigma_{11}^2 + F_{22}\,\sigma_{22}^2 + F_{33}\,\sigma_{33}^2 + F_{44}\,\sigma_{23}^2 + F_{55}\,\sigma_{13}^2 \\ &+ F_{66}\,\sigma_{12}^2 + 2F_{12}\,\sigma_{11}\sigma_{22} + 2F_{13}\,\sigma_{11}\sigma_{33} + 2F_{23}\,\sigma_{22}\sigma_{33} = 1 \end{aligned} \tag{98}$$

that contains 12 coefficients. Nine of the coefficients can be determined from unidirectional tests

$$F_1 = \frac{1}{X_t} - \frac{1}{X_c}, \quad F_2 = \frac{1}{Y_t} - \frac{1}{Y_c}, \quad F_3 = \frac{1}{Z_t} - \frac{1}{Z_c},$$

$$F_{11} = \frac{1}{X_t X_c}, \quad F_{22} = \frac{1}{Y_t Y_c}, \quad F_{33} = \frac{1}{Z_t Z_c}, \tag{99}$$

$$F_{44} = \frac{1}{S_{23}^2}, \quad F_{55} = \frac{1}{S_{13}^2}, \quad F_{44} = \frac{1}{S_{12}^2}$$

The three coefficients F_{12}, F_{13}, and F_{23} can have a significant effect on the shape of the failure surface and they are best determined through biaxial tests that are difficult to perform. When this information is not available, one can use the fact the three terms $\frac{1}{2}F_{11}\sigma_{11}^2 + \frac{1}{2}F_{22}\sigma_{22}^2 + 2F_{12}\sigma_{11}\sigma_{22}$ form the perfect square $\frac{1}{2}\left(\sqrt{F_{11}}\,\sigma_{11} - \sqrt{F_{22}}\,\sigma_{22}\right)^2$ if

$$F_{12} = -\frac{1}{2}\sqrt{F_{11}F_{22}} \tag{100}$$

In addition, if we also take

$$F_{13} = -\frac{1}{2}\sqrt{F_{11}F_{33}} \quad \text{and} \quad F_{23} = -\frac{1}{2}\sqrt{F_{33}F_{22}} \tag{101,102}$$

then, the criterion can be written as

$$\begin{gathered}\frac{1}{2}\left(\sqrt{F_{11}}\,\sigma_{11} - \sqrt{F_{22}}\,\sigma_{22}\right)^2 + \frac{1}{2}\left(\sqrt{F_{33}}\,\sigma_{33} - \sqrt{F_{22}}\,\sigma_{22}\right)^2 + \frac{1}{2}\left(\sqrt{F_{11}}\,\sigma_{11} - \sqrt{F_{33}}\,\sigma_{33}\right)^2 \\ + F_1\sigma_{11} + F_2\sigma_{22} + F_3\sigma_{33} + F_{44}\sigma_{23}^2 + F_{55}\sigma_{13}^2 + F_{66}\sigma_{12}^2 = 1\end{gathered} \tag{103}$$

a form similar to the criterion proposed by Caddell et al (1973) (Equation 89). Combining Equations (99-103) gives expressions that are used by many authors

$$F_{12} = -\frac{1}{2\sqrt{X_t X_c Y_t Y_c}}, \quad F_{23} = -\frac{1}{2\sqrt{Y_t Y_c Z_t Z_c}}, \quad F_{13} = -\frac{1}{2\sqrt{X_t X_c Z_t Z_c}} \tag{104}$$

In the case of a transversely isotropic material where the 23 plane is the plane of symmetry, the number of constants are reduced (1971) since F_2=F_3, F_{22}=F_{33},

F_{55}=F_{66}, F_{12}=F_{13}, and $F_{44}=2(F_{22}-F_{23})$. DeTeresa and Larsen (2003) also suggested that $F_{12}=-F_{11}/4$ and $F_{23}=-F_{22}$. Wu and Stachurski (1984) proposed to calculate F_{12} using

$$F_{12}=-\frac{1}{X_tX_c+Y_tY_c} \tag{105}$$

These last references [53-55] (Tsai and Wu (1971), DeTeresa and Larsen (2003) and Wu and Stachurski (1984)) indicate that the form of the three stress interaction factors is not well established and that authors make different assumptions in the absence of direct experimental data.

4.10- Hoffman's Criterion

Hoffman (1967) extended Hill's criterion (Equation 86) by adding linear functions of the three normal stress components as in Caddell et al (1973, 1981) and Raghawa and Caddell (1974) (see Equation 89). With his notation the failure criterion is written as

$$\begin{aligned}&C_1(\sigma_{22}-\sigma_{33})^2+C_2(\sigma_{33}-\sigma_{11})^2+C_3(\sigma_{11}-\sigma_{22})^2+C_4\sigma_{11}+C_5\sigma_{22}+C_6\sigma_{33}\\&+C_7\sigma_{23}^2+C_8\sigma_{31}^2+C_9\sigma_{12}^2=1\end{aligned} \tag{106}$$

with nine parameters

$$C_1=\frac{1}{2}\left(\frac{1}{Y_tY_c}+\frac{1}{Z_tZ_c}-\frac{1}{X_tX_c}\right),\quad C_2=\frac{1}{2}\left(\frac{1}{Z_tZ_c}+\frac{1}{X_tX_c}-\frac{1}{Y_tY_c}\right),$$

$$C_3=\frac{1}{2}\left(\frac{1}{X_tX_c}+\frac{1}{Y_tY_c}-\frac{1}{Z_tZ_c}\right),\ C_4=\frac{1}{X_t}-\frac{1}{X_c},\ C_5=\frac{1}{Y_t}-\frac{1}{Y_c}, \tag{107}$$

$$C_6=\frac{1}{Z_t}-\frac{1}{Z_c},\qquad C_7=1/S_{23}^2,\ C_8=1/S_{13}^2,\ C_9=1/S_{12}^2$$

Hoffman's criterion can be written in the same form as the Tsai-Wu criterion (Equation 98) with the first nine parameters as in Equation 99. The normal stress interaction parameters are

$$F_{12} = -\frac{1}{2}\left(\frac{1}{X_t X_c} + \frac{1}{Y_t Y_c} - \frac{1}{Z_t Z_c}\right), \quad F_{13} = -\frac{1}{2}\left(\frac{1}{Z_t Z_c} + \frac{1}{X_t X_c} - \frac{1}{Y_t Y_c}\right),$$

$$F_{13} = -\frac{1}{2}\left(\frac{1}{Y_t Y_c} + \frac{1}{Z_t Z_c} - \frac{1}{X_t X_c}\right) \tag{108}$$

These interaction parameters are different from those in Eqs. 101-103. This example shows that criteria expressed in different forms and often developed in different ways can be quite similar. A number of authors use Eqs. 108 to calculate the three interaction parameters.

4.11- Three-Dimensional Tsai-Hill Criterion

In the Tsai-Hill criterion, the tensile and compressive strengths are assumed to be the same. Therefore, the σ_{11}, σ_{22}, and σ_{33} terms are no longer needed

$$\begin{aligned} &F_{11}\sigma_{11}^2 + F_{22}\sigma_{22}^2 + F_{33}\sigma_{33}^2 + F_{44}\sigma_{23}^2 + F_{55}\sigma_{13}^2 + F_{66}\sigma_{12}^2 + 2F_{12}\sigma_{11}\sigma_{22} \\ &+ 2F_{13}\sigma_{11}\sigma_{33} + 2F_{23}\sigma_{22}\sigma_{33} = 1 \end{aligned} \tag{109}$$

with

$$F_{11} = \frac{1}{X^2}, \quad F_{22} = \frac{1}{Y^2}, \quad F_{33} = \frac{1}{Z^2}, \quad F_{44} = \frac{1}{S_{23}^2}, \quad F_{55} = \frac{1}{S_{13}^2}, \quad F_{44} = \frac{1}{S_{12}^2} \tag{110}$$

A 2D version of the Tsai-Hill criterion can be found in several texts (e.g. Kedwards et al (1990), Kaw (2005))

$$\left(\frac{\sigma_{11}}{X_t}\right)^2 + \left(\frac{\sigma_{22}}{Y_t}\right)^2 + \left(\frac{\sigma_{12}}{S_{12}}\right)^2 - \frac{\sigma_{11}\sigma_{22}}{X_t^2} = 1 \tag{111}$$

while Naik et al (2000) used

$$\left(\frac{\sigma_{11}}{X_t}\right)^2 + \left(\frac{\sigma_{22}}{Y_t}\right)^2 + \left(\frac{\sigma_{12}}{S_{12}}\right)^2 - \left(\frac{\sigma_{11}}{X_t}\right)\left(\frac{\sigma_{22}}{Y_t}\right) = 1 \tag{112}$$

This illustrates the fact that the strength parameters accounting for the normal stress interactions are not very well defined. For the three dimensional case, Singh and Kumar (1998, 1999) and Ochoa and Reddy (1992) used

$$F_{12} = -\frac{1}{2}\left(\frac{1}{X^2}+\frac{1}{Y^2}-\frac{1}{Z^2}\right), \qquad F_{13} = -\frac{1}{2}\left(\frac{1}{Z^2}+\frac{1}{X^2}-\frac{1}{Y^2}\right), \qquad (113)$$

$$F_{23} = -\frac{1}{2}\left(\frac{1}{Y^2}+\frac{1}{Z^2}-\frac{1}{X^2}\right)$$

Eqs. 113 can be obtained from Eqs. 108 when compressive and tensile strengths are identical.

5- Criteria for Delamination Propagation

It is generally accepted that the onset of debonding between adjacent plies can be predicted using one of the criteria discussed in Section 4.3. Using a finite element approach, many researchers track the progression of the delamination front as predicted by the failure criterion as the loading increases, taking into account the debonding of the interface. This approach does not accurately account for the stress singularity at the delamination front. To account for the presence of the delamination crack, two approaches are used: the fracture mechanics approach and the cohesive element approach. In the following we discuss mixed-mode fracture criteria used for this type of problems and provide an overview of the cohesive element approach.

5.1- Fracture Mechanics Approach

Failure criteria described in Section 4.3 can be used to predict the onset of delamination. Once initiated, the propagation of a delamination is a dynamic fracture event. Li et al (2002) used a Griffith type criteria to determine whether an existing delamination extends or not based on the value of the total strain energy release rate: the delamination does not extend if $G \le G_c$ and it extends if $G > G_c$. This approach requires a detailed stress analysis near the crack tip and the ability to calculate the change in strain energy as the delamination increases by a small amount. However it does not require knowledge of the individual strain energy release rates G_I, G_{II}, G_{III} (Rinderknecht and Kroplin (1997)).

When the stress analysis provides the individual strain energy release rates, a criterion for mixed-mode fracture can be used. Camanho et al (2003), Jiang et al (2007), and Johnson et al (2001, 2003, 2006) used a power law criterion

$$\left(\frac{G_I}{G_{IC}}\right)^{\alpha}+\left(\frac{G_{II}}{G_{IIC}}\right)^{\alpha}=1 \qquad (114)$$

first introduced by Wu and Reuter (1965). Szekrényes (2007) used a generalized the power law criterion by introducing two independent parameters α_1 and α_2

$$\left(\frac{G_I}{G_{IC}}\right)^{\alpha_1}+\left(\frac{G_{II}}{G_{IIC}}\right)^{\alpha_2}=1 \tag{115}$$

and what is called Williams' criterion

$$\left(\frac{G_I}{G_{IC}}-1\right)\left(\frac{G_{II}}{G_{IIC}}-1\right)-I_i\left(\frac{G_I}{G_{IIC}}\right)\left(\frac{G_{II}}{G_{IIC}}\right)=0 \tag{116}$$

where I_i is an interaction parameter to be determined from experiments. Curve fits of experimental results on a particular delaminated beam specimen by these two expressions show very little differences.

The criterion proposed by Benzeggagh and Kenane (1996) and Pereira et al (2004)

$$G_T=G_I+(G_{II}-G_I)\left(\frac{G_{II}}{G_T}\right)^{\alpha} \tag{117}$$

where $G_T=G_I+G_{II}$ and α is a parameter was used by Camanho et al (2003), Ducept et al. (1997). The bilinear failure criterion proposed by Reeder (1993),

$$G_I\geq\xi G_{II}+G_{IC}\ \text{ for }\ \frac{G_I}{G_{II}}\geq 1.0\quad\text{or}\quad G_I\geq\xi(G_{II}-G_{IIC})\ \text{ for }\ \frac{G_I}{G_{II}}\geq 1.0 \tag{118}$$

was used by Davidson (2007). Reeder (1992) and Iannucci (2006) used

$$\left(\frac{G_I}{G_{IC}}\right)^{\lambda}+\left(\frac{G_{II}}{G_{IIC}}\right)^{\lambda}+\left(\frac{G_{III}}{G_{IIIC}}\right)^{\lambda}=1 \tag{119}$$

Lammerant and Verpoest (1996) used Equation 110 with $\lambda=1$. Elder et al (2004) and Zou (2002-a,b) used

$$\left(\frac{G_I}{G_{IC}}\right)^{\alpha_I}+\left(\frac{G_{II}}{G_{IIC}}\right)^{\alpha_{II}}+\left(\frac{G_{II}}{G_{IIC}}\right)^{\alpha_{III}}=1 \tag{121}$$

This brief overview indicates that the development of adequate fracture criteria is still an area of active research. The use of these criteria requires the accurate determination of the critical energy release rates from experiments and an accurate calculation of the individual strain energy release rates as the delamination grows.

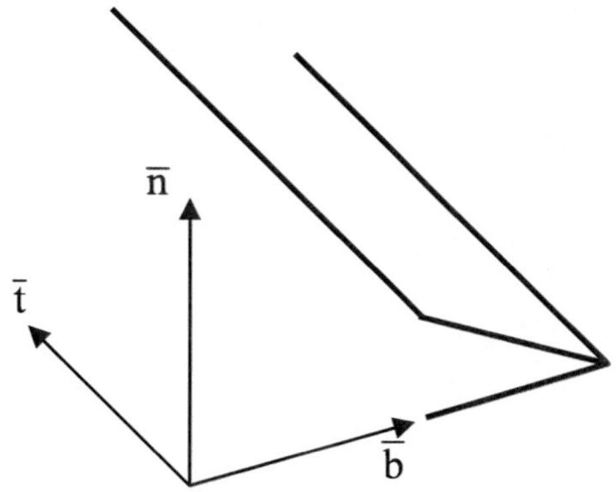

Figure 10. Coordinate system at the interface between adjacent plies.

5.2- Cohesive Element Approach

The general approach consisting of using volume elements to model the bulk of the material and a cohesive model to simulate the spontaneous creation of failure surfaces has been called the CVFE (Cohesive Volume Finite Element) scheme by Geubelle and Baylor (1998). In that reference, the basic idea of the CVFE scheme is traced back to the pioneering work of Dugdale and Barrenblatt in early 1960s. In this approach, the surface tractions and the relative displacements at an interface are referred to a btn coordinate system in which n is the normal direction, b is perpendicular to the crack front, and t is parallel to the crack front (Figure 10). Surface tractions T_n, T_b and T_t and the relative displacements Δ_n, Δ_b, and Δ_t correspond to fracture modes I, II, and III respectively. The interface can be thought of as a layer of zero thickness connecting the nodes from the elements located above and below it.

In general, the relationship between the surface tractions and the relative displacements at the interface are taken as shown in Figure 11 for the normal component. Initially, the cohesive traction T_n increases linearly with the relative displacement until a maximum value T_{max} is reached. As the two layers separate further, damage is introduced and the cohesive traction decreases linearly. If unloading occurs, unloading occurs along a line like the dotted line in Figure 11. Once the load exceeds T_{max} and the relative displacement exceeds the corresponding value Δ_o, the interface is damaged and the stiffness is degraded until the critical value of the relative displacement Δ_{nc} is reached and complete delamination occurs. Δ_{nc} is determined by requiring that the area under the decreasing portion of the curve be equal to the critical energy release rate under Mode I. The same approach is followed for the three surface tractions at the interface.

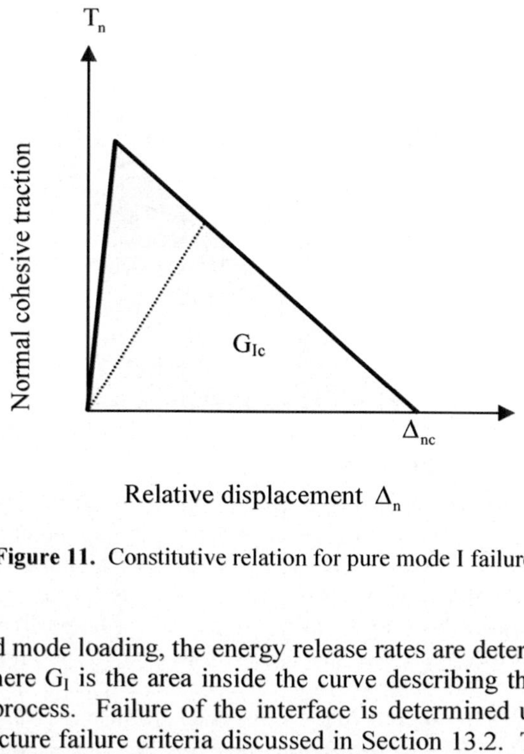

Figure 11. Constitutive relation for pure mode I failure

Under mixed mode loading, the energy release rates are determined as shown in Figure 12 where G_I is the area inside the curve describing the entire loading and unloading process. Failure of the interface is determined using one of the mixed mode fracture failure criteria discussed in Section 13.2. This approach is discussed in the review article by Elder et al (2004) and is used in several articles including by Borg et al (2000, 2002, 2004), Guedes (2008), Suemasu (2008), Aymerich et al (2008), and Yoshimura et al (2008). Nishikawa et al (2007) proposed a different model in which the cohesive law is as shown in Figure 13 and the damage criterion is in terms of the relative displacements

$$\left(\frac{\Delta_{I}}{\Delta_{Ic}}\right)^{2}+\left(\frac{\Delta_{II}}{\Delta_{IIc}}\right)^{2}+\left(\frac{\Delta_{III}}{\Delta_{IIIc}}\right)^{2}=1 \qquad (121)$$

In this model, damage is introduced while the surface traction remains constant at its maximum value.

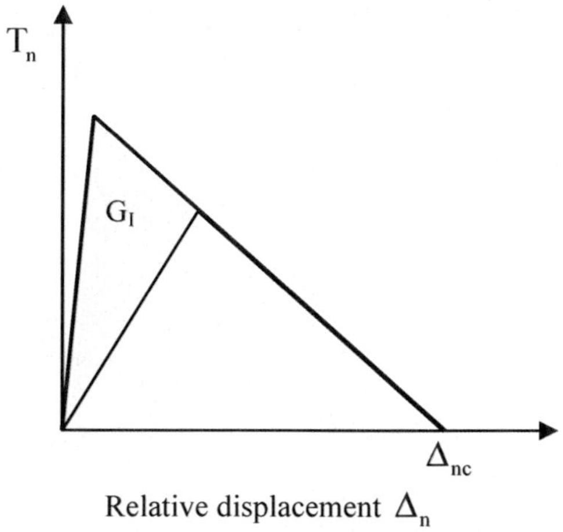

Figure 12. Calculation on the strain energy release rate

6- Summary

This chapter presented a brief introduction to the concepts of stress and strains. Basic stress-strain relations for isotropic and orthotropic materials have been introduced and the failure criteria used in most analysis of composite structures have been presented in details. These criteria are used to predict the onset of failure inside a layer or the debonding between two layers. Special criteria for predicting the extension of existing delaminations are also discussed. It must be noted that, in general, failure predicted by one of the criteria described in Section 4 does not mean that this particular ply has completely failed and cannot carry anymore load. In a progressive failure analysis some scheme for property degradation must be used. A comprehensive review of that area by Garnish and Akula (2009) provides detailed descriptions of the various approaches available.

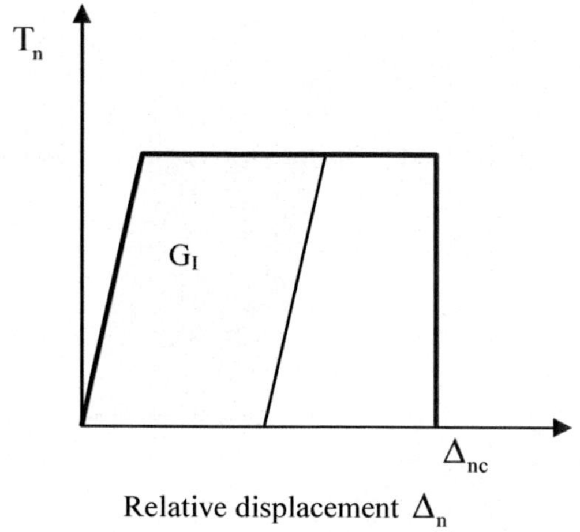

Figure 13. Constitutive relation for the interface used by Nishikawa et al [89]

References

Aymerich F., Dore F. and Priolo P. (2008). Prediction of impact-induced delamination in cross-ply composite laminates using cohesive interface elements. *Composites Science and Technology* 68 (12): 2383-2390.

Benzeggagh M.L., Khellil K. and Chotard T. (1995). Experimental determination of Tsai failure tensorial terms F_{ij} for unidirectional composite materials. *Composites Science and Technology* 55 (2): 145-156.

Benzeggagh M.L. and Kenane M. (1996). Measurement of mixed-mode delamination fracture toughness of unidirectional glass/epoxy composites with mixed-mode bending apparatus. *Composites Science and Technology* 56 (4) : 439-449.

Borg R., Nilsson L. and Simonsson K. (2004). Simulation of low velocity impact on fiber laminates using a cohesive zone based delamination model. *Composites Science and Technology*, 64, 2, 279-288.

Borg R., Nilsson L. and Simonsson K. (2002). Modeling delamination using discretized cohesive zone and damage formulation. *Composites Science and Technology* 62: 1299–314.

Borg R., Nilsson L. and Simonsson K. (2000). Simulation in fibre composites with a discrete cohesive failure model. *Composites Science and Technology* 61: 667–77.

Brewer J.C. and Lagace P.A. (1988). Quadratic stress criterion for initiation of delamination. *Journal of Composite Materials* 22(12): 1141-1155.

Bridgman P.W. (1947). The effect of hydrostatic pressure on the fracture of brittle substances. *Journal of Applied Physics* 18(2): 246-258.
Caddell R.M. and Kim J.W. (1981). Influence of hydrostatic pressure on the yield strength of anisotropic polycarbonate. *International Journal of Mechanical Sciences* 23(2), 99-104.
Caddell R.M., Raghava R.S. and Atkins A.G. (1973). Yield criterion for anisotropic and pressure dependent solids such as oriented polymers. *Journal of Materials Science* 8 (11): 1641-1646.
Camanho P. P., Davila C. G., and de Moura M. F. (2003). Numerical Simulation of Mixed-Mode Progressive Delamination in Composite Materials. *Journal of Composite Materials* 37: 1415-1438.
Cesari F., Dal Re V., Minak G. and Zucchelli A. (2007). Damage and residual strength of laminated carbon–epoxy composite circular plates loaded at the centre. *Composites Part A: Applied Science and Manufacturing* 38 (4): 1163-1173
Chan S., Fawaz Z., Behdinan K. and Amid R. (2007). Ballistic limit prediction using a numerical model with progressive damage capability. *Composite Structures* 77 (4): 466-474.
Chen G., Li Z., Kou C., and Gui L. (2004). Finite Element Analysis of Low-Velocity Impact Damage of Stitched Laminates. *Journal of Reinforced Plastics and Composites* 23 (9): 987-995.
Choi H.Y. and Chang F.K. (1992). A model for predicting damage in graphite/ epoxy laminated composites resulting from low-velocity point impact. *Journal of Composite Materials* 26: 2134–69.
Christensen R.M. and DeTeresa S.J. Delamination failure investigation for out-of-plane loading in laminates. *Journal of Composite Materials* 38 (24): 2231-2238, 2004
Davidson B.D. and Zhao W. (2007). An Accurate Mixed-Mode Delamination Failure Criterion for Laminated Fibrous Composites Requiring Limited Experimental Input. *Journal of Composite Materials* 41(6): 679-702.
DeTeresa S.J. and Larsen G.J. (2003). Reduction in the Number of Independent Parameters for the Tsai-Wu Tensor Polynomial Theory of Strength for Composite Materials. *Journal of Composite Materials* 37: 1769-1785.
Ducept F., Davies P., Gamby D. (1997). An experimental study to validate tests used to determine mixed mode failure criteria of glass/epoxy composites. *Composites Part A: Applied Science and Manufacturing* 28 (8): 719-729.
Elder D.J., Thomson R.S., Nguyen M.Q. and Scott M.L. (2004). Review of delamination predictive methods for low speed impact of composite laminates. *Composite Structures* 66 (1-4): 677-683.
Fenske M.T. and Vizzini A.J. (2001). The Inclusion of In-Plane Stresses in Delamination Criteria. *Journal of Composite Materials* 35 (15): 1325 – 1342.

Foo CC, Chai GB and Seah LK. (2008). A model to predict low-velocity impact response and damage in sandwich composites. *Composites Science and Technology* 68 (6): 1348-1356, 2008

Fuoss E., Straznicky P.V. and Poon C. (1998). Effects of stacking sequence on the impact resistance in composite laminates - Part 1: parametric study. *Composite Structures* 41 (1): 67-77.

Garnich M.R. and Akula V.M.K. (2009). Review of degradation models for progressive failure analysis of fiber reinforced polymer composites. *Applied Mechanics Reviews* 62 (1): 010801.

Geubelle P.H. and Baylor J.S. (1998). Impact-induced delamination of composites: a 2D simulation. *Composites Part B: Engineering* 29 (5): 589–602.

Gómez-del Río T., Zaera R., Barbero E. and Navarro C. (2005). Damage in CFRPs due to low velocity impact at low temperature. *Composites Part B: Engineering* 36(1): 41-50.

Green E.R., Morrison C.J., and Luo R.K. (2000). Simulation and experimental investigation of impact damage in composite plates with holes. *Journal of Composite Materials* 34 (6): 502-521.

Guedes R.M., de Moura M.F.S.F. and Ferreira F.J. (2008). Failure analysis of quasi-isotropic CFRP laminates under high strain rate compression loading. *Composite Structures* 84 (4): 362-368.

Hashin Z. (1980). Failure criteria for unidirectional fibre composites. *Journal of Applied Mechanics* 47: 329–334.

Hatami-Marbini H. and Pietruszczak S. (2007). On inception of cracking in composite materials with brittle matrix. *Computers and Structures* 85 (15-16): 1177-1184.

Her S.C. and Liang Y.C. (2004). The finite element analysis of composite laminates and shell structures subjected to low velocity impact. *Composite Structures* 66 (1-4), 277-285.

Hill R. (1950). *The Mathematical Theory of Plasticity*. Oxford University Press, Oxford.

Hill R. (1948). A theory of the yielding and plastic flow of anisotropic metals, *Proceedings of the Royal Society of London*, Series A, 193(1033): 281-297.

Hoffman O. (1967). The Brittle Strength of Orthotropic Materials. *Journal of Composite Materials* 1 (2): 200-206.

Hou J.P., Petrinic N. and Ruiz C. (2001). A delamination criterion for laminated composites under low-velocity impact. *Composites Science and Technology* 61 (14): 2069-2074.

Hou J.P., Petrinic N., Ruiz C. and Hallett S.R. (2000). Prediction of impact damage in composite plates. *Composites Science and Technology* 60 (2): 273-281.

Huang C.H. and Lee Y.J. (2003). Experiments and simulation of the static contact crush of composite laminated plates. *Composite Structures* 61 (3): 265-270.
Iannucci L. (2006). Dynamic delamination modelling using interface elements. *Computers and Structures* 84 (15-16) : 1029-1048.
Jiang W.G., Hallett S.R., Green B.G., and Wisnom M.R. (2007). A concise interface constitutive law for analysis of delamination and splitting in composite materials and its application to scaled notched tensile specimens. *International Journal for Numerical Methods in Engineering* 69 (9) : 1982-1995.
Johnson A.F., Pickett A.K. and Rozycki P. (2001). Computational methods for predicting impact damage in composite structures. *Composites Science and Technology* 61 (15): 2183-2192.
Johnson A.F. and Holzapfel M. (2006). Influence of delamination on impact damage in composite structures. *Composites Science and Technology* 66 (6): 807-815.
Johnson A.F. and Holzapfel M. (2003). Modelling soft body impact on composite structures. *Composite Structures* 61 (1-2): 103-113.
Kaw A.K. (2005). *Mechanics of Composite Materials*. CRC Press, Boca Raton, FL, USA.
Kedward K.T., Whitney J.M., Carlsson L.A. and Gillespie J.W. (1990) *Delaware Composites Design Encyclopedia: Design studies.* CRC Press, Boca Raton, FL, USA.
Kim J.S. and Chung S.K. (2007). A study on the low-velocity impact response of laminates for composite railway bodyshells. *Composite Structures* 77 (4): 484-492.
Kim S.J., Goo N.S. and Kim T.W. (1997). The effect of curvature on the dynamic response and impact-induced damage in composite laminates. *Composites Science and Technology* 57 (7): 763-773.
Krishnamurthy S., Mahajan P. and Mittal R.K. (2001). A parametric study of the impact response and damage of laminated cylindrical composite shells. *Composites Science and Technology* 61 (12) : 1655-1669, 2001
Lammerant L. and Verpoest I. (1996). Modelling of the interaction between matrix cracks and delaminations during impact of composite plates. *Composites Science and Technology* 56 (10): 1171-1178.
Lee J.D. (1982). Three dimensional finite element analysis of damage accumulation in composite laminate. *Computers and Structures* 15: 335–350.
Lee Y.J. and Huang C.H. (2003). Ultimate strength and failure process of composite laminated plates subjected to low-velocity impact. *Journal of Reinforced Plastics and Composites* 22(12): 1059-1081.
Li C.F., Hu N., Yin Y.J., Sekine H. and Fukunaga H. (2002). Low-velocity impact-induced damage of continuous fiber-reinforced composite laminates. Part

I. An FEM numerical model. *Composites Part A: Applied Science and Manufacturing* 33 (8): 1055-1062, 2002
Li S., Reid S.R. and Zou Z. (2006). Modelling damage of multiple delaminations and transverse matrix cracking in laminated composites due to low velocity lateral impact. *Composites Science and Technology* 66(6): 827-836.
Li X., Hallett S.R. and Wisnom M.R. (2008). Predicting the effect of through-thickness compressive stress on delamination using interface elements. *Composites Part A - Applied Science and Manufacturing* 39 (2): 218-230.
Liu X. and Wang G. (2007). Progressive failure analysis of bonded composite repairs. *Composite Structures* 81 (3) : 331-340.
Luo R.K., Green E.R. and Morrison C.J. (2001). An approach to evaluate the impact damage initiation and propagation in composite plates. *Composites Part B: Engineering* 32 (6): 513-520.
Luo R.K., Green E.R. and Morrison C.J. (1999). Impact damage analysis of composite plates. *International Journal of Impact Engineering* 22 (4): 435-447.
Mahanta B.B., Chakraborty D. and Dutta A. (2004). Accurate prediction of delamination in FRP composite laminates resulting from transverse impact. *Composites Science and Technology* 64 (15): 2341-2351.
Naik N.K., Meduri S. and Chandrasekher Y. (2001). Polymer matrix woven fabric composites subjected to low velocity impact: Part III - Effect of incident impact velocity and impactor mass. *Journal of Reinforced Plastics and Composites* 20 (9): 720-743.
Naik N.K., Sekher Y.C. and Meduri S. (2000). Damage in woven-fabric composites subjected to low-velocity impact. *Composites Science and Technology* 60 (5): 731-744.
Naik N.K., Chandra Sekher Y., and Meduri S. (2000). Polymer Matrix Woven Fabric Composites Subjected to Low Velocity Impact: Part I- Damage Initiation Studies. *Journal of Reinforced Plastics and Composites* 19 (12): 912 – 954.
Nishikawa M., Okabe T., Takeda N. (2007). Numerical simulation of interlaminar damage propagation in CFRP cross-ply laminates under transverse loading. *International Journal of Solids and Structures* 44 (10): 3101-3113.
Ochoa O.O. and Reddy J.N. (1992). *Finite Element Analysis of Composite Laminates.* Springer, Dordrecht, The Netherlands.
Pereira A.B., de Morais A.B., Marques A.T., de Castro P.T. (2004). Mode II interlaminar fracture of carbon/epoxy multidirectional laminates. *Composites Science and Technology* 64 (10-11) : 1653-1659.
Pradhan B. and Kumar S. (2000). Finite element analysis of low-velocity impact damage in composite laminates. *Journal of Reinforced Plastics and Composites* 19 (4): 322-339.
Raghava R.S. and Caddell R.M. (1974). Yield locus studies of oriented polycarbonate: an anisotropic and pressure-dependent solid. *International Journal of Mechanical Sciences* 16(11): 789-792, 1974

Rahul S.G., Chakraborty D., and Dutta A. (2006). Multi-objective optimization of hybrid laminates subjected to transverse impact. *Composite Structures* 73 (3): 360-369.

Rahul C.D. and Dutta A. (2005). Optimization of FRP composites against impact induced failure using island model parallel genetic algorithm. *Composites Science and Technology* 65 (13), 2003-2013.

Reeder J.R. (1993). A Bilinear Failure Criterion for Mixed-Mode Delamination, In: Camponeschi, Jr, E.T. (ed.), *Composite Materials: Testing and Design*, Eleventh Volume, ASTM STP 1206, pp. 303–322.

Reeder J.R. (1992). Evaluation of mixed-mode delamination failure criteria. *NASA report TM 104210.*

Rinderknecht S. and Kroplin B. (1997). Computational method for the analysis of delamination growth in composite plates. *Computers and Structures* 64 (1-4) : 359-374.

Singh S.B. and Kumar A. (1998). Postbuckling response and failure of symmetric laminates under in-plane shear. *Composites Science and Technology* 58 (12): 1949-1960.

Singh S.B. and Kumar A. (1999). Postbuckling response and strength of laminates under combined in-plane loads. *Composites Science and Technology* 59 (5): 727-736.

Suemasu H., Sasaki W., Ishikawa T. and Aoki Y. (2008). A numerical study on compressive behavior of composite plates with multiple circular delaminations considering delamination propagation. *Composites Science and Technology* 68 (12): 2562-2567.

Sun C.T. and Yamada S.E. (1978). Strength distribution of a unidirectional fiber composite. *Journal of Composite Materials* 12: 169-176.

Sun CT, Quinn BJ, Tao J, and Oplinger DW. (1996). Comparative evaluation of failure analysis methods for composite laminates. *Report DOT/FAA/AR-95/109.*

Szekrényes A. (2007). Delamination fracture analysis in the G_{II}–G_{III} plane using prestressed transparent composite beams. *International Journal of Solids and Structures* 44 (10): 3359-3378.

Tsai S.W. and Wu E.M. (1971). General Theory of Strength for Anisotropic Materials. *Journal of Composite Materials* 5 : 58-80.

Wagner W., Gruttmann F. and Sprenger W. (2001). A finite element formulation for the simulation of propagating delaminations in layered composite structures. *International Journal for Numerical Methods in Engineering* 51 (11): 1337-1359.

Wu E.M. and Reuter Jr. R.C. (1965). Crack Extension in Fiberglass Reinforced Plastics, *T. & AM Report No. 275*, University of Illinois.

Wu R.Y. and Stachurski Z. (1984). Evaluation of the Normal Stress Interaction Parameter in the Tensor Polynomial Strength Theory for Anisotropic Materials. *Journal of Composite Materials* 18: 456-463.

Yamada S E. and Sun C T. (1978). Analysis of Laminate Strength and Its Distribution. *Journal of Composite Materials* 12: 275-284.
Yeh H.Y. and Kim C.H. (2004). The Yeh–Stratton criterion for composite materials. *Journal of Composite Materials* 28 (10): 926–39.
Yoshimura A., Nakao T., Yashiro S. and Takeda N. (2008). Improvement on out-of-plane impact resistance of CFRP laminates due to through-the-thickness stitching. *Composites Part A: Applied Science and Manufacturing*, 39 (9): 1370-1379.
Zhang Z., Shen J., Zhong W. and Sun, Z. (2002). A dynamic model of ceramic/fibre-reinforced plastic hybrid composites under projectile striking. *Proceedings of the Institution of Mechanical Engineers, Part G: Journal of Aerospace Engineering* 216 (6): 325-331.
Zhao G.P. and Cho C.D. (2007). Damage initiation and propagation in composite shells subjected to impact. *Composite Structures* 78 (1): 91-100.
Zhao G.P., Cho C.D. (2004). On Impact Damage of Composite Shells by a Low-velocity Projectile. *Journal of Composite Materials* 38 (14): 1231 – 1254.
Zheng C., Ren M., Zhao W. and Chen H. (2006). Delamination prediction of composite filament wound vessel with metal liner under low velocity impact. *Composite Structures* 75 (1-4): 387-392.
Zou Z., Reid S.R., Li S. and Soden P.D. (2002-a). Modelling Interlaminar and Intralaminar Damage in Filament-Wound Pipes under Quasi-Static Indentation. *Journal of Composite Materials* 36(4): 477-499.
Zou Z, Reid SR, Li S, et al. (2002-b). Application of a delamination model to laminated composite structures. *Composite Structures* 56(4): 375-389.

Mechanics of Plates

Serge Abrate

Department of Mechanical Engineering and Energy Processes, Southern Illinois University, Carbondale, IL, USA

In most structural applications, composite structures can be idealized as beams, plates or shells. The analysis is reduced from a three-dimensional elasticity problem to a one-dimensional or two-dimensional problem based on certain simplifying assumptions that can be made because the structure is thin. This chapter presents the basic assumptions made in the development of several commonly used plate theories. The simplifying assumptions deal with various ways of approximating the transverse normal strain and the transverse shear strains. The present discussion also applies to beam theories and shell theories where parallel developments take place. Knowledge of the basic elements of these structural theories is required for the development of mathematical models for modeling the impact event.

1- Derivation of the Equations of Motion in terms of Force and Moment Resultants

The equations of motions of a plate are derived by integration of the equations of motion for a three dimensional elastic solid. The definitions of force and moment resultants are introduced in the derivation and the equations obtained are valid for all three of the theories considered here. As discussed in Chapter 1, the equations of motion for linear elastic solid are

$$\frac{\partial \sigma_{xx}}{\partial x} + \frac{\partial \sigma_{yx}}{\partial y} + \frac{\partial \sigma_{zx}}{\partial z} + q_x = \rho \frac{\partial^2 u}{\partial t^2}$$

$$\frac{\partial \sigma_{xy}}{\partial x} + \frac{\partial \sigma_{yy}}{\partial y} + \frac{\partial \sigma_{zy}}{\partial z} + q_y = \rho \frac{\partial^2 v}{\partial t^2} \qquad (1)$$

$$\frac{\partial \sigma_{xz}}{\partial x} + \frac{\partial \sigma_{yz}}{\partial y} + \frac{\partial \sigma_{zz}}{\partial z} + q_z = \rho \frac{\partial^2 w}{\partial t^2}$$

where q_x, q_y, and q_z are the forces/unit volume (body forces) and u, v, and w are the body forces in the x, y, and z directions. Integrating with respect to z gives

$$\frac{\partial}{\partial x}\int_{z^-}^{z^+}\sigma_{xx}dz+\frac{\partial}{\partial y}\int_{z^-}^{z^+}\sigma_{xy}dz+\int_{z^-}^{z^+}\frac{\partial\sigma_{xz}}{\partial z}dz+\int_{z^-}^{z^+}q_x dz=\int_{z^-}^{z^+}\rho\frac{\partial^2 u}{\partial t^2}dz$$

$$\frac{\partial}{\partial x}\int_{z^-}^{z^+}\sigma_{xy}dz+\frac{\partial}{\partial y}\int_{z^-}^{z^+}\sigma_{yy}dz+\int_{z^-}^{z^+}\frac{\partial\sigma_{yz}}{\partial z}dz+\int_{z^-}^{z^+}q_y dz=\int_{z^-}^{z^+}\rho\frac{\partial^2 v}{\partial t^2}dz \qquad (2)$$

$$\frac{\partial}{\partial x}\int_{z^-}^{z^+}\sigma_{xz}dz+\frac{\partial}{\partial y}\int_{z^-}^{z^+}\sigma_{yz}dz+\int_{z^-}^{z^+}\frac{\partial\sigma_{zz}}{\partial z}dz+\int_{z^-}^{z^+}q_z dz=\int_{z^-}^{z^+}\rho\frac{\partial^2 w}{\partial t^2}dz$$

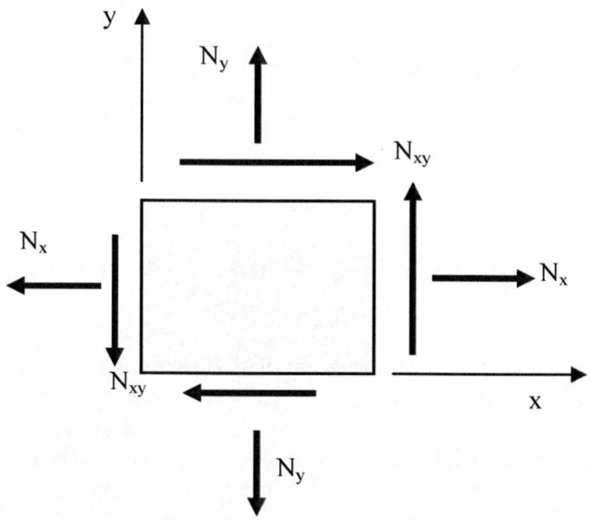

Figure 1. Inplane force resultants

In Equations 2, we recognize the force resultants shown in Figure 1 that are defined by

$$N_x=\int_{-h/2}^{h/2}\sigma_{xx}dz,\quad N_y=\int_{-h/2}^{h/2}\sigma_{yy}dz,\quad N_{xy}=\int_{-h/2}^{h/2}\sigma_{xy}dz \qquad (3)$$

and the transverse shear forces in Figure 2 defined by

$$\begin{Bmatrix} Q_x \\ Q_y \end{Bmatrix} = \int_{z^-}^{z^+} \begin{Bmatrix} \sigma_{xz} \\ \sigma_{yz} \end{Bmatrix} dz \tag{4}$$

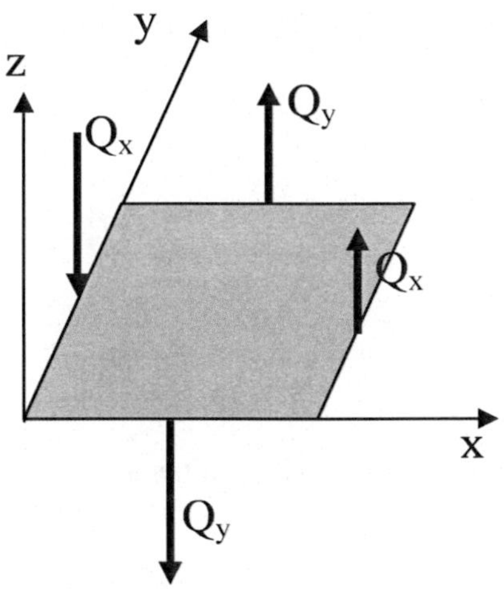

Figure 2. Transverse shear forces

After substitution, the equations of motion (Equation 2) become

$$\frac{\partial N_x}{\partial x} + \frac{\partial N_{xy}}{\partial y} + p_x = \int_{z^-}^{z^+} \rho \frac{\partial^2 u}{\partial t^2} dz, \quad \frac{\partial N_{xy}}{\partial x} + \frac{\partial N_{yy}}{\partial y} + p_y = \int_{z^-}^{z^+} \rho \frac{\partial^2 v}{\partial t^2} dz \tag{5}$$

and

$$\frac{\partial Q_x}{\partial x} + \frac{\partial Q_y}{\partial y} + p_z = \int_{z^-}^{z^+} \rho \frac{\partial^2 w}{\partial t^2} dz$$

where the applied forces per unit area are given by

$$p_x = (\sigma_{xz})^+ - (\sigma_{xz})^- + \int_{z^-}^{z^+} q_x dz, \quad p_y = (\sigma_{yz})^+ - (\sigma_{yz})^- + \int_{z^-}^{z^+} q_y dz \tag{6}$$

and $$p_z = (\sigma_{zz})^+ - (\sigma_{zz})^- + \int_{z^-}^{z^+} q_z dz$$

With the present notation, $(\sigma_{zz})^+$, $(\sigma_{xz})^+$ and $(\sigma_{yz})^+$ are the stresses acting on the top surface of the plate. $(\sigma_{zz})^-$, $(\sigma_{xz})^-$ and $(\sigma_{yz})^-$ are the stresses acting on the bottom surface of the plate. The integrals in Equations 6 account the effect of the body forces applied on the plate. Multiplying Equations (1.a, b) by z and integrating through the thickness gives

$$\frac{\partial}{\partial x}\int_{z^-}^{z^+} \sigma_{xx} z dz + \frac{\partial}{\partial y}\int_{z^-}^{z^+} \sigma_{xy} z dz + \int_{z^-}^{z^+} \frac{\partial \sigma_{xz}}{\partial z} z dz + \int_{z^-}^{z^+} q_x z dz = \int_{z^-}^{z^+} \rho \frac{\partial^2 u}{\partial t^2} z dz \tag{7}$$

$$\frac{\partial}{\partial x}\int_{z^-}^{z^+} \sigma_{xy} z dz + \frac{\partial}{\partial y}\int_{z^-}^{z^+} \sigma_{yy} z dz + \int_{z^-}^{z^+} \frac{\partial \sigma_{yz}}{\partial z} z dz + \int_{z^-}^{z^+} q_y z dz = \int_{z^-}^{z^+} \rho \frac{\partial^2 v}{\partial t^2} z dz$$

In Equations 7 we recognize the moment resultants

$$\begin{Bmatrix} M_x \\ M_y \\ M_{xy} \end{Bmatrix} = \int_{z^-}^{z^+} \begin{Bmatrix} \sigma_{xx} \\ \sigma_{yy} \\ \sigma_{xy} \end{Bmatrix} z dz \tag{8}$$

that are shown in Figure 3. Using the expressions

$$\frac{\partial}{\partial z}(\sigma_{xz} z) = \frac{\partial \sigma_{xz}}{\partial z} z + \sigma_{xz} \quad \text{and} \quad \frac{\partial}{\partial z}(\sigma_{yz} z) = \frac{\partial \sigma_{yz}}{\partial z} z + \sigma_{yz} \tag{9}$$

we can evaluate the integrals

$$\int_{z^-}^{z^+} \frac{\partial \sigma_{xz}}{\partial z} z dz = \sigma_{xz} z \Big|_{z^-}^{z^+} - \int_{z^-}^{z^+} \sigma_{xz} dz = (\sigma_{xz} z)^+ - (\sigma_{xz} z)^- - Q_x \tag{10.a}$$

$$\int_{z^-}^{z^+} \frac{\partial \sigma_{yz}}{\partial z} z dz = \sigma_{yz} z \Big|_{z^-}^{z^+} - \int_{z^-}^{z^+} \sigma_{yz} dz = (\sigma_{yz} z)^+ - (\sigma_{yz} z)^- - Q_y \tag{10.b}$$

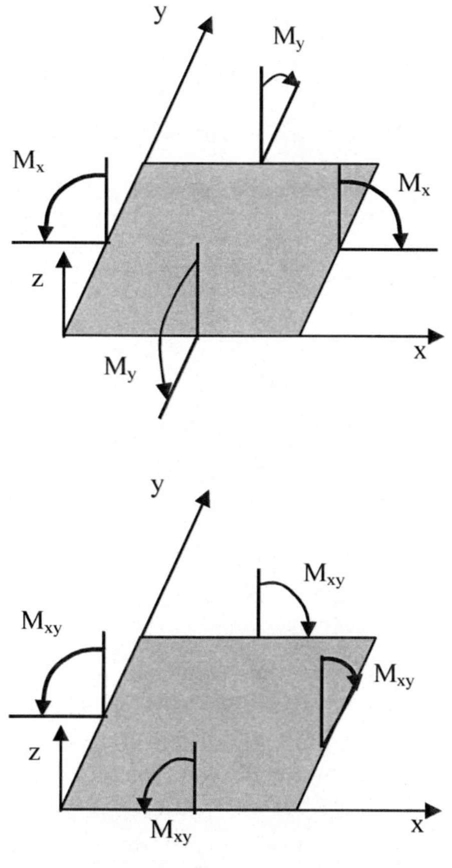

Figure 3. Bending moments acting on a unit square element of the plate

In Equations 10, the quantities in parentheses are the distributed moments created by the shear stresses acting on the top (+) and bottom (-) surfaces. Using Equations 8 and 10, Equations 7 can be written as

$$\frac{\partial M_x}{\partial x}+\frac{\partial M_{xy}}{\partial y}-Q_x+m_x=\int_{z^-}^{z^+}\rho\frac{\partial^2 u}{\partial t^2}z\,dz \tag{11}$$

$$\frac{\partial M_{xy}}{\partial x}+\frac{\partial M_y}{\partial y}-Q_y+m_y=\int_{z^-}^{z^+}\rho\frac{\partial^2 v}{\partial t^2}zdz \qquad (12)$$

with

$$m_x=(\sigma_{xz}z)^+-(\sigma_{xz}z)^++\int_{z^-}^{z^+}q_x zdz,\quad m_y=(\sigma_{yz}z)^+-(\sigma_{yz}z)^++\int_{z^-}^{z^+}q_y zdz \qquad (13)$$

Equations 5, 11, 12 are the equations of motion of the plate in terms of the force and moment resultants. Note that in general m_x and m_y are zero or can be neglected. The inertia terms on the right-hand sides will be made explicit in each plate theory once the kinematics of the plate deformations are made explicit.

2- Kinematics Assumptions

The Classical Plate Theory (CPT), the First order Shear Deformation Theory (FSDT), and Reddy's Shear Deformation Theory (RSDT) can be derived from two basic assumptions. The first assumption (H1) is that the transverse normal strain is zero

$$\varepsilon_{zz}=\frac{\partial w}{\partial z}=0 \qquad (14)$$

which implies that the transverse displacement w is independent of z or $w=w_o(x,y)$. Sometimes this assumption is stated backwards by saying that the transverse displacement is uniform through the thickness.
The second assumption (H2) concerns the variation of transverse shear strains

$$\varepsilon_{xz}=\frac{\partial u}{\partial z}+\frac{\partial w}{\partial x}\qquad \varepsilon_{yz}=\frac{\partial v}{\partial z}+\frac{\partial w}{\partial y} \qquad (15)$$

through the thickness. Different assumptions are made: in the classical plate theory (CPT) the transverse shear strains are assumed to be zero; in the first order shear plate theory (FSDT) they are assumed to be constant through the thickness of the plate; in Reddy's third order shear deformation theory (RSDT) the transverse shear strains are assumed to vary parabolically through the thickness and vanish on the top and bottom surfaces.

3- Classical Plate Theory

In the following we show how the assumption that the transverse normal strains are negligible leads to what are called the kinematic assumptions of the classical plate theory (CPT) from which we derive the strain-displacement relations, the constitutive relations, and the equations of motion.

3.1- Kinematics of the deformation

When the transverse shear strains are neglected $(\varepsilon_{xz} = \varepsilon_{yz} = 0)$, Equations 15 give

$$\frac{\partial u}{\partial z} = -\frac{\partial w}{\partial x} \quad \text{and} \quad \frac{\partial v}{\partial z} = -\frac{\partial w}{\partial y} \tag{16}$$

which, after integration, yield the kinematic relations for the Classical Plate Theory (CPT)

$$u = u_o - z\frac{\partial w_o}{\partial x}, \quad v = v_o - z\frac{\partial w_o}{\partial y}, \quad w = w_o(x,y) \tag{17}$$

where u_o, v_o, and w_o are the displacements of the reference surface (xy-plane). Equations (17.a,b) express the concept that a line segment initially normal to the reference surface remains perpendicular to that surface after deformation. Often, this statement or Equation 17 are taken as the point of departure for the CPT which is also called the Love-Kirchhoff plate theory. The kinematics of the CPT are illustrated in Figure 4. With these assumptions, the inplane strains are

$$\varepsilon_{xx} = \frac{\partial u}{\partial x} = \frac{\partial u_o}{\partial x} - z\frac{\partial^2 w}{\partial x^2}, \quad \varepsilon_{yy} = \frac{\partial v}{\partial y} = \frac{\partial v_o}{\partial y} - z\frac{\partial^2 w}{\partial y^2} \tag{18}$$

$$\varepsilon_{xy} = \frac{\partial u}{\partial y} + \frac{\partial v}{\partial x} = \frac{\partial u_o}{\partial y} + \frac{\partial v_o}{\partial x} - 2\frac{\partial^2 w}{\partial x \partial y}$$

Defining the total strain, the midplane strain, and the curvature as

$$\{\varepsilon\} = \begin{Bmatrix} \varepsilon_{xx} \\ \varepsilon_{yy} \\ \varepsilon_{xy} \end{Bmatrix}, \quad \{\varepsilon_o\} = \begin{Bmatrix} \dfrac{\partial u_o}{\partial x} \\ \dfrac{\partial v_o}{\partial y} \\ \dfrac{\partial u_o}{\partial y} + \dfrac{\partial v_o}{\partial x} \end{Bmatrix}, \quad \{K\} = \begin{Bmatrix} -\dfrac{\partial^2 w}{\partial x^2} \\ -\dfrac{\partial^2 w}{\partial y^2} \\ -2\dfrac{\partial^2 w}{\partial x \partial y} \end{Bmatrix} \tag{19}$$

respectively, the strain at an arbitrary point can be expressed as

$$\{\varepsilon\}=\{\varepsilon_o\}+z\{K\} \tag{20}$$

In this theory, transverse shear deformations are considered to be negligible and the corresponding strains are set to zero. However, the shear forces Q_x and Q_y are not zero even though from their definition (Equation 4) it would appear that they should be zero if the corresponding strains are uniformly zero. This is sometimes called the paradox of the classical beam or plate theories.

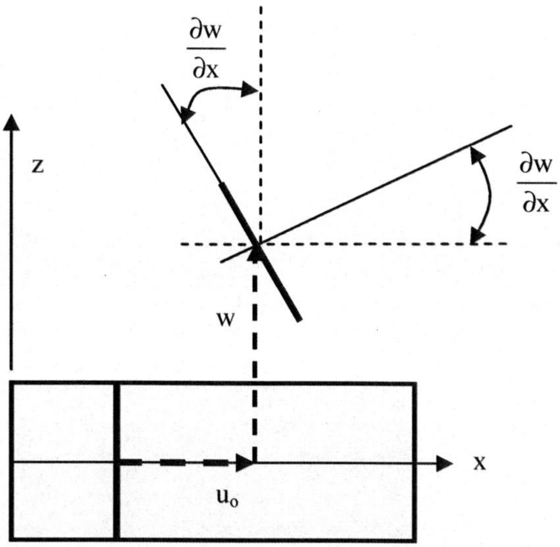

Figure 4. Kinematics of the deformation according to the classical plate theory

3.2- Constitutive Equations

From Chapter 1, we recall that the constitutive equations for a single layer can be written as

$$\begin{Bmatrix}\sigma_{xx}\\ \sigma_{yy}\\ \sigma_{xy}\end{Bmatrix}=[\overline{Q}]\begin{Bmatrix}\varepsilon_{xx}\\ \varepsilon_{yy}\\ \varepsilon_{xy}\end{Bmatrix} \tag{21}$$

The force resultants defined in Equation 3 can be written as

$$\begin{Bmatrix} N_x \\ N_y \\ N_{xy} \end{Bmatrix} = \int_{z^-}^{z^+} \begin{Bmatrix} \sigma_{xx} \\ \sigma_{yy} \\ \sigma_{xy} \end{Bmatrix} dz = \int_{z^-}^{z^+} [\overline{Q}] \begin{Bmatrix} \varepsilon_{xx} \\ \varepsilon_{yy} \\ \varepsilon_{xy} \end{Bmatrix} dz = \int_{z^-}^{z^+} [\overline{Q}] \{\varepsilon^o\} dz + \int_{z^-}^{z^+} [\overline{Q}] \{K\} z dz \quad (22)$$

Similarly, the moment resultants in Equation (8) are written as

$$\begin{Bmatrix} M_x \\ M_y \\ M_{xy} \end{Bmatrix} = \int_{z^-}^{z^+} \begin{Bmatrix} \sigma_{xx} \\ \sigma_{yy} \\ \sigma_{xy} \end{Bmatrix} z dz = \int_{z^-}^{z^+} [\overline{Q}] \begin{Bmatrix} \varepsilon_{xx} \\ \varepsilon_{yy} \\ \varepsilon_{xy} \end{Bmatrix} z dz = \int_{z^-}^{z^+} [\overline{Q}] \{\varepsilon^o\} z dz + \int_{z^-}^{z^+} [\overline{Q}] \{K\} z^2 dz \quad (23)$$

These constitutive equations (Equations 22, 23) can be written in compact form as

$$\begin{Bmatrix} N \\ M \end{Bmatrix} = \begin{bmatrix} A & B \\ B & D \end{bmatrix} \begin{Bmatrix} \varepsilon_o \\ K \end{Bmatrix} \quad (24)$$

where

$$A_{ij} = \sum_{i=1}^{N} \overline{Q}_{ij}^{k} [z_k - z_{k-1}], \quad B_{ij} = \frac{1}{2} \sum_{i=1}^{N} \overline{Q}_{ij}^{k} [z_k^2 - z_{k-1}^2], \quad D_{ij} = \frac{1}{3} \sum_{i=1}^{N} \overline{Q}_{ij}^{k} [z_k^3 - z_{k-1}^3]$$

The A, B, and D matrices are 3x3 matrices that represent the inplane stiffness, the inplane-bending coupling terms, and the bending rigidities respectively. If [B] is not zero, an inplane load will produce both inplane and transverse deformations. When the layup is symmetric, B=0 and there is no elastic coupling between the inplane and transverse deformations. The A and D matrices are usually written with the following notation

$$[A] = \begin{bmatrix} A_{11} & A_{12} & A_{16} \\ A_{12} & A_{22} & A_{16} \\ A_{16} & A_{26} & A_{66} \end{bmatrix} \qquad [D] = \begin{bmatrix} D_{11} & D_{12} & D_{16} \\ D_{12} & D_{22} & D_{16} \\ D_{16} & D_{26} & D_{66} \end{bmatrix} \quad (25)$$

When [B]=0, we can further notice that the A_{16} and A_{26} terms couple the extensional deformations measured by ε_{xx} and ε_{yy} and the shear deformation defined by ε_{xy}. When $A_{16} = A_{26} = 0$, the extension-shear coupling has been eliminated. Similarly, the D_{16} and D_{26} terms represent the bending-twisting coupling of the out-of-plane deformations.

3.3- Equations of Motion for the CPT

With the kinematic assumptions of the classical plate theory, the inertia terms on the right hand side of Equations (5, 11, 12) can be written as

$$\int_{z^-}^{z^+} \rho \frac{\partial^2 u}{\partial t^2} dz = \frac{\partial^2}{\partial t^2} \int_{z^-}^{z^+} \rho \left(u_o - z \frac{\partial w}{\partial x} \right) dz = I_o \frac{\partial^2 u_o}{\partial t^2} + I_1 \frac{\partial^2}{\partial t^2} \left(\frac{\partial w}{\partial x} \right) \qquad (26a\text{-}e)$$

$$\int_{z^-}^{z^+} \rho \frac{\partial^2 v}{\partial t^2} dz = \frac{\partial^2}{\partial t^2} \int_{z^-}^{z^+} \rho \left(v_o - z \frac{\partial w}{\partial y} \right) dz = I_o \frac{\partial^2 v_o}{\partial t^2} + I_1 \frac{\partial^2}{\partial t^2} \left(\frac{\partial w}{\partial y} \right)$$

$$\int_{z^-}^{z^+} \rho \frac{\partial^2 w}{\partial t^2} dz = I_o \frac{\partial^2 w}{\partial t^2}, \quad \int_{z^-}^{z^+} \rho \frac{\partial^2 u}{\partial t^2} z dz = I_1 \frac{\partial^2 u_o}{\partial t^2} - I_2 \frac{\partial^2}{\partial t^2} \left(\frac{\partial w}{\partial x} \right)$$

$$\int_{z^-}^{z^+} \rho \frac{\partial^2 v}{\partial t^2} z dz = I_1 \frac{\partial^2 v_o}{\partial t^2} - I_2 \frac{\partial^2}{\partial t^2} \left(\frac{\partial w}{\partial y} \right)$$

where the inertia terms are

$$(I_o, I_1, I_2) = \int_{z^-}^{z^+} \rho (1, z, z^2) dz \qquad (27)$$

If I_1=0, there is no inertial coupling between inplane and transverse motion. I_2 is the rotary inertia and is neglected in the CPT since the effects of shear deformations have been neglected and the effect of rotary inertia is usually of the same order of magnitude. The five equations of motion can then be written as

$$\frac{\partial N_x}{\partial x} + \frac{\partial N_{xy}}{\partial y} + p_x = I_o \frac{\partial^2 u_o}{\partial t^2}, \quad \frac{\partial N_{xy}}{\partial x} + \frac{\partial N_{yy}}{\partial y} + p_y = I_o \frac{\partial^2 v_o}{\partial t^2},$$

$$\frac{\partial Q_x}{\partial x} + \frac{\partial Q_y}{\partial y} + p_z = I_o \frac{\partial^2 w}{\partial t^2} \qquad (28\ a\text{-}e)$$

$$\frac{\partial M_x}{\partial x} + \frac{\partial M_{xy}}{\partial y} - Q_x + m_x = 0, \quad \frac{\partial M_{xy}}{\partial x} + \frac{\partial M_y}{\partial y} - Q_y + m_y = 0$$

In most practical cases, $m_x = m_y$ =0. Solving for Q_x and Q_y from Equations (28.d,e) and substituting into Equation 27.c gives

$$\frac{\partial^2 M_x}{\partial x^2}+2\frac{\partial^2 M_{xy}}{\partial x\partial y}+\frac{\partial^2 M_y}{\partial y^2}+p_z=I_o\frac{\partial^2 w}{\partial t^2} \tag{29}$$

In this form the equation of motion is the same for a plate made of single isotropic material or for a laminated plate. For a symmetric laminate, [B]=0 and using Equation 24, Equation 29 gives the governing equation for the transverse motion of the plate

$$\begin{aligned}&D_{11}\frac{\partial^4 w}{\partial x^4}+2(D_{12}+2D_{66})\frac{\partial^4 w}{\partial x^2\partial y^2}+D_{22}\frac{\partial^4 w}{\partial y^4}+4D_{16}\frac{\partial^2 w}{\partial x^3\partial y}+\\&4D_{26}\frac{\partial^2 w}{\partial x\partial y^2}+I_o\frac{\partial^2 w}{\partial t^2}=p_z\end{aligned} \tag{30}$$

D_{16} and D_{26} couple the bending and twisting deformations. It is not possible to eliminate those coupling terms unless each layer is oriented at either 0 or 90° (specially orthotropic plates). For a laminate with a large number of layers the effect of these terms becomes negligible. The equation of motion becomes

$$D_{11}\frac{\partial^4 w}{\partial x^4}+2(D_{12}+2D_{66})\frac{\partial^4 w}{\partial x^2\partial y^2}+D_{22}\frac{\partial^4 w}{\partial y^4}+I_o\frac{\partial^2 w}{\partial t^2}=p_z \tag{31}$$

This equation of motion is similar to that for an isotropic plate when in that case $D_{11} = D_{22} = D_{16} + 2D_{16}$. Exact solutions can be found for rectangular plates that are simply supported on two opposite edges.

4- First Order Shear Deformation Theory

The first order shear deformation theory (FSDT) also called Mindlin's plate theory is based on the assumptions that the transverse shear strains are constant which leads to a particular kinematics of the deformation. The constitutive equations and the derivation of the equations of motion in terms of displacement variables are discussed.

4.1- Kinematics of the Deformation

With the First order Shear Deformation Theory (FSDT), the transverse shear strains are assumed to be constant through the thickness but not necessarily zero.

$$\varepsilon_{xz} = \frac{\partial u}{\partial z} + \frac{\partial w}{\partial x} = c_1(x, y, t) \qquad \varepsilon_{yz} = \frac{\partial v}{\partial z} + \frac{\partial w}{\partial y} = c_2(x, y, t) \tag{32}$$

After integration, the inplane displacements can be written as

$$u = u_o - z\left(-c_1 + \frac{\partial w}{\partial x}\right), \qquad v = v_o - z\left(-c_2 + \frac{\partial w}{\partial y}\right) \tag{33}$$

The quantities $\psi_x = -c_1 + \frac{\partial w}{\partial x}$ and $\psi_y = -c_2 + \frac{\partial w}{\partial y}$ are the rotations of a line segment initially perpendicular to the midplane and they are different from $\frac{\partial w}{\partial x}$ and $\frac{\partial w}{\partial y}$, rotations of line elements initially oriented in the x and y directions. Using these newly defined rotations, Equations 33 give the kinematic relations of the FSDT

$$u = u_o - z\psi_x, \quad v = v_o - z\psi_y, \quad w = w_o(x, y) \tag{34}$$

Equations 34 state that, line segments perpendicular to the reference surface remain straight but not necessarily normal to the reference surface after deformation. This plate theory is often called the Reissner-Mindlin plate theory since it was developed by Reissner (1945) and Mindlin (1951). It can be considered as an extension of the work of Timoshenko (1921) who was the first to examine both the effects of shear deformation and rotary inertia in the analysis of beams. With Equations 34, the inertia terms in the constitutive equations become

$$\int_{z^-}^{z^+} \rho \frac{\partial^2 u}{\partial t^2} dz = \frac{\partial^2}{\partial t^2} \int_{z^-}^{z^+} \rho (u_o - z\psi_x) dz = I_o \frac{\partial^2 u_o}{\partial t^2} - I_1 \frac{\partial^2 \psi_x}{\partial t^2} \tag{35}$$

$$\int_{z^-}^{z^+} \rho \frac{\partial^2 v}{\partial t^2} dz = \frac{\partial^2}{\partial t^2} \int_{z^-}^{z^+} \rho (v_o - z\psi_y) dz = I_o \frac{\partial^2 v_o}{\partial t^2} - I_1 \frac{\partial^2 \psi_y}{\partial t^2}$$

$$\int_{z^-}^{z^+} \rho \frac{\partial^2 w}{\partial t^2} dz = I_o \frac{\partial^2 w}{\partial t^2}$$

$$\int_{z^-}^{z^+} \rho \frac{\partial^2 u}{\partial t^2} z dz = I_1 \frac{\partial^2 u_o}{\partial t^2} - I_2 \frac{\partial^2 \psi_x}{\partial t^2} \quad \int_{z^-}^{z^+} \rho \frac{\partial^2 v}{\partial t^2} z dz = I_1 \frac{\partial^2 v_o}{\partial t^2} - I_2 \frac{\partial^2 \psi_y}{\partial t^2}$$

Inertial coupling between the inplane and transverse deformations is eliminated if $I_1=0$ which is generally the case for composite since the density of the different layers is usually the same even if the ply orientation changes from ply to ply. Note that in this theory the effects of rotary inertia are included so the I_2 terms are not neglected in Equations 35.

4.2- Constitutive Equations

For the FSDT the inplane forces and the moments are related to the midplane strains and the curvatures by Equation 24. With the present kinematic assumptions (Equation 34), the curvatures and the transverse shear strains are

$$\{\kappa\} = \left[-\frac{\partial \psi_x}{\partial x}, -\frac{\partial \psi_y}{\partial y}, -\left(\frac{\partial \psi_x}{\partial y} + \frac{\partial \psi_y}{\partial x} \right) \right]^T \quad (36)$$

and

$$\begin{Bmatrix} \varepsilon_{xz} \\ \varepsilon_{yz} \end{Bmatrix} = \begin{Bmatrix} \frac{\partial w}{\partial x} - \psi_x \\ \frac{\partial w}{\partial y} - \psi_y \end{Bmatrix} \quad (37)$$

The transverse shear forces can be written as

$$\begin{Bmatrix} Q_x \\ Q_y \end{Bmatrix} = \begin{bmatrix} A_{55} & A_{45} \\ A_{45} & A_{44} \end{bmatrix} \begin{Bmatrix} \varepsilon_{xz} \\ \varepsilon_{yz} \end{Bmatrix} \quad (38)$$

where $\begin{bmatrix} A_{55} & A_{45} \\ A_{45} & A_{44} \end{bmatrix} = \int_{z^-}^{z^+} \begin{bmatrix} \overline{Q}_{55} & \overline{Q}_{45} \\ \overline{Q}_{45} & \overline{Q}_{55} \end{bmatrix} dz$.

The equations of motion of the plate in terms of the displacements are obtained by substituting Equations 24, 35 and 38 into Equations (5, 11, 12).

The FSDT is expressed in terms of 5 displacement functions: u_o, v_o, w, ψ_x, and ψ_y. For symmetrically laminated plates, [B]=0 and the transverse motion is governed by three equations in terms of w, ψ_x, and ψ_y.

5- Limitation of the CPT and the FSDT

The CPT neglects shear deformation in its kinematics assumptions. The FSDT assumes that shear strains remain constants through the thickness. This can only be an approximation since we know that on the top and bottom surfaces shear stresses will generally be zero. Then, the question becomes: what is the shear stress distribution through the thickness of the plate? Assuming constitutive equations of the form

$$\begin{Bmatrix} \sigma_{xx} \\ \sigma_{yy} \\ \sigma_{zz} \\ \sigma_{yz} \\ \sigma_{xz} \\ \sigma_{xy} \end{Bmatrix} = \begin{bmatrix} c_{11} & c_{12} & c_{13} & 0 & 0 & 0 \\ c_{21} & c_{22} & c_{23} & 0 & 0 & 0 \\ c_{31} & c_{32} & c_{33} & 0 & 0 & 0 \\ 0 & 0 & 0 & c_{44} & 0 & 0 \\ 0 & 0 & 0 & 0 & c_{55} & 0 \\ 0 & 0 & 0 & 0 & 0 & c_{66} \end{bmatrix} \begin{Bmatrix} \varepsilon_{xx} \\ \varepsilon_{yy} \\ \varepsilon_{zz} \\ \varepsilon_{yz} \\ \varepsilon_{xz} \\ \varepsilon_{xy} \end{Bmatrix} \tag{39}$$

the displacement field of the FSDT (Equations 34) implies that, during bending deformations $(u_o = v_o = 0)$, the “in plane” stresses are

$$\sigma_{xx} = c_{11}\left(-z\frac{\partial \psi_x}{\partial x}\right) + c_{12}\left(-z\frac{\partial \psi_y}{\partial y}\right) \quad \text{and} \quad \sigma_{xy} = -zc_{66}\left(\frac{\partial \psi_x}{\partial y} + \frac{\partial \psi_y}{\partial x}\right) \tag{40}$$

For linear elasticity, equilibrium in the x direction is governed by

$$\frac{\partial \sigma_{xx}}{\partial x} + \frac{\partial \sigma_{yx}}{\partial y} + \frac{\partial \sigma_{zx}}{\partial z} = 0 \tag{41}$$

Substituting Equations 40 into Equation 41 gives

$$\frac{\partial \sigma_{zx}}{\partial z} = z\left\{c_{11}\frac{\partial^2 \psi_x}{\partial x^2} + c_{12}\frac{\partial^2 \psi_y}{\partial x \partial y} + c_{66}\left(\frac{\partial^2 \psi_x}{\partial y^2} + \frac{\partial^2 \psi_y}{\partial x \partial y}\right)\right\} \tag{42}$$

The expression inside the brackets does not depend on z if the c_{ij}'s are constant through the thickness. Therefore, for a material that is homogeneous through the thickness, integrating Equation 42 with respect to z gives

$$\sigma_{zx} = \sigma_{zx}^{o} + z^2 f(x,y,t) \tag{43}$$

That is, the transverse shear stress σ_{zx} follows a parabolic distribution. It consists of a constant term σ_{zx}^{o} plus a z^2 term. So, when material properties remain constant through the thickness, both ε_{xz} and ε_{yz} follow a parabolic distribution through the thickness. Several higher order plate theories, including the RSDT that will discussed in the next section, assume that the shear strain follows a parabolic stress distribution through the thickness.

6- Reddy Shear Deformation Theory

The RSDT is one of many so-called higher order theories but it is widely used. From the outset it is expected to lead to an improved approximation of the state of stress in the plate. However, we shall see that it also leads to increased complexity of the equations.

6.1- Kinematics of the Deformation

Reddy (1984) presented a shear deformation plate theory (RSDT) in which the transverse shear strains vary parabolically. In addition to Hypothesis I, transverse shear strains are assumed to follow a parabolic variation through the thickness ($\varepsilon_{xz} = a_o + a_1 z + a_2 z^2$, $\varepsilon_{yz} = b_o + b_1 z + b_2 z^2$) and to vanish on the top and bottom surfaces. Therefore,

$$\varepsilon_{xz} = a_o\left(1 - 4\frac{z^2}{h^2}\right), \quad \varepsilon_{yz} = b_o\left(1 - 4\frac{z^2}{h^2}\right) \tag{44}$$

which leads to the following kinematics relations

$$u = u_o - z\psi_x\left(1 - \frac{4}{3}\frac{z^2}{h^2}\right) - \frac{4}{3}\frac{z^3}{h^2}\frac{\partial w_o}{\partial x} \tag{45}$$

$$v = v_o - z\psi_y\left(1-\frac{4}{3}\frac{z^2}{h^2}\right)-\frac{4}{3}\frac{z^3}{h^2}\frac{\partial w_o}{\partial y}, \qquad w = w_o(x,y)$$

and the strain-displacement relations

$$\begin{Bmatrix}\varepsilon_x\\ \varepsilon_y\\ \varepsilon_{xy}\end{Bmatrix}=\begin{Bmatrix}\frac{\partial u_o}{\partial x}\\ \frac{\partial v_o}{\partial y}\\ \frac{\partial u_o}{\partial y}+\frac{\partial v_o}{\partial x}\end{Bmatrix}+z\begin{Bmatrix}-\frac{\partial \psi_x}{\partial x}\\ -\frac{\partial \psi_y}{\partial y}\\ -\frac{\partial \psi_x}{\partial y}-\frac{\partial \psi_y}{\partial x}\end{Bmatrix}+\frac{4}{3}\frac{z^3}{h^2}\begin{Bmatrix}\frac{\partial \psi_x}{\partial x}-\frac{\partial^2 w}{\partial x^2}\\ \frac{\partial \psi_y}{\partial y}-\frac{\partial^2 w}{\partial y^2}\\ \frac{\partial \psi_x}{\partial y}+\frac{\partial \psi_y}{\partial x}-2\frac{\partial^2 w}{\partial x\partial y}\end{Bmatrix} \tag{46}$$

In these strain-displacement relations, the first two terms are the same as in the FSDT, the third term is new and can be though of as a correction on the previous theory. With Equations 34 the inertia terms in the constitutive equations become

$$\int_{z^-}^{z^+}\rho\frac{\partial^2 u}{\partial t^2}dz = I_o\frac{\partial^2 u_o}{\partial t^2}-I_1\frac{\partial^2 \psi_x}{\partial t^2}+I_4\frac{\partial^2}{\partial t^2}\left(\psi_x-\frac{\partial w_o}{\partial x}\right) \tag{47}$$

$$\int_{z^-}^{z^+}\rho\frac{\partial^2 v}{\partial t^2}dz = I_o\frac{\partial^2 v_o}{\partial t^2}-I_1\frac{\partial^2 \psi_y}{\partial t^2}+I_4\frac{\partial^2}{\partial t^2}\left(\psi_y-\frac{\partial w_o}{\partial y}\right)$$

$$\int_{z^-}^{z^+}\rho\frac{\partial^2 w}{\partial t^2}dz = I_o\frac{\partial^2 w}{\partial t^2}$$

$$\int_{z^-}^{z^+}\rho\frac{\partial^2 u}{\partial t^2}zdz = I_1\frac{\partial^2 u_o}{\partial t^2}-I_2\frac{\partial^2 \psi_x}{\partial t^2}+I_5\frac{\partial^2}{\partial t^2}\left(\psi_x-\frac{\partial w_o}{\partial x}\right)$$

$$\int_{z^-}^{z^+}\rho\frac{\partial^2 v}{\partial t^2}zdz = I_1\frac{\partial^2 v_o}{\partial t^2}-I_2\frac{\partial^2 \psi_y}{\partial t^2}+I_5\frac{\partial^2}{\partial t^2}\left(\psi_y-\frac{\partial w_o}{\partial y}\right)$$

The two additional inertia terms introduced here are $(I_4, I_5)=\int_{z^-}^{z^+}\rho(z^4,z^5)dz$.

6.2- Constitutive Equations

The constitutive equations for this plate theory are

$$\begin{Bmatrix} N \\ M \end{Bmatrix} = \begin{bmatrix} A & B & E \\ B & D & F \end{bmatrix} \begin{Bmatrix} \varepsilon^0 \\ K \\ K^{(2)} \end{Bmatrix} \tag{48}$$

where $\{\varepsilon^o\}$ and {K} are given by Equations (19.b, 37) and

$$\{K^{(2)}\} = \begin{Bmatrix} \dfrac{\partial\psi_x}{\partial x} - \dfrac{\partial^2 w}{\partial x^2} \\ \dfrac{\partial\psi_y}{\partial y} - \dfrac{\partial^2 w}{\partial y^2} \\ \dfrac{\partial\psi_x}{\partial y} + \dfrac{\partial\psi_y}{\partial x} - 2\dfrac{\partial^2 w}{\partial x \partial y} \end{Bmatrix} \tag{49}$$

In addition to the usual A, B, D matrices, we define

$$[E] = \frac{4}{3h^2}\int_{z^-}^{z^+} [\bar{Q}] z^3 dz, \quad [F] = \frac{4}{3h^2}\int_{z^-}^{z^+} [\bar{Q}] z^4 dz \tag{50}$$

With these strain-displacements relations and constitutive equations, substitution into Equations (5.c, 11, 12) give the equations of motion of the RSDT in terms of the kinematic variables u_o, v_o, w, ψ_x, and ψ_y. Note that this theory that the same number of variables as the FSDT. The RSDT was used by many authors including Meunier and Shenoi (2001) to develop finite elements for the analysis of sandwich plates with composite facings.

7- Limitation of the CPT, FSDT and RSDT

The three most commonly used plate theories are all based on the hypothesis that the stress in the transverse direction is negligible. From the Equations 39 we see that, due to Poisson's effect, the transverse normal strain calculated from

$$0 = c_{31}\varepsilon_{xx} + c_{32}\varepsilon_{yy} + c_{33}\varepsilon_{zz} \tag{51}$$

is not zero even if the transverse normal stress is zero. For the FSDT, using Equations 34,

$$\varepsilon_{zz} = \frac{\partial w}{\partial z} = z\left[c_{31}\frac{\partial \psi_x}{\partial x} + c_{32}\frac{\partial \psi_x}{\partial x}\right] / c_{33} \tag{52}$$

Therefore, the transverse normal strain is expected to vary linearly through the thickness. If the material properties remain constant through the thickness, integrating Equation 51 with respect to z gives

$$w = w_o + z^2\,\varphi_z(x, y, t) \tag{53}$$

The transverse displacement consists of a constant term w_o and a z^2 term multiplied by a function of x, y and t. The transverse displacement is expected to vary according to Equation 53 in a number of higher order theories.

8- Theories Including Deformations in the Transverse Direction

Several theories allow for deformation in the transverse direction. In the theory presented by Reissner (1944), assuming that "in plane" and bending deformations are uncoupled,

$$u = z\psi_x + z^3\varphi_x \qquad v = z\psi_y + z^3\varphi_y \qquad w = w_o + z^2\varphi_z \tag{54}$$

Hanna and Leissa (1994) started with the same kinematics assumption and required that the transverse shear strains vanish on the top and bottom surfaces of the plate. This way, ϕ_x and ϕ_y can be expressed in terms of the other variables and the displacements can be written as

$$u = \left(z - \frac{4z^3}{3h^2}\right)\psi_x - \frac{4z^3}{3h^2}\frac{\partial w_o}{\partial x} - \frac{z^3}{3}\frac{\partial \varphi_x}{\partial x} \tag{55}$$

$$v = \left(z - \frac{4z^3}{3h^2}\right)\psi_y - \frac{4z^3}{3h^2}\frac{\partial w_o}{\partial y} - \frac{z^3}{3}\frac{\partial \varphi_x}{\partial y}, \qquad w = w_o + z^2\varphi_z$$

Taking $\varphi_z = 0$ in Equation 54 we recover the kinematic assumptions made by Pandya and Kant (1988). The same theory was used by Roque et al (2007). When $u_o = v_o = 0$, Equations 55 reduce to the kinematics relation of the RSDT. As more complicating factors are taken into account, the theories become more complex and the equations become more involved.

9- Dynamic Response

In the most general case, the dynamic response of a composite plate is best determined numerically to account for the geometry of the plate, its boundary conditions, and the coupling between the different types of motion introduced by the anisotropy of the material. However, for most impact test conditions, the geometry is simple (rectangular or circular), the plate is simply supported along the edges, and the plate is symmetrically laminated. For such a plate with a rectangular planform, a simple analytical solution can be obtained that leads to a simple spring-mass model used by many investigators.

9.1- Free Vibrations

In this section we consider the free vibration of plates in the transverse direction according to the classical plate theory for cases where the motion is governed by Equation 31 (classical plate theory). For a rectangular plate with simple supports along the edges, the boundary conditions are

$$\begin{aligned} &w(0,y,t) = w(a,y,t) = w(x,0,t) = w(x,b,t) = 0 \\ &M_x(0,y,t) = M_x(a,y,t) = M_y(x,0,t) = M_y(x,b,t) = 0 \end{aligned} \qquad (56)$$

In this case, the mode shapes can be taken in the form

$$\varphi_{ij}(x,y) = \sin\frac{i\pi x}{a}\sin\frac{j\pi y}{b} \qquad (57)$$

The function $w = \varphi_{ij}(x,y)\sin\omega_{ij}t$ satisfies the equation of motion (Equation 31) and the boundary conditions provided that the natural frequencies are given by

$$\omega_{ij}^2 = \frac{\pi^4}{I_o a^4}\left[D_{11}i^4 + 2(D_{12} + 2D_{66})i^2 j^2 r^2 + D_{22} j^4 r^4\right] \quad (58)$$

where r = a/b is the aspect ratio of the plate. It can be shown that $\int_\Omega \phi_{ij}\phi_{mn}\, d\Omega = 0$ when $\phi_{ij} \neq \phi_{mn}$ and it is said that the mode shapes are orthogonal.

9.2- Modal Superposition

Modal superposition is a technique in which the displacements of the structure are expanded into a linear combination of the mode shapes in order to obtain a set of uncoupled second order ordinary differential equations from the partial differential equation governing the motion of the system. The resulting ODE can be solved separately and the results can be combined to give the response of the structure. Using the modal superposition approach, the transverse displacements are expanded in the double series

$$w(x,y,t) = \sum_{i=1}^{m}\sum_{j=1}^{n} \alpha_{ij}(t)\, \varphi_{ij}(x,y) \quad (59)$$

where the α's are called modal participation factors. Substituting into the equation of motion and using the orthogonality property of the mode shapes yields a set of m x n uncoupled modal equations of the form

$$\ddot{\alpha}_{ij} + \omega_{ij}^2 \alpha_{ij} = q_{ij} \quad (60)$$

where

$$q_{ij}(t) = \frac{4}{mab}\int_0^a\int_0^b q(x,y,t)\, \phi_{ij}(x,y)\, dx dy \quad (61)$$

For a concentrated force F applied at (x_o, y_o), $q = F\delta(x - x_o)\delta(y - y_o)$ where δ is the Dirac delta function. Solving each modal equation and substituting into Equation 59 gives the total response to the loading. In general only a few modes contribute to the overall response and often the overall response is dominated by that of the first mode.

9.3- SDOF Model

In most cases, impact tests are conducted on rectangular specimens impacted in the center. In this case the equation of motion for the first mode is

$$m\ddot{\alpha}_{11} + k\alpha_{11} = F \qquad (62)$$

where $m = I_o ab/4$ and $k = m\omega_{11}^2$ are the modal mass and the modal stiffness. F is the external force. This equation is that of a single degree of freedom (SDOF) system of mass m and stiffness k. This type of model is often used in the analysis of composite structures subjected to impacts.

10- Conclusion

This chapter provides an introduction to the three most commonly used plate theories starting from three-dimensional elasticity. The derivation of the equations of motion emphasized the commonality between the theories and their differences based on various assumptions made. These theories are part of what are called "equivalent single layer theories". Other plate theories such as layer-by-layer theories or zig-zag theories are also available but are not discussed here. The various theories described in this chapter are used in modeling low velocity impacts on composite structures and finite element models based on these theories are available in finite element codes. Therefore it is important to be familiar with these theories when studying the dynamic impact response. The formulation of beam theories and shells theories follow the development of plate theories quite closely so the material in this chapter should be helpful in understanding these various theories.

11- Acknowledgements

Chapters 1-3 in this book were written with the support of a grant from the Office of Naval Research. The help and support of Dr. Y.D.S. Rajapakse, manager of the Solid Mechanics program is greatly appreciated.

References

Meunier M. and Shenoi R.A. (2001) Dynamic analysis of composite sandwich plates with damping modelled using high-order shear deformation theory. *Composite Structures* 54 (2-3): 243-254.

Mindlin R.D. (1951). Influence of rotatory inertia and shear on flexural motions of isotropic, elastic plates. *Journal of Applied Mechanics* 73: 31–38.

Reddy J.N. (1984). A simple higher-order theory for laminated composite plates. *Journal of Applied Mechanics* 51, 745-752.

Reissner E. (1945). The effect of transverse shear deformation on the bending of elastic plates. *Journal of Applied Mechanics*, 67: A-69–A-77.

Reissner E. (1944). On the theory of bending of elastic plates. *Journal of Mathematical Physics*, **23**, pp. 184–191.
Timoshenko S.P. (1921). On the correction for shear of the differential equation for transverse vibration of prismatic bars. *Philosophical Magazine Series* 6, 41: 744-746.
Hanna N.F. and Leissa A.W. (1994). A higher order shear deformation theory for the vibration of thick plates. *Journal of Sound and Vibration* 170 (4): 545-555.
Pandya B.N., Kant T. (1988) Higher-order shear deformable theories for flexure of sandwich plates-finite element evaluations. *International Journal of Solids and Structures* 24: 419–451.
Roque C.M.C., Ferreira A.J.M. and Jorge R.M.N. (2007). A radial basis function approach for the free vibration analysis of functionally graded plates using a refined theory. *Journal of Sound and Vibration* 300 (3-5): 1048-1070.

Impact Dynamics

Serge Abrate

Department of Mechanical Engineering and Energy Processes
Southern Illinois University, Carbondale, IL, USA

1- Introduction

This chapter provides an introduction to the dynamic response of composite structures subjected to low velocity impacts and the issues related to the development of accurate models for the analysis of such events. Impact due to tool drops or flying debris on a runway can introduce significant damage in composite structures. When subjected to the same impact composite structures are more likely to be damaged than similar structures made out of metals. Usually, damage is internal and cannot be detected by visual inspection. It can grow under load and can significantly reduce the load carrying capacity of the structure. Consequently, the effects of impacts on composite structures have been studied extensively and the literature on that subject has been reviewed in Abrate (1991, 1994, 1997, 1998).

This chapter starts with a review of the rigid body dynamics approach to impact which is based on the conservation of linear momentum and the definition of a coefficient of restitution with values between 0 and 1. This approach has the merit of being simple and of providing accurate estimates of the final velocities once the coefficient of restitution is known. This approach is shown not to be useful in this case because it does not provide the information required for the present problem: the contact force history. Then, we investigate the impact dynamics using simple spring-mass models to determine the effect of various parameters and the various phenomena that must be include in the analysis to obtain accurate predictions of the contact force history. The development of complete models that fully account for the dynamics of the projectile, the dynamics of the structure, and their interaction in the contact zone is discussed next. Finally, we address the problem of impacts on composite sandwich structures.

2- Rigid Body Impacts

In the study of the dynamics of rigid bodies, the direct central impact between two bodies A and B (Figure 1) is governed by the principle of conservation of energy

$$m_A v_A + m_B v_B = m_A v_A' + m_B v_B' \tag{1}$$

which provides a relationship between the velocities of the two bodies before and after impact.

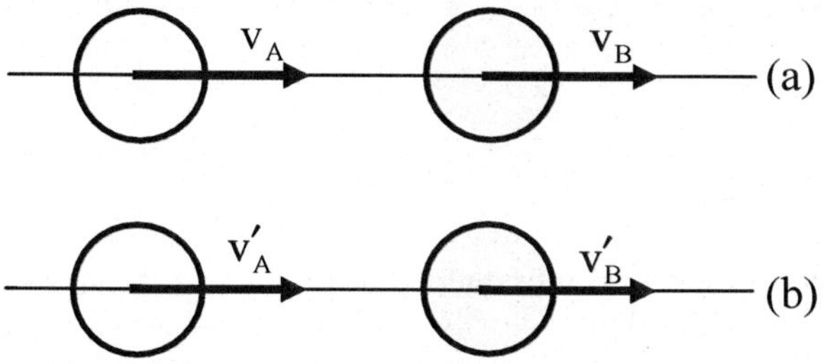

Figure 1. Impact between two rigid bodies A and B. (a): before impact; (b) after impact

In order to determine the velocities after impact, another equation is needed. In what is often called Newton's law of impact, the coefficient of restitution e is introduced as the ratio between the separation velocity and the approach velocity. That is

$$e = \frac{v_B' - v_A'}{v_A - v_B} \tag{2}$$

Solving those two equations, gives the velocities after impact

$$v_A' = [v_A(m_A - e m_B) + m_B v_B (1+e)]/(m_A + m_B) \tag{3}$$

$$v_B' = [m_A v_A (1+e) + v_B (m_B - e m_A)]/(m_A + m_B) \tag{4}$$

These two expressions can be used to calculate the kinetic energy after impact. Since the kinetic energy after impact cannot be larger than the initial kinetic energy, it can be shown that the coefficient of restitution must be less than or equal to one. When e=1, there is no loss of kinetic energy and the impact is said to be elastic.

If the positions of the two masses are defined by the coordinates x_A and x_B, the indentation x_A-x_B is maximum when the relative velocity $\dot{x}_A - \dot{x}_B = 0$ or, in other words, when the two masses have the same velocity: $\dot{x}_A = \dot{x}_B = v$. If the contact force is an increasing function of the indentation, it reaches a maximum value when the velocity of both masses is equal to v. The impact is divided into two phases: the loading phase (or compression phase) and the restitution phase. The loading phase starts from time t=0 and ends at time $t=t_{max}$ when the velocities of both masses are equal. Applying the impulse momentum relations to each body during that phase gives

$$m_A v_A - I_1 = m_A v \qquad m_B v_B + I_1 = m_B v \tag{5, 6}$$

where $I_1 = \int_0^{t_{max}} F dt$ is the impulse applied during the loading phase. Adding Equations 5, gives

$$v = (m_A v_A + m_B v_B)/(m_A + m_B) \tag{7}$$

Substituting into Equation 5 gives

$$I_1 = \frac{m_A m_B}{m_A + m_B}(v_A - v_B) \tag{8}$$

The restitution phase starts at $t= t_{max}$ and ends at $t=t_c$, the end of the contact phase. Applying the impulse momentum relations to the restitution phase gives

$$m_A v - I_2 = m_A v'_A \qquad m_B v + I_2 = m_B v'_B \tag{9, 10}$$

where $I_2 = \int_{t_{max}}^{t_c} F dt$ is the impulse applied during the loading phase. From Equations (9, 10)

$$I_2 = \frac{m_A m_B}{m_A + m_B}(v'_A - v'_B) \tag{11}$$

From Equations (2, 8, 11) we find that the ratio of the two impulses is equal to coefficient of restitution. That is,

$$I_2 / I_1 = e \tag{12}$$

Equation 12 is the definition of the coefficient of restitution originally introduced by Poisson: e is the ratio of the impulse applied during the restitution phase and the impulse applied during the contact phase. During the impact the contact force F is always positive so I_2 is non-negative. The coefficient or restitution must then be in the range

$$0 \leq e \leq 1 \tag{13}$$

When e=0, the two bodies have the same velocity after impact and the impact is said to be plastic. In general, e is a number between zero and one. For two given bodies, the coefficient of restitution must be determined from experiments and is assumed to be constant in a certain velocity range.

With this approach, the final velocities of the two bodies can be easily predicted (Equations 3, 4). However, it cannot predict the contact force history, the contact duration, or the maximum contact force. Therefore, it is of little use in investigating the effects of impacts on composite structures.

3- Quasi-Static Approximation

In order to get some insight into the impact dynamics it is useful to consider a simple model in which the structure is assumed to respond in a quasi-static manner and is modeled by a spring which is then impacted by a rigid body of mass m. This minimal model provides an estimate of the contact force history and gives some insight into the effect of several parameters: impact velocity, mass of the impactor, and stiffness of the projectile. Then, we consider the effect of several complicating factors: (1) the weight of projectile which is often neglected because it is expected to be small compared to the contact force; (2) the mass of the structure which is often small in laboratory test conditions but not necessarily so in actual applications; (3) large deflections; (4) internal damping; (5) impact damage which is usually neglected at this stage of the analysis. This type of model can be directly applicable in some situations and the insights it provides are valuable.

3.1- Single Degree of Freedom Model

Considering the impact of a projectile of mass m on a linear elastic structure, assume that the structure deforms in a quasi-static fashion. At the point of impact, the deflection of the structure will be directly proportional to the displacement so that the structure can be modeled as a spring with a stiffness k that can be determined from experiments or numerically. The equation of motion of the single degree of freedom spring-mass system in Figure 2 is

$$m\ddot{x} + kx = 0 \tag{14}$$

with initial conditions

$$x(0) = 0 \qquad \text{and} \qquad \dot{x}(0) = V \tag{15}$$

Solving this initial value problem gives the contact force history as

$$F = kx = V\sqrt{mk}\,\sin \omega t \tag{16}$$

for $0 \leq t \leq T_c$. The contact duration and the maximum contact force are given by

$$T_c = \pi\sqrt{\frac{m}{k}} \qquad \text{and} \qquad F_{max} = V\sqrt{km} \tag{17}$$

respectively. The contact force history is a sine pulse (Equation 16). This model predicts that the maximum contact force increases linearly with the velocity of the projectile (Equation 17.b). It also increases with the square root of the mass of the projectile and the square root of the stiffness. The contact duration is proportional to the square root of the mass and inversely proportional to the square root of the stiffness (Equation 17.a).

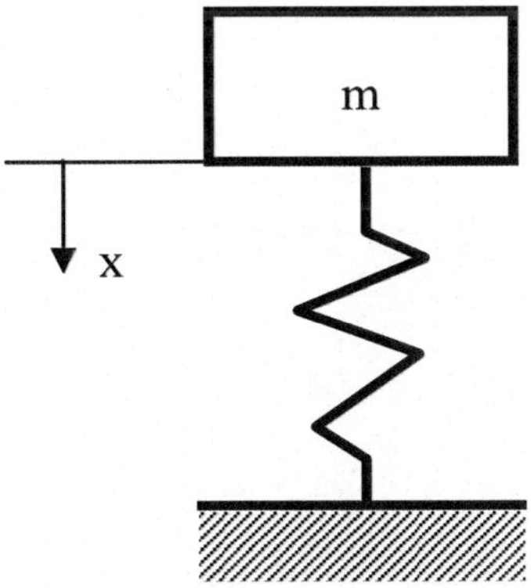

Figure 2. Single degree of freedom spring-mass model for impact dynamics analysis

Often, a series of experiments are conducted in which the initial velocity of the projectile is varied and all other parameters are kept constant. Equation 17 predicts that, when the maximum contact force is plotted versus the initial velocity, results should plot on a straight line. Deviation from linearity indicates the onset of significant damage that reduces the stiffness of the structure.

Recalling that the initial kinetic energy of the projectile is given by $U = \frac{1}{2}mV^2$, the maximum contact force can be written as

$$F_{max} = \sqrt{2kU} \tag{18}$$

Therefore, the maximum contact force is not a linear function of the kinetic energy and when F_{max} is plotted versus U, as it is often done in the literature, the resulting curve is a parabola opening towards the positive U axis.

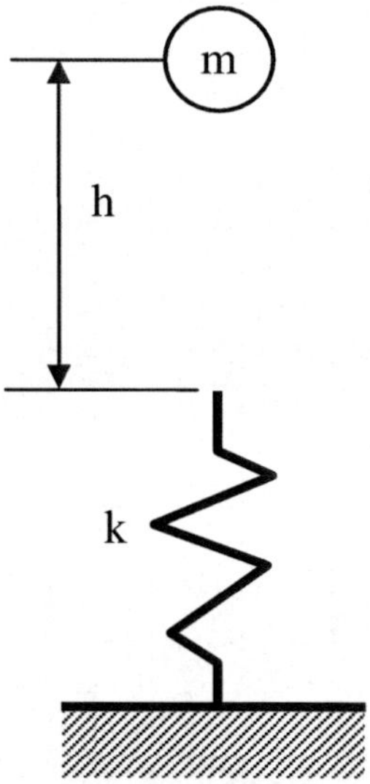

Figure 3. Drop-weight impact

3.2- Effect of the Weight of the Projectile

Often, the weight of the projectile is neglected in the analysis as in section 3.1. Adding the weight mg to the right hand side of Equation 14, the contact force history becomes

$$F = kx = V\sqrt{mk}\sin\omega t + mg(1 - \cos\omega t) \tag{19}$$

In order to assess the importance of the second term in the contact force history, consider the case of a mass dropped from a distance h (Figure 3). Using the work-energy principle, we find that, at the end of the free flight of the projectile,

$$V = \sqrt{2gh} \tag{20}$$

Then,

$$\frac{V\sqrt{km}}{mg} = \sqrt{2\frac{kh}{mg}} = \sqrt{2\frac{h}{\delta_{st}}} \tag{21}$$

where $\delta_{st} = mg/k$ is the static deflection of the spring under the weight mg. Therefore, the effect of the weight of the projectile is negligible when h is large compared with δ_{st}. Figure 4 shows an example where $h/\delta_{st} = 100$. The non-dimensional force $F/\left(V\sqrt{mk}\right)$ is plotted versus the non-dimensional time ωt. Curve a is the sine term in Equation 19, curve b represents the $(1-\cos\omega t)$ term, and curve c is the sum of the two. The weight of the projectile has a significant effect on the contact duration and the maximum contact force even if h is large. Including this effect is often necessary in order to match experimental results.

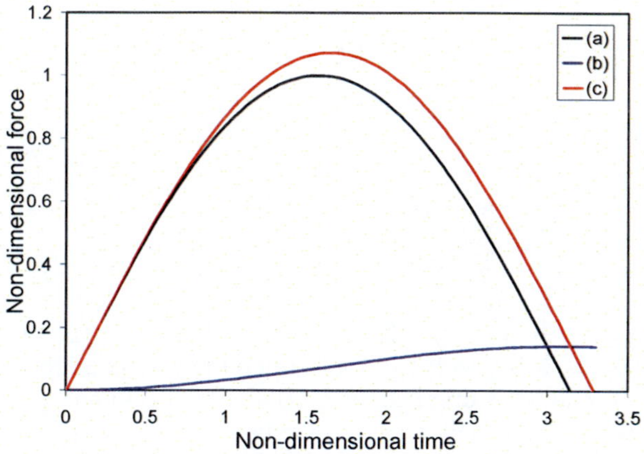

Figure 4. Effect of the weight of the projectile. (a): first term in Equation 19; (b): 2nd term in Equation 19; (c): Equation 19

3.3- Effect of the Mass of the Structure

In the previous model the mass of the structure was neglected assuming that is was much smaller than that of the projectile. Using a one term modal expansion, motion of the structure is governed by

$$m_s \ddot{x} + k_s x = V\sqrt{km}\ \sin\omega t \tag{22}$$

where m_s and k_s are the modal mass and the modal stiffness of the structure, x is the displacement of the beam at the point of impact, and the right hand side of Equation 22 is the contact force history obtained in Section 3.1 (Equation 16). The solution

$$x = \frac{V\sqrt{km}}{k_s\left(1 - \dfrac{\omega^2}{\Omega^2}\right)}\left[\sin\omega t - \frac{\omega}{\Omega}\sin\Omega t\right] \tag{23}$$

where $\Omega = \sqrt{k_s/m_s}$. When $\dfrac{\omega}{\Omega} << 1$, the mass of the structure is negligible. For the special case of a simply-supported beam impacted at mid-span with a bending rigidity EI, a length L, and a total mass M,

$$k = \frac{48EI}{L^3}, \qquad k_s = \frac{\pi^4 EI}{2L^3}, \qquad m_s = M/2 \tag{24}$$

In this case, the ratio

$$\frac{\omega}{\Omega} = \sqrt{\frac{48}{\pi^4}\frac{M}{m_B}} \tag{25}$$

indicates that the effect of the mass of the beam is negligible when it is small compared with the mass of the projectile. This is often the case in a drop-weight impact test where a large mass is dropped on a small size specimen. However, the inertia effects cannot be neglected when analyzing impacts by a small mass.

3.4- Effect of geometrical nonlinearities

When deflections become large, the behavior of the structure becomes nonlinear as membrane stiffening effects become significant. The force-displacement relationship can be written as

$$P = k_{bs} x + k_m x^3 \tag{26}$$

where k_{bs} is the stiffness including the effect of bending and shear deformation and k_m is the membrane stiffness. The equation of motion becomes

$$m\ddot{x} + k_{bs} x + k_m x^3 = 0 \tag{27}$$

or, in non-dimensional form,

$$\bar{x}'' + \bar{x} + \lambda \bar{x}^3 = 0 \tag{28}$$

where $\bar{x} = x/b$, $\tau = \omega t$, $\omega = \sqrt{k_{bs}/m}$, $b = V/\omega$ and $\lambda = mV^2 k_m / k_{bs}^2$. The initial condition (Equation 15) becomes $\bar{x}(0) = 0$ and $\bar{x}'(0) = 1$. Figure 5 shows the effect of the cubic nonlinearity on the contact force history. As λ increases, the contact duration decreases and the maximum contact force becomes larger.

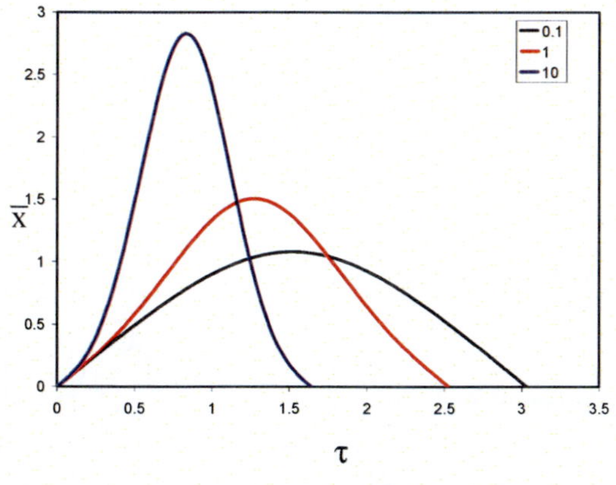

Figure 5. Non-dimensional contact force history for SDOF with cubic nonlinearity for three values of the parameter λ

3.5- Effect of damping

Composite materials generally have a certain amount of damping. For single degree of freedom systems with damping, the equation of motion is Peters (1997)

$$m\ddot{x} + c\dot{x} + kx = 0 \tag{29}$$

with initial conditions $x(0)=0$ and $\dot{x}(0)=V$. Defining the undamped natural frequency $\omega=\sqrt{k/m}$ and the damping ratio η by $c/m=2\omega\eta$, the solution is given by

$$x=\frac{V}{\omega}e^{-\omega\eta t}\sin\left(\omega\sqrt{1-\eta^2}\,t\right) \tag{30}$$

The contact force history $F=c\dot{x}+kx$ can be written as

$$F=V\sqrt{km}\,e^{-\omega\eta t}\left[\left(1-2\eta^2\right)\sin\left(\omega_d t\right)+2\eta\sqrt{1-\eta^2}\cos\left(\omega_d t\right)\right] \tag{31}$$

where $\omega_d=\omega\sqrt{1-\eta^2}$ is the damped natural frequency.

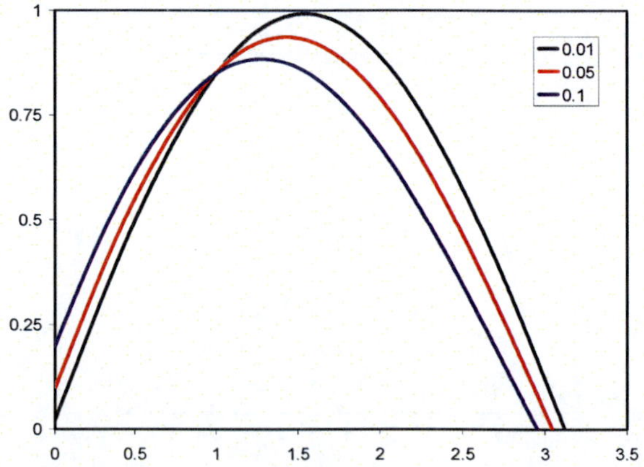

Figure 6. Non-dimensional contact force history for a SDOF system with damping for three values of the damping ratio

Figure 6 shows the non-dimensional contact force history $\overline{F}=F/\left(V\sqrt{km}\right)$ versus the non-dimensional time $\overline{t}=\omega t$ for three values of the damping ratio. As the damping ratio increases the contact duration and the maximum contact force decrease. We should also note that at t=0 the contact force jumps immediately from zero to

$$F(0) = V\sqrt{km}\left[2\eta\sqrt{1-\eta^2}\right] \qquad (32)$$

That initial jump is not physically realistic, it is not observed during experiments, and it increases with damping. The present results indicate that small amounts of damping have little effect of the contact force history. For larger amounts of damping other models are needed to provide realistic results.

3.6- Effect of damage

Most of the time analyses of the dynamic response of a composite structure to low velocity impacts are performed without accounting for the introduction of damage during the impact. This approach is convenient because it does not require the ability to predict the onset and growth of damage during the impact event. On the other hand it is logical to inquire whether or not this approach is sufficiently accurate to predict the contact force history. Following Sutherland and Guedes Soares (2004), the effect impact induced damage is introduced in the analysis assuming that force-deflection behavior of the structure is bilinear. That is, $P = k_o x$ when $x \le x_1$ and $P = k_1 x + (k_o - k_1)x_1$ when $x \ge x_1$. The motion is governed by

$$m\ddot{x} + k_o x = 0 \qquad \text{when} \qquad x \le x_1 \qquad (33)$$

$$m\ddot{x} + k_1 x + (k_o - k_1)x_1 = 0 \quad \text{when} \quad x \ge x_1 \qquad (34)$$

In non-dimensional form,

$$\bar{x}'' + \bar{x} = 0 \ \text{ when } \ \bar{x} \le \bar{x}_1 \qquad (35)$$

and

$$\bar{x}'' + \frac{k_1}{k_o}x + \left(1 - \frac{k_1}{k_o}\right)\bar{x}_1 = 0 \ \text{ when } \ \bar{x} \ge \bar{x}_1 \qquad (36)$$

Figure 7 shows the effect of stiffness reduction on the contact force history. In that figure, damage occurs when $\bar{x}_1 = 0.5$ and three values of the k_1 / k_o ratio were considered: 0.4, 0.6, and 0.8. The introduction of damage results in lower maximum contact forces and longer contact durations. These results also indicate that, in order to have a significant effect on the contact force history, damage must cause significant reduction in the overall stiffness of the composite which is often not the case with low velocity impacts. Therefore, when damage is limited to a small area near the point of impact, neglecting damage development is acceptable.

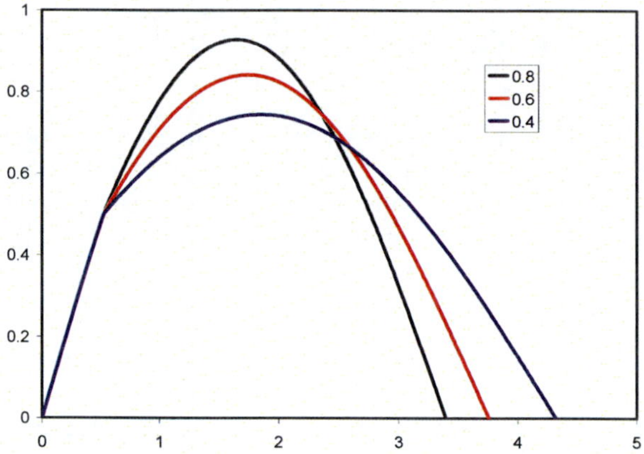

Figure 7. Contact force history including the effect of damage

4- Two Degree of Freedom Model

The previous section showed that the SDOF model is adequate when the mass of the projectile is large compare to that of the target. However it is expected to be lacking for small size projectiles. The previous model also did not account for the relative motion between the projectile and the target that is due to local indentation. This is acceptable for the type of problems treated by this model where the overall deflections of the target are much larger than the local indentation. In this section we consider a two-degree-of-freedom model that accounts for the mass of the target and the local indentation in the contact zone.

The two degree of freedom model in Figure 8 starts with a one mode approximation for the dynamics of the structure. This is represented by the modal mass m_1 and the spring S_1. The mass of the projectile m_2 is connected to the mass m_1 by a spring S_2. The indentation of the structure by the projectile is defined as the difference between the displacement of the projectile and that of the structure

$$\alpha = x_2 - x_1 \tag{37}$$

Relationships between the contact force P and the indentation are called contact laws. The mechanics of contact between bodies of different shapes has been studied extensively by many authors for the static case. Many authors have shown experimentally and though numerical analyses that the same static contact laws can be used to study impact problems.

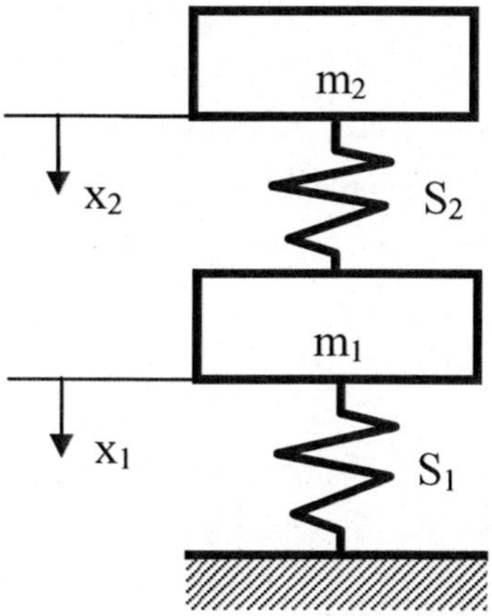

Figure 8: Two degree of freedom spring-mass model for impact

4.1- Contact Laws

Most investigators model local deformations using Hertz's contact law or the modified contact law proposed by Sun and Yang (see Abrate 1998). According to Hertz's law, the contact force P is related to the indentation α as follows

$$P = k_c\,\alpha^{3/2} \tag{38}$$

where the contact stiffness k_c is given by

$$k_c = \frac{4}{3} E\sqrt{R} \tag{39}$$

where

$$\frac{1}{R} = \frac{1}{R_1} + \frac{1}{R_2} \quad \text{and} \quad \frac{1}{E} = \frac{1-\nu_1^2}{E_1} + \frac{1-\nu_2^2}{E_2} \tag{40}$$

where R_1 and R_2 are the radii of curvature of the two bodies, E_1 and E_2 are the two elastic moduli, and ν_1 and ν_2 are the two Poisson ratios. With Hertz's contact law, Equation (38) is used for both the loading and the unloading phases. However, it is known that the unloading path is usually different from that followed during the loading phase. In 1952, Crook proposed the following relationship for the unloading phase (Abrate1998)

$$P = P_m \left[(\alpha - \alpha_o)/(\alpha_m - \alpha_o)\right]^{5/2} \tag{41}$$

where P_m is the maximum force reached before unloading, α_m is the maximum indentation and α_o is the permanent indentation. $\alpha_o = 0$ when α_m is less than a critical value α_{cr}. When $\alpha_m > \alpha_{cr}$, the permanent indentation is given by

$$\alpha_o = \alpha_m \left[1 - (\alpha_{cr} / \alpha_m)^{2/5}\right] \tag{42}$$

In 1982, Yang and Sun proposed what is now called the modified Hertz's law which consists of Equation 38 for the loading phase, Equation 41 for the unloading phase, and

$$P = P_m \left[(\alpha - \alpha_o)/(\alpha_m - \alpha_o)\right]^{3/2} \tag{43}$$

for subsequent reloading. These contact laws are widely used.

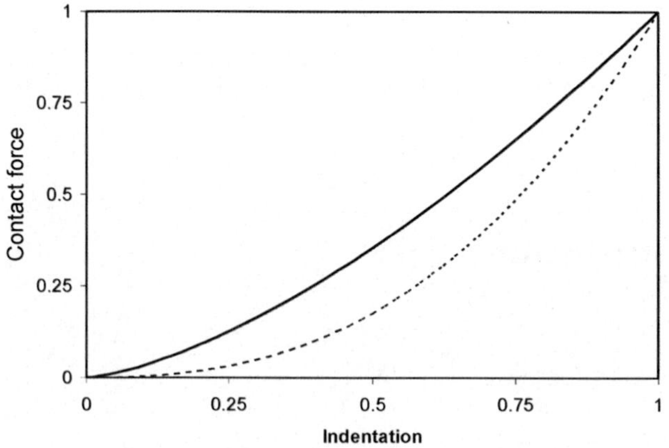

Figure 9: Non-dimensional contact law (P/P_m versus α_{cr} / α_m): thick line= loading, dashed line= unloading.

Figure 9 shows the contact law defined by Equations (38, 41) assuming that there is no permanent indentation. The area under the solid line is the energy absorbed during the loading phase and it is sometimes called the contact energy

E_c. The area under the dashed curve is the energy restituted during the unloading phase. From Equations (38, 41) it can be shown that the restituted energy is $\frac{5}{7}E_c$ so only 71.4 % of the energy is recovered.

In the static case, for a beam under three point bending, the energy stored in the beam under bending deformations is $U_b = P^2L^2/(96EI)$ and the energy absorbed by local indentation is $U_c = \frac{2}{5}k_c^{-2/3}P^{5/3}$. For a rectangular beam of width b and height h, using Equation 39, we find that

$$\frac{U_c}{U_b} = \frac{16}{5}\left(\frac{3}{4}\right)^{2/3} \frac{b}{P^{1/3}R^{1/3}} \frac{h^3}{L^3} \tag{44}$$

This expression shows several trends:

1. when the ratio h/L becomes small (thin beams), U_c becomes small. This explains why in many analyses the effect of local indentations is neglected entirely. In some analyses, the same contact law (Equation 38) is used for both the loading and the unloading phases. This is possible when U_c is not negligible but it remains relatively small so that the difference between U_c and the energy restituted during unloading is negligible.
2. Local indentation is always important when P is small but becomes small as P increases.
3. Local indentation effects are important for small size impactors and become less so when the radius R increases
4. Increasing the width of the beam increases the bending rigidity and decreases the amount of energy stored in the beam by bending deflections.

4.2- Impact on a Thick Laminate

If the indentation is much larger than the overall deflection of the structure, the motion of the structure is governed by

$$m\ddot{x} + k_c x^{3/2} = 0 \tag{45. a}$$

or, in non-dimensional form as

$$\bar{x}'' + \bar{x}^{3/2} = 0 \tag{45.b}$$

where $\bar{x} = x/b$, $\tau = \omega t$, $\omega = k_c^{2/5} m^{-2/5} V^{1/5}$, $b = V/\omega$. The initial conditions becomes $\bar{x}(0) = 0$ and $\bar{x}'(0) = 1$ again. This model is used to study cases that are often called "impact on a half-space". Figure 10 shows the non-dimensional contact force history predicted by this model.

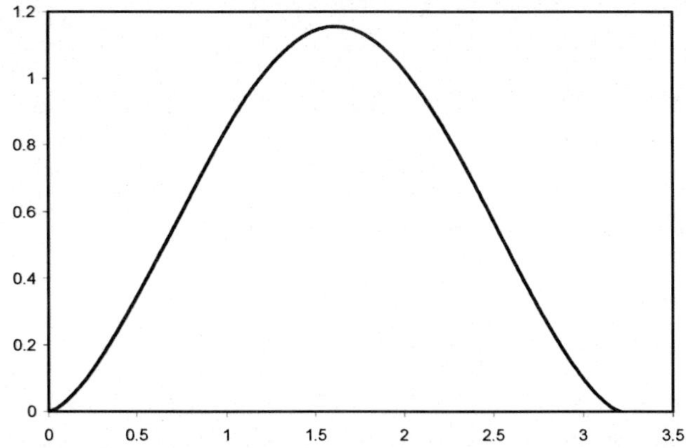

Figure 10. Nondimensional contact force versus time for impacts on a thick laminate

The maximum contact force can be determined using the work-energy principle. The work done by the contact force as the indentation increases from zero to its maximum value α_{max} is

$$E_c = \int_0^{\alpha_{max}} P\, d\alpha = \frac{2}{5} k_c\, \alpha_{max}^{5/2} \tag{46}$$

The initial kinetic energy is equal to E_c when the contact force reaches its maximum

$$\frac{1}{2} m V^2 = \frac{2}{5} k_c\, \alpha_{max}^{5/2} \tag{47}$$

Solving for the maximum indentation and substituting into the contact law (Equation 38), gives the maximum contact force

$$F_{max} = \left(\frac{5}{4}\right)^{3/5} k_c^{2/5} m^{3/5} V^{6/5} \tag{48}$$

This equation shows the influence of the contact stiffness, the mass of the impactor, and the impact velocity.

4.3- Two degree of freedom models

Many investigators use a two degree of freedom model (Figure 8) in which S_1 is a linear spring with a stiffness k_1 and S_2 is a nonlinear spring that represents the contact stiffness. k_1 accounts for the geometry of the structure, the support conditions, bending and shear deformations and be a linear spring if deformations remain small. Often, Hertz's contact law is used to model the indentation process so the force in S_2 is given by

$$P = k_c \left(x_2 - x_1\right)^{3/2} \tag{49}$$

The equations of motion of the system are

$$m_1\ddot{x}_1 + k_1 x_1 - k_c \left(x_2 - x_1\right)^{3/2} = 0 \tag{50}$$

and

$$m_2\ddot{x}_2 + k_c \left(x_2 - x_1\right)^{3/2} = 0 \tag{51}$$

These equations of motion are nonlinear because of the nonlinear contact behavior. Cheon et al (1999) and Lee et al (2000) used this model but neglected the mass of the structure (m_1 =0). Gong, Shim and Toh (1998) and Toh et al (1995) used a two degree of freedom spring-mass model to study impacts on composite shells. Both springs were assumed to be linear with spring constants k_1 and k_2. The motion of the system is governed by

$$m_1\ddot{x}_1 + k_1 x_1 + k_2(x_1 - x_2) = 0 \quad \text{and} \quad m_2\ddot{x}_2 + k_2(x_2 - x_1) = 0 \tag{52, 53}$$

The two natural frequencies of the system are

$$\omega_{1,2}^2 = \frac{1}{2}\frac{k_2}{m_2}\left\{\left[1 + \frac{m_2}{m_1}\left(1 + \frac{k_1}{k_2}\right)\right] \pm \sqrt{\left[1 + \frac{m_2}{m_1}\left(1 + \frac{k_1}{k_2}\right)\right] - 4\frac{k_1}{k_2}\frac{m_2}{m_1}}\right\} \tag{54}$$

The contact force history takes the form

$$P = A \sin \omega_1 t + B \sin \omega_2 t \tag{55}$$

In terms of the nondimensional time $\tau = t\sqrt{k_2/m_2}$ and the nondimensional displacements $\overline{x}_1 = x_1/\left(V\sqrt{m_2/k_2}\right)$ and $\overline{x}_2 = x_2/\left(V\sqrt{m_2/k_2}\right)$, the equations of motion (Equations 52) become

$$\overline{x}_1'' + \frac{m_2}{m_1}\left[\left(1+\frac{k_1}{k_2}\right)\overline{x}_1 - \overline{x}_2\right] = 0 \qquad \text{and} \qquad \overline{x}_2'' + \overline{x}_2 - \overline{x}_1 = 0 \qquad (56)$$

Figs. (11, 12) show the results for a particular case where $m_2/m_1=10$ and $k_1/k_2=1$. The displacement of the projectile depends essentially on the first mode of vibration while for the mass m_1 both modes make a significant contribution in this case. The contact force being proportional to the difference between the two displacements is seen to have significant oscillations. The natural frequencies calculated using Equation 54 show that, after non-dimensionalization, the half period for the first mode is 4.499 which corresponds to the time when the displacement of the projectile becomes negative. This example was selected to illustrate how dynamic interactions between the projectile and the structure can be important and require the use of two degree of freedom models. Similar results are obtained using the modified Hertz contact model. The present analysis showed that the oscillations observed in the contact force history are not due to nonlinearity but are due to the dynamics of the system. This subject is discussed further in Abrate (2001).

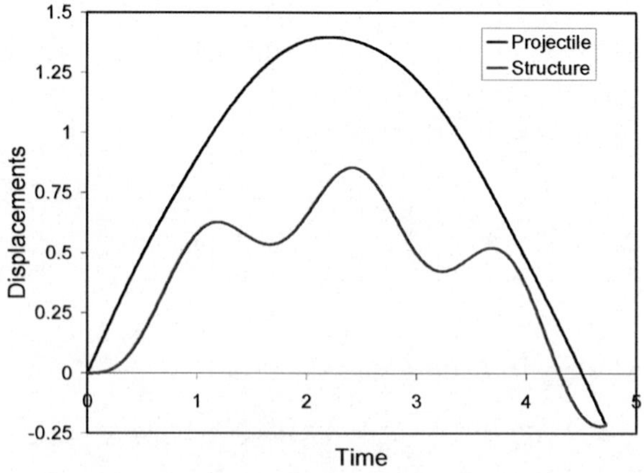

Figure 11: Non-dimensional displacements versus time during impact with $m_2/m_1 = 10$ and $k_1/k_2 = 1$

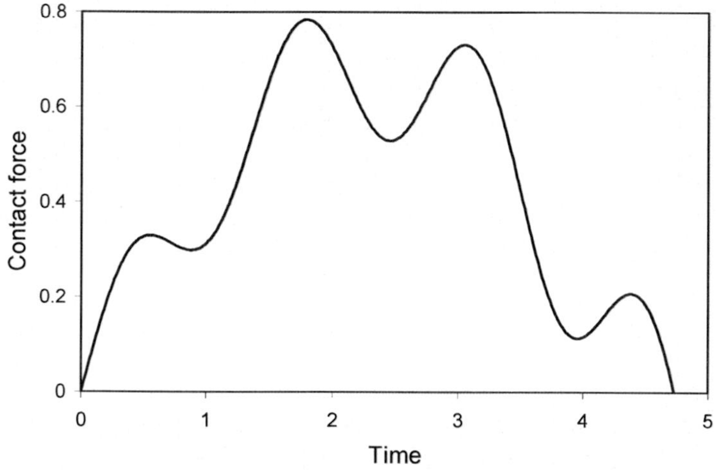

Figure 12: Non-dimensional contact force versus time for an impact with $m_2 / m_1 = 10$ and $k_1 / k_2 = 1$

5- Complete Models

For studying the impact dynamics, a complete model is one that fully accounts for the dynamic behavior of the structure, the contact behavior, and the dynamics of the projectile. In most cases, the structure is modeled using a beam, plate, or shell theory but in a number of studies a three dimensional elasticity approach is used. Then, as with any concentrated external force, the transient response is determined using either a closed form solution (if possible), a variational approximation method, the finite element method or some other numerical approach. Either way, the equations of motion of the structure can be written as

$$[M]\{\ddot{X}\}+[K]\{X\}=\{F\} \tag{57}$$

where $\{X\}$ is a vector with n degrees of freedom, [M] and [K] are n x n matrices, and $\{F\}$ is and external force vector in which the components are nonlinear functions of the displacements. The projectile is usually modeled as a rigid body and its motion is governed by

$$m\ddot{X}_{n+1} = -P \tag{58}$$

where m is the mass of the projectile, X_{n+1} is its displacements and P is the contact force. The contact force is modeled using an appropriate contact law, usually the modified Hertz law (Equation 38, 41). Equations (57, 58) form a set of n+1 nonlinear differential equations with n+1 unknowns.

If the structure is linear, the nonlinear coupling between the equations comes from the nonlinearity of the contact laws. Nonlinear behavior is also introduced by large deflections, and by material nonlinear behavior. For example, Meo et al (2005) included the effect of nonlinear stress-strain behavior of the composite in shear through the equation

$$\varepsilon_{12} = \frac{1}{G_{12}}\tau_{12} + \alpha\tau_{12}^3 \qquad (59)$$

Nonlinear behavior of a lamina is also observed under transverse loading (σ_{22}) and that phenomenon can be modeled by an expression similar to Equation 59. The stress-strain behavior in the fiber direction is usually linear. The effect of this type of material nonlinearity is usually limited in a laminate that is dominated by the properties of the material in the fiber direction.

Finally, another complicating factor is the fact that impacts introduce damage in composite structures: matrix cracks, delaminations, fiber fracture. Usually, the analysis of the impact dynamics does not include the effects of damage development. The contact force history is used as an input for a detailed stress analysis conducted to determine the onset and growth of damage. The uncoupling of the two analyses is justified because typical impact damage has little effect on the overall dynamic behavior of the structure.

6- Impact on Sandwich Structures

In the analysis of impacts on sandwich structures the new aspect to be included in the analysis is the response to local indentation. While the projectile still indents the top facesheets during the impact, a much larger deflection is produced due to the deformation of the core. The indentation in the case of of sandwich structures is defined as the difference between the displacement of the indentor and that of the midplane of the structure. This type of deformation is not accounted for in beam, plate or shell theories and should be included in the analysis through the use of a contact law. This aspect will be discussed next along with simple impact models that are used in the literature. Again, the goal is not to present an exhaustive review but to give an overview of the issues involved in modeling impacts on composite sandwich structures.

6.1- Contact Behavior

With sandwich structures, the projectile makes contact with one of the facesheets and indents that facesheet. In addition, because the core material is usually much softer, a much larger local deformation of the facesheet takes place. This second type of local deformation relative to the overall deformation of the structure is what is to be characterized by a contact law.

Experimental Results

For sandwich structures with foam and aluminum cores under low velocity impact, Akil Hazizan and Cantwell (2002, 2003) conducted experiments and determined coefficients Meyer's contact law

$$P = k_L \alpha^n \tag{60}$$

by curve fitting and showed that the exponent n had an average value of 1.2. The tests also showed no loading rate dependency. Anderson (2000, 2005) also modeled the indentation of sandwich structures using Meyer's law and found the exponent n to be between 1.1 and 1.2. Equation 60 is adequate in cases where the contact forces remain small. Some tests show a linear behavior when the contact force is small and a different behavior when the contact force exceeds a certain value. In the experiments conducted by Hasebe and Sun (2000) the loading curve was modeled by

$$P = \begin{cases} K_1 \alpha & \text{when} \quad \alpha < \alpha_Y \\ K_2 \alpha^n + K_3 & \text{when} \quad \alpha > \alpha_Y \end{cases} \tag{61}$$

where α_Y is the apparent yielding point in the loading curve. We note that the initial portion of the curve is linear and that nonlinear behavior is allowed after yielding. For unloading

$$P = F_m \left(\frac{\alpha - \alpha_o}{\alpha_m - \alpha_o} \right)^q \tag{62}$$

This equation is similar to the unloading curve used by Yang and Sun for composite structures (Equation 41).

Beam on elastic foundation model for the indentation of a sandwich beam

In Abot et al (2002), the initial linear portion is modeled as a beam resting on an elastic foundation. According to Hetenyi's solution for a finite beam on an elastic foundation subjected to a concentrated load at the center, the deflection at the center can be calculated as

$$\alpha = \frac{P\lambda}{2kb}\frac{\cosh(\lambda L)+\cos(\lambda L)+2}{\sinh(\lambda L)+\sin(\lambda L)} \tag{63}$$

where

$$\lambda = \left(\frac{bk}{4D_f}\right)^{1/4} \tag{64}$$

k is the modulus of the foundation defined as stress per unit deflection, P the applied load, L the beam length and b the beam width. The modulus k of the foundation can be calculated as

$$k = 0.28E_c\left(\frac{bE_c}{D_f}\right)^{1/3} \tag{65}$$

in terms of E_c, the modulus of elasticity of the core, D_f, the bending rigidity of the facing, and the width of the beam. Equation 65 was shown to be inaccurate, so that k must be determined experimentally. However, in order to determine the effect of the various parameters, Equations 63-65 can be combined to give

$$\alpha = 0.9185\frac{P}{E_c}\frac{\cosh(\lambda L)+\cos(\lambda L)+2}{\sinh(\lambda L)+\sin(\lambda L)} \tag{66}$$

Therefore, the indentation of the sandwich beam increases linearly with the contact force and it is inversely proportional to the compressive modulus of the core material.

6.2- SDOF impact models

When the contact behavior follows Meyer's law and the overall deflections are negligible, a SDOF model is governed by

$$m\ddot{w} + kw^n = 0 \tag{67}$$

Including damping effects with a dashpot in parallel with the spring (Figure 13), the equation of motion becomes

$$m\ddot{w} + c\dot{w} + k w^n = 0 \qquad (68)$$

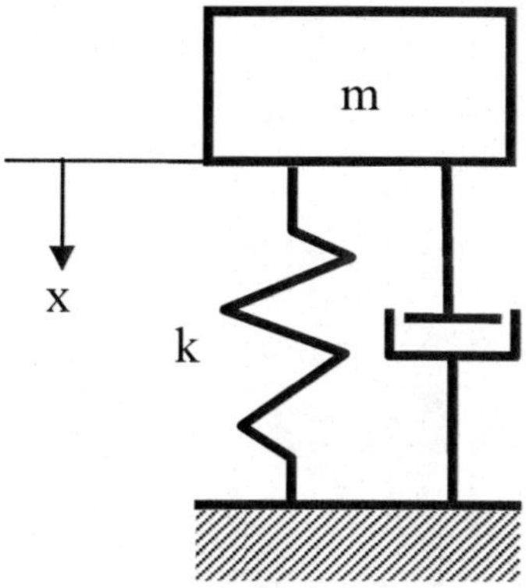

Figure 13. Single degree of freedom model with damping

Anderson (2005) and Olsson (2002) added the dashpot in series with the nonlinear spring (Figure 14).

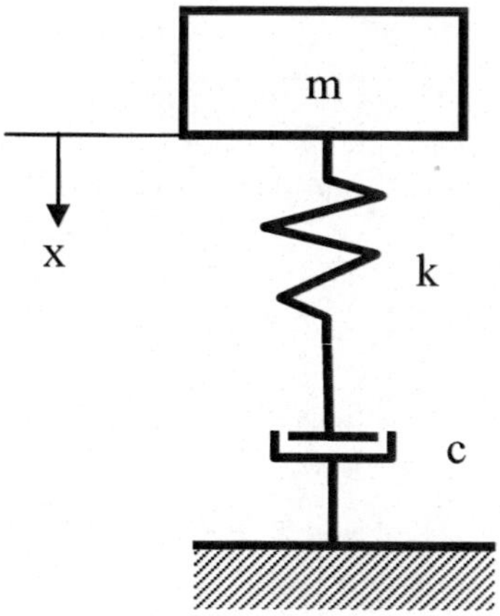

Figure 14. Single degree of freedom model with damping

6.3- Two Degree of Freedom Models

In the two degree of freedom model (Figure 15) adopted by Zhou and Stronge (2006), Olsson (2002) and others, m_1 represents the mass of the plate, m_2 the mass of the projectile, k_b is the bending stiffness and k_m is the membrane stiffness. In general, k_m is a nonlinear spring. Sometime this nonlinear behavior is not considered and k_m is omitted. S_2 is a nonlinear spring representing the contact stiffness and the force in S_2 is the contact force P.

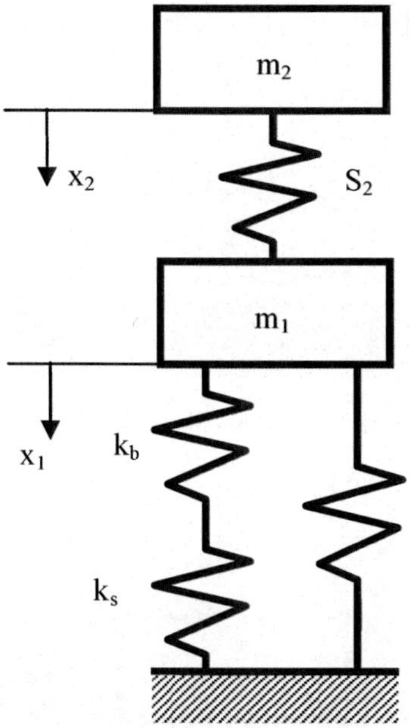

Figure 15. Two degree of freedom spring-mass model for impact dynamics

The equations of motion of the system are

$$m_2\ddot{x}_2 + P = 0 \tag{69}$$

$$m_1\ddot{x}_1 + k_{bs}\, x_1 + k_m\, x_1^3 - P = 0 \tag{70}$$

According to Zhou and Stronge, the contact law is given by

$$P = \chi \sqrt{\alpha + \beta\alpha^3} \tag{71}$$

where the indentation is defined as $\alpha = x_2 - x_1$ and χ and β are parameters. However, any contact law including Meyer's law (Equation 60), Hasebe and Sun's law (Equation 61) and other contact laws can be used.

7- Conclusion

This chapter presents an introduction to the impact dynamics problem based on simple analytical models that are used by many investigators and that allow to investigate the effects of various factors. The insight gained through this exercise is helpful in the development of numerical models for more complicated geometries and support conditions.

References

Abot J.L., Daniel I.M. and Gdoutos E.E. (2002). Contact Law for Composite Sandwich Beams. *Journal of Sandwich Structures and Materials* 4(2): 157-173.

Abrate S. (2001). Modeling of impacts on composite structures. *Composite Structures* 51(2): 129-138.

Abrate S. (1998). *Impact on Composite Structures.* Cambridge University Press.

Abrate S. (1997). Localized impact on sandwich structures with laminated facings. *Applied Mechanics Reviews* 50(2): 69-82.

Abrate S. (1994). Impact on Laminated Composites: Recent Advances. Applied Mechanics Reviews 47(11): 517-544.

Abrate S. (1991). Impact on Laminated Composite Materials. Applied Mechanics Reviews 44(4): 155-189.

Akil Hazizan Md. and Cantwell W.J. (2003). The low velocity impact response of an aluminium honeycomb sandwich structure. *Composites Part B: Engineering* 34(8): 679-687.

Akil Hazizan Md. and Cantwell W.J. (2002). The low velocity impact response of foam-based sandwich structures. *Composites Part B: Engineering* 33(3): 193-204.

Anderson T.A. (2005). An investigation of SDOF models for large mass impact on sandwich composites. *Composites Part B: Engineering* 36(2): 135-142.

Anderson T. and Madenci E. (2000). Experimental investigation of low-velocity impact characteristics of sandwich composites. *Composite Structures* 50(3): 239-247.

Cheon S.S., Lim T.S. and Lee D.G. (1999). Impact energy absorption characterristics of glass fiber hybrid composites. *Composite Structures* 46(3): 267-278.
Gong S.W., V.P.W. Shim and S. L. Toh. (1998). Determining effective contact stiffness between striker and composite shell. *Composite Structures* 43(2): 137-145.
Hasebe R.S. and Sun C.T. (2000) Performance of Sandwich Structures with Composite Reinforced Core. *Journal of Sandwich Structures and Materials* 2 (1): 75-100.
Lee D.G., Lim T.S. and Cheon S.S. Impact energy absorption characteristics of composite structures. *Composite Structures* 50(4): 381-390, 2000
Meo M., Achard F. and Grassi M. (2005). Finite element modelling of bridging micro-mechanics in through-thickness reinforced composite laminates. *Composite Structures* 71(3-4): 383-387.
Olsson R. (2002). Engineering Method for Prediction of Impact Response and Damage in Sandwich Panels. *Journal of Sandwich Structures and Materials* 4: 3-29.
Peters D.A. (1997). Optimum spring-damper design for mass impact. *SIAM Review* 30(1): 118-122.
Sutherland L.S. and Guedes Soares C. (2004). Effect of laminate thickness and of matrix resin on the impact of low fibre-volume, woven roving E-glass composites. *Composites Science and Technology* 64: 1691–1700.
Toh S. L., Gong S.W. and Shim V.P.W. (1995). Transient stresses generated by low velocity impact on orthotropic laminated cylindrical shells. *Composite Structures*, 31(3): 213-228.
Zhou D.W. and Stronge W.J. (2006). Low velocity impact denting of HSSA lightweight sandwich panel. *International Journal of Mechanical Sciences* 48(10): 1031-1045.

IMPACT RESPONSE OF LAMINATED AND SANDWICH COMPOSITES

Uday K. Vaidya
Department of Materials Science & Engineering, University of Alabama at Birmingham
Birmingham, AL 35294

ABSTRACT

With increasing use of composites in transportation, military ground and air vehicles, ship structures, infrastructure, sporting goods and the power industry, the understanding of impact and dynamic behavior is critical to composite designers and end-users. A wealth of knowledge has been published on dynamic impact response of composite materials and structures. Yet, with continually emerging materials and processes, there is a lack of systematic structure-property- performance relationships that provide guidelines on dynamic impact behavior of composites.

This chapter deals with the impact response of composites for the regimes of low, intermediate and high velocity. Material models used in finite element modeling of impact problems in composites are described. Using LS-DYNA as a finite element modeling platform, the modeling of progressive damage in laminated and sandwich composite, and its correlation to experiments for quasi-static shear punch and high velocity impact is described in detail. High strain rate impact behavior and nondestructive evaluation (NDE) of impact damage are also addressed. For each of these topics, details of test equipment, test methodology, instrumentation and test parameters, past and recent work are covered. This chapter is interrelated to two other chapters in this book, Chapter 3, and Chapter 7, and the information in the present Chapter complements the other two.

1.0 INTRODUCTION

The impact response of materials is generally categorized into low (large mass) velocity, intermediate velocity, high/ballistic (small mass) velocity, and hyper velocity regimes. As illustrated in Figure 1 - (i) Large mass impact, also known as low velocity impact (LVI) results from conditions arising from tool drops, which typically occur at velocities below 10 m s^{-1}. Global plate motion is established in LVI; (ii) Secondary blast debris, hurricane and tornado debris, and foreign object debris on roads and runways are considered to fall under the intermediate velocity impact regime. Intermediate impact events occur in the 10 m/s to 50 m/s range, and are characteristic of both low and high velocity impact; (iii) High velocity (ballistic) impact is usually a result of small arms fire or explosive warhead fragments. High-velocity impact response is dominated by stress wave propagation through the thickness of the material, in which the structure does not have time to respond, leading to a localized damage. Boundary condition effects can be ignored because the impact event passes before the stress waves reach the boundary. High velocity impacts range from 50 m/s to 1000 m/s, and (iv) In hyper velocity impact >2-5 km/s, the projectile is moving at very high velocities and the target material behaves like a fluid. This type of impact is most frequently studied in the context of developing protection against micrometeorites of objects and personnel in low earth orbit.

High strain rate of composite materials includes foreign object damage, blast loading, structural impacts and terminal ballistics. When a composite material is subjected to impact, the material undergoes large amounts of strain depending upon the magnitude of the impact, temperature range and strain rates. The composite material must, therefore, be characterized at the strain rates and temperatures of the intended application.

The mechanisms, progression and extent of damage need assessment to establish critical damage state and repair procedures. Nondestructive testing (NDE) procedures such as X-Ray radiography, ultrasonics, acoustic emission and vibration are promising NDE techniques for composites.

This chapter is organized as follows. Test conditions, equipment, methodology, past and recent work on impact and NDE on laminated and sandwich composites are provided for each of the following (i) low velocity impact (LVI) - The basic theory of LVI, test equipment, damage mechanisms in composites subjected to LVI and structural response in relation to material type (fiber, matrix, laminate construction etc), fiber length scales, impactor type and geometry, and residual strength is discussed. These discussions complement the Chapter 3 which covers the impact mechanics in greater detail; (ii) intermediate velocity impact (IVI) – the conditions of testing that fall under the IVI impact regime, test equipment and recent work in IVI of composites using blunt and sharp impactor on laminates and sandwich composites is discussed; (iii) The test equipment and recent work in ballistic impact, also referred to as high velocity impact (HVI) is discussed; (iv) Material models for simulating quasi-static penetration and ballistic impact are covered. The commonality of quasi-static shear punch and ballistic impact is presented. Chapter 7 complements the ballistic impact in the experimental aspects; (iv) high strain rate impact (HSR) – theory of HSR impact, typical test equipment, and HSR response of laminated and sandwich composites is described; (v) Nondestructive testing (NDE) methods to evaluate impact damage for various conditions such as barely visible damage, perforation and delamination in both laminated and sandwich composites are presented.

2.0 LOW VELOCITY IMPACT (LVI)

2.1 BACKGROUND

Polymer matrix composites are susceptible to internal damage caused by LVI. In many instances, damage is not apparent on the surface; however internal damage can be significant, thereby reducing the residual strength and service life of a composite structure. Studies on impact damage of composites have focused on impact dynamics, damage mechanics, post impact residual property characterization, and damage resistance[1-3].

LVI can be treated as a quasi-static event, the upper limit of which can vary from one to tens of meters per second depending on the target stiffness, material properties and the impactor mass and stiffness[4]. When impact velocities are below 5 m/s, the response type is controlled by impactor/plate mass ratio rather than impact velocity[5]. In low-velocity impact, the dynamic structural response of the target is important because the contact duration is long enough for the entire structure to respond to the impact and in consequence more energy is absorbed elastically.

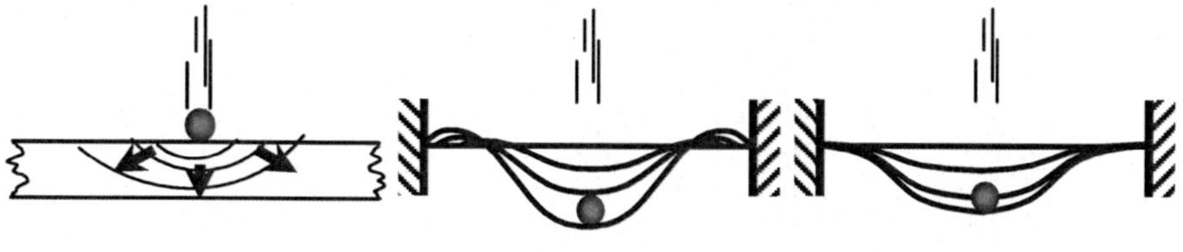

a b c

Figure 1. Impact Regimes; (i) Ballistic impact, very short impact times with dilatational wave dominated response; (ii) Intermediate velocity impact, short impact times with flexural and shear wave dominated response; and (iii) Low velocity impact, long impact times with quasi-static response. (Adapted from Olsson (2000))[5].

impact, the dynamic structural response of the target is important because the contact duration is long enough for the entire structure to respond to the impact and in consequence more energy is absorbed elastically.

Cantwell and Morton[6] classify low velocity as up to 10 m/s by considering test techniques including Charpy, Izod and instrumented falling weight impact testing. Liu and Malvern[7] suggest that the type of impact can be classified according to the damage incurred. Abrate[1] and Davies and Robinson[8] define a low-velocity impact as being one in which the through-thickness stress wave plays no significant part in the stress distribution, and suggest a model to determine the transition to high velocity.

A cylindrical zone under the impactor is considered to undergo a uniform strain as the stress wave propagates through the plate, resulting in a compressive strain ε_c as[3]:

$$\varepsilon_c = \frac{V_i}{V_s} \tag{1}$$

where V_i is the impact velocity and V_s is the speed of sound in the material. For failure strains between 0.5 and 1%, this results in transition to stress wave dominated events at 10-20 m/s for epoxy composites.

The failure modes depend on the loading condition. For LVI, the failure mode and energy absorption is highly dependent on the specimen size, stiffness, and boundary conditions[1-3]. The impact energy for a compliant specimen subjected to LVI is absorbed by primarily in the form of strain energy, in addition to energy for microcracking, fiber breakage and delamination, not accounting for loss from supports and boundary.

2.2 TEST EQUIPMENT

LVI testing of composites can be conducted using several types of equipment arrangements. Generally the impact is imparted through the use of a swinging pendulum, dropping weight, a rotating flywheel, or a gas-gun driven projectile[9,10]. The load may be applied in flexure, tension, compression, or torsion. The most typical equipment for LVI studies is the Izod and Charpy Impact Testers and Instrumented Falling Weight Impact Testing[9,10].

2.2.1 Izod and Charpy Impact Testing. Both these methods impart impact from a swinging pendulum mass as illustrated in Fig.2[11]. As the velocity and kinetic energy of a striking mass are varied, energy is transferred and work is done on the specimen. The energy is absorbed in the form of elastic and plastic deformation, hysteresis effects, friction between the specimen and test fixture and specimen acquiring kinetic energy. The effect of an impact load in producing stress depends on the extent to which the energy is expended in causing deformation.

The Charpy and Izod impact tests are frequently used for comparing the impact response of isotropic materials that have different compositions or that are fabricated from different processing conditions[12, 13]. Both Charpy and Izod tests employ a swinging pendulum (the pendulum is identified as 'striking edge' in Figure 2). In the Charpy test, the specimen is supported as a simple beam, while in the Izod test it is supported as a cantilever (ASTM E 23 for metals and ASTM D 256 for plastics).

For homogenous isotropic materials, the tests are usually conducted on specimens with a notch on the tension side. The notch acts as a stress concentration site to minimize the energy required for initiation of fracture. Thus the total measured energy required to fracture is essentially the energy required for the propagation for fracture. When applied to polymer composites in which the fracture phenomenon is more complex, these tests may not be adequate to represent a realistic impact condition[13,14].

2.2.2 Instrumented Falling Weight Impact Testing. This test is more representative of physical scenarios that produce transverse impact to a laminated or sandwich composite. The instrumented impact apparatus illustrated in Figure 3 consists of a drop tower equipped with an instrumented indentor (tup or dart). The tup houses a force transducer (Figure 4). Weights are added to the tup holder to provide the desired impact energy. Two optical flags separated by a distance of about 75 mm (3") to 150 mm (6") are placed in the path of the striker before it strikes the specimen. The velocity of impact is measured based on separation distance of the flags and time of flight of travel.

The force-time history is measured from the point of initial contact with the specimen, and as the striker traverses through the thickness of the specimen. Energy is calculated from integration of the force-time signal. The load-displacement, force-time and energy-time history are some of the parameters that are recorded.

Typical force-time histories – one representing minimal impact damage and the other with significant impact damage is shown in Figure 5a and 5b respectively. A force-time history provides good indication of the progression of damage in the composite[9-10, 15]. In Figure 5a, the force-time history is symmetrical in the ascending and descending portions of the force-time curve. The energy is returned to the impactor as evidenced by the descending nature of the energy-time curve past peak load. In Figure 5b, the force-time history is asymmetrical. The oscillations represent progressive impact damage in the specimen as the impactor penetrates, as subsequently all the impact energy is absorbed by the specimen, as seen by the constant energy profile past the peak load.

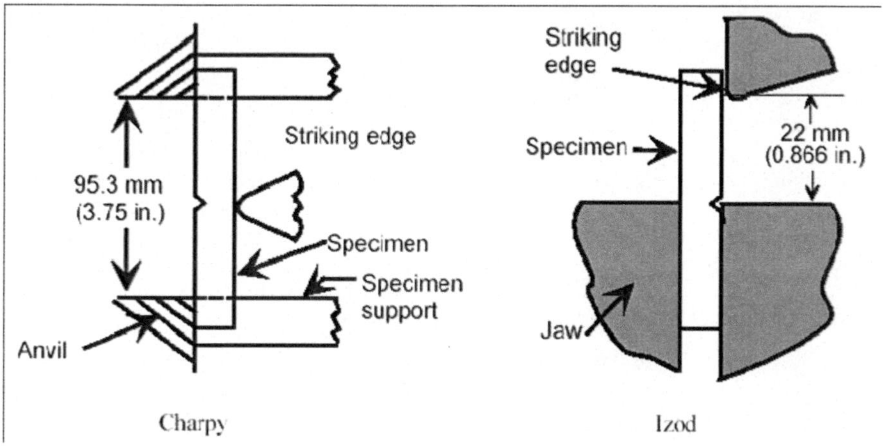

Figure 2. Charpy and Izod test configurations for low velocity impact testing[11].

In several studies conducted on composites, adjustable falling weights range from 1 (2.2 lb) to 15 kg (30 lbs), drop height of few centimeters to 3-4 m, velocity less than 10 m/s, impact energy ranges from 1 to 150 J and tup diameter of 12 mm (0.5") to 19 mm (0.75"). Sharp tip and flat area impactors have also been used[16,17].

For an impact condition in which the striker rebounds from the specimen, repeated (or multiple) impacts can occur which can cause excessive damage that is not representative of a single impact[18]. To avoid repeated impacts, (Figure 6) the impact device is equipped with two rebound arrestors on either side of the specimen. The rebound arrestors are pneumatically actuated, and spring up and separate the striker from the specimen, after the first impact.

The instrumented falling weight apparatus can have various specimen fixtures for loading flat laminate, cylindrical and other shape specimens. In general, a flat laminate is typically used for investigation of material response to impact. A specimen size of 150 mm x 150 mm (6" x 6"), 100 mm x 100 mm (4" x 4") and 150 mm x 100 mm (6" x 4") is typical. The 150 mm x 100 mm (6" x 4") specimen is used when the compression-after-impact (CAI) tests are to be conducted on the impacted specimen. In CAI, the specimen is first subjected to a LVI event that imparts controlled impact damage, followed by an in-plane compression test to evaluate residual strength (also see Section 2.8.2 for CAI). For LVI testing the specimen is fixed in a fully clamped support condition with a 50-75 mm (2-3") diameter opening that allows the tup to strike the specimen as illustrated in Fig. 4. The output of the load (force) transducer can be recorded by an oscilloscope or fed directly to a computer's signal processing board.

The various parameters from the LVI test include the incident impact energy, impact velocity, and incipient energy (E_i), total energy absorbed (E_t), total deflection (l_t), incipient damage point (P_i), maximum load (P_{max}), failure load point (P_f), total load point (P_t), energy at maximum load (E_m), deflection at maximum load (l_m), and energy ($E_p = E_t - E_m$) and deflection ($l_p=l_t-l_m$) after maximum load[11] (Figure 7). The point of incipient damage (P_i and E_i) is the first significant deviation or break from the initial portion of the load-time curve. This point signifies discernible matrix microcracking, fiber damage or onset of debonding and/or impact side facesheet rupture (in a sandwich). The point of maximum load (P_m) and energy (E_m) signifies the maximum penetration of the impactor and beginning of its rebound. In many cases, this incipient damage point coincides with the maximum load point. The failure load (P_t) and energy (E_t) points represent the specimen response to the end of the rebound phase of the impactor, and subsequently the end of event is represented by P_t and E_t respectively. Each of these parameters are influenced by number of factors including material thickness and geometry, boundary conditions, progression and sequence of damage accumulation, fiber orientation, interface variations and impactor geometry (sharp, blunt, spherical).

2.3 IMPACT ENERGY

The maximum energy E_o in the hammer (for Charpy, Izod) or the instrumented tup assembly prior to impact of the specimen is given by[11]:

$$E_0 = \frac{1}{2} m * V^2{}_0 \tag{2}$$

where, V_o is the hammer velocity just prior to impact and m is the mass of the impactor.

If w is the total weight of the system (impactor (with tup) and test specimen) then, for pendulum impact testing

$$w = w_h + (\frac{1}{3}) w_b \tag{3}$$

where w_h is the impactor weight and w_b is the beam weight. The impactor can be regarded as a free-falling body, $V_o = \sqrt{2gh_0}$ where h_o is the drop height. When the impactor tup makes contact with a test specimen, the impactor energy is reduced by an amount ΔE_0

$$\Delta E_0 = E_0 - E_f = (1/2) m (V^2{}_0 - V^2{}_f) \tag{4}$$

where, E_f is the kinetic energy of the time τ after initial contact between specimen and the tup, and V_f is the corresponding velocity.

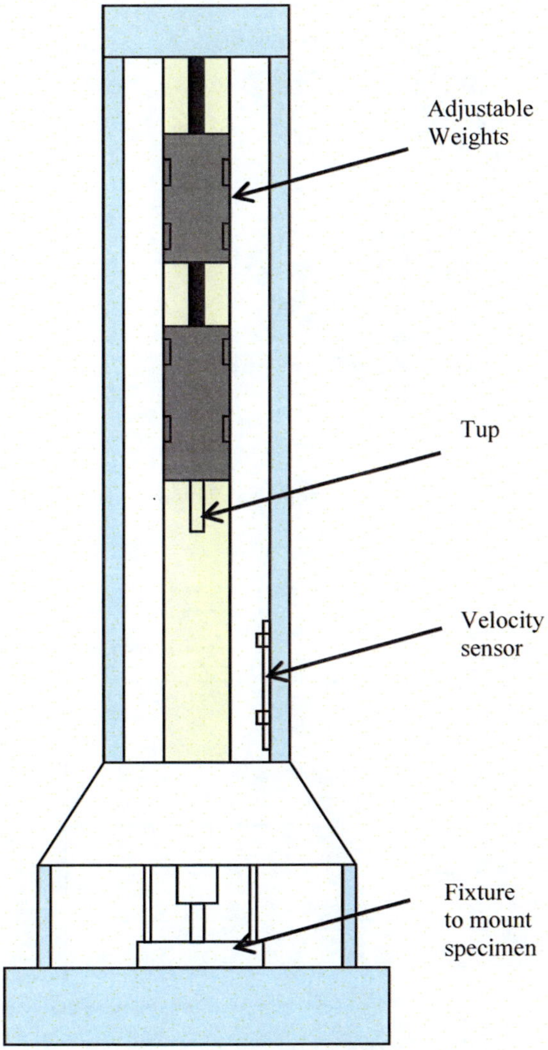

Figure 3. Low velocity instrumented falling weight impact testing equipment.

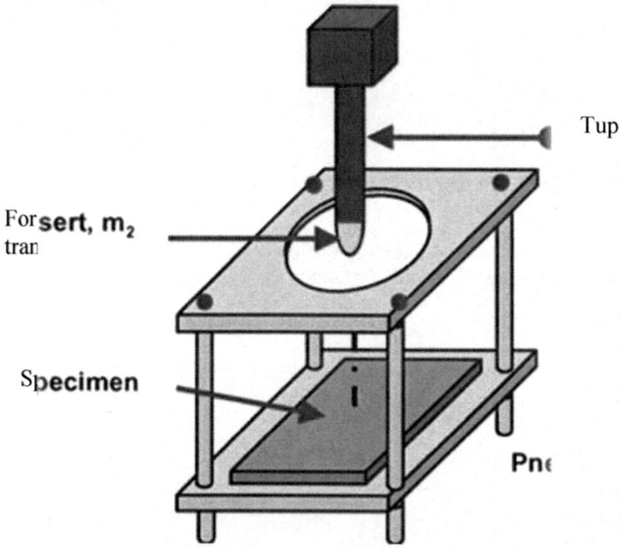

Figure 4. Typical specimen fixture with circular opening for tup to make contact with specimen.

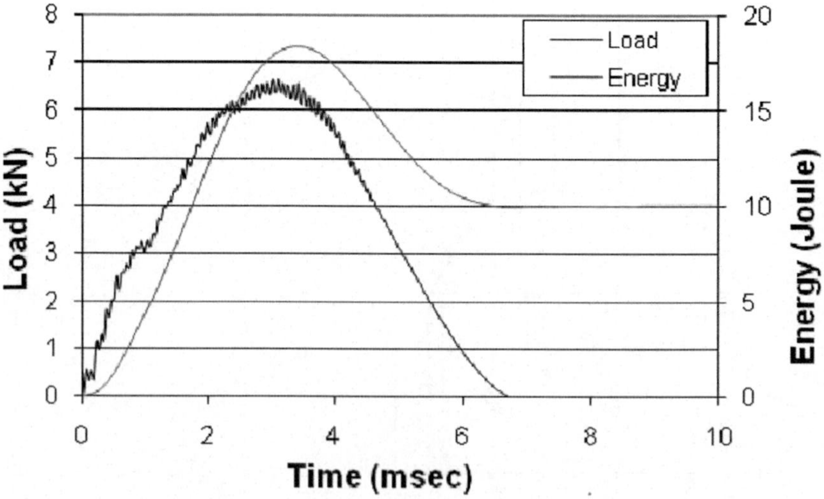

(a) continued on next page

(b) continued from previous page

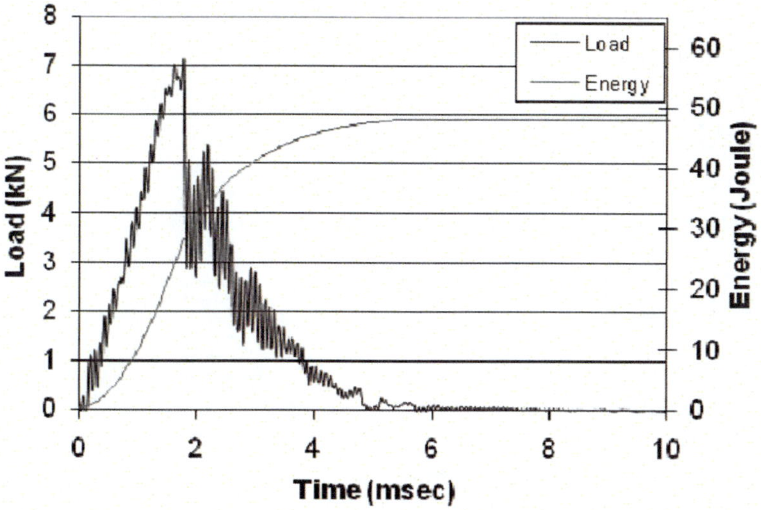

Figure 5. Typical load-time and energy-time curves; (a) laminate with minimal impact, (b) laminate with perforation damage[31].

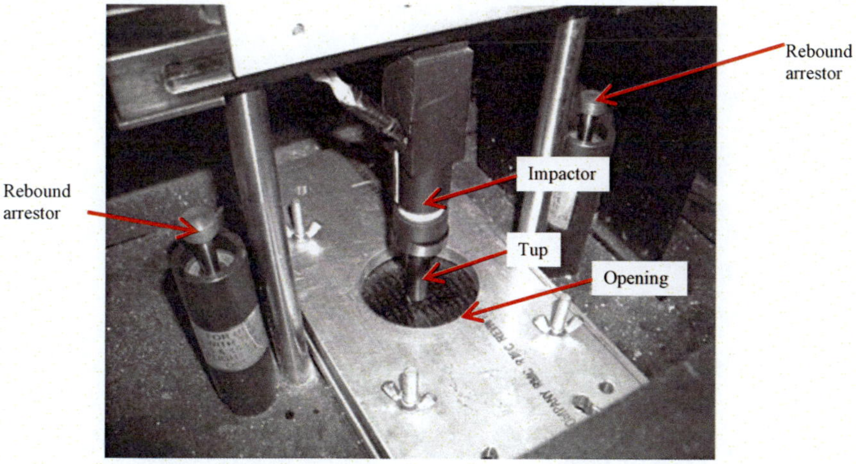

Figure 6. Rebound arrestors spring back after the first impact.

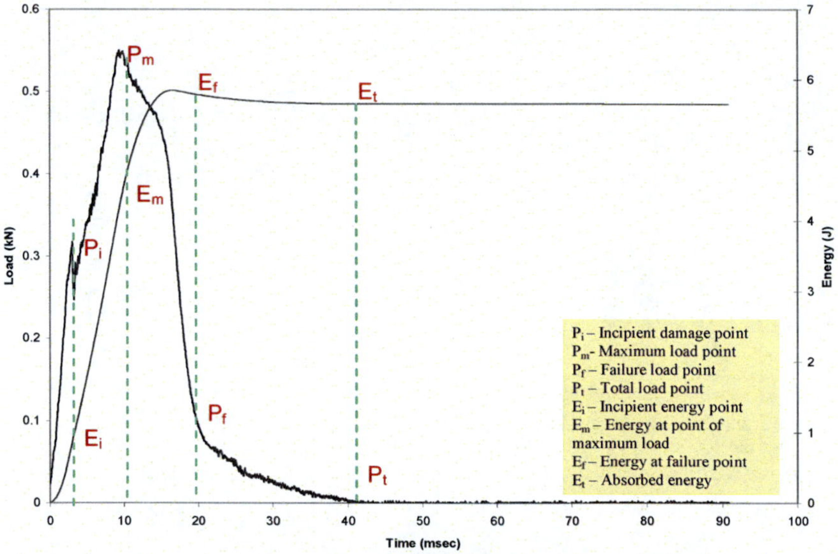

Figure 7. Typical Load and Energy versus Time curve and characteristic points for post impact analysis.

From the relationship between the impulse and momentum, it can be shown that

$$\int_0^{\tau} P dt = m(V_0 - V_f) \tag{5}$$

where P is the load at time τ.

Equation (5) in combination with Equation 4 yields:

$$\Delta E_0 = E_a[1 - E_a/(4E_0)] \tag{6}$$

where

$$E_a = V_0 \int_0^{\tau} P dt \tag{7}$$

Equation 6 can be shown to be equivalent to:

$$\Delta E_0 = \overline{V}\int_0^{\tau} P dt \tag{8}$$

where, by definition,

$$\overline{V} = (1/2)(V_0 + V_f) \tag{9}$$

2.4 MODES OF FAILURE IN LOW-VELOCITY IMPACT

The heterogeneous and anisotropic nature of fiber reinforced plastic (FRP) laminates contributes to different modes of failure. In most cases, these include; (1) matrix cracking that occurs parallel to the fibers due to tension, compression or shear; (2) debonding and/or delamination between the plies as a result of interlaminar stresses; (3) fiber breakage and buckling; and (4) penetration. The interaction between failure modes influences damage mode initiation and propagation.

2.4.1 Matrix damage. Matrix damage is induced by transverse LVI, and usually takes the form of matrix cracking, fiber/matrix debonding and delamination initiation. Barely visible or minimal damage occurs at low impact energy levels (1 to 5 J). Matrix cracks are usually oriented in planes parallel to the fiber direction in unidirectional fiber composites. The matrix cracks in the upper layers initiate at the contact edges of the impactor. Shear cracks are formed by the very high transverse shear stress through the material. Transverse shear stresses are related to the contact force and contact area[6].

The crack on the tensile side (opposite side of the impact) is termed a bending crack because it is induced by high tensile bending stresses and is characteristically perpendicular to the laminate. The bending stress is closely related to the flexural deformation of the laminate. The type of matrix cracking is dependent on the global characteristics of the impacted specimens. For long thin specimens bending cracks in the lower layers occur due to excessive transverse deflection and subsequent membrane effects predominate, whereas short thick specimens are stiffer and so higher peak contact forces induce transverse shear cracks under the impactor for impact side plies.

2.4.2 Delamination. A delamination is a separation of plies which progresses in the resin rich area between the plies. Delamination is a result of the bending stiffness mismatch between adjacent layers, i.e. the different fiber orientations between the layers[11,12]. The delamination area is generally oblong shaped with its major axis being coincident with the fiber orientation of the layer below the interface. For 0/90 laminates a peanut shape damage is typically formed. The bending-induced stresses are the major cause of delamination. Some studies have defined a bending mismatch coefficient between two adjacent laminates[7]. The greater the mismatch, the greater is the delamination area. This is also affected by material properties, stacking sequence and laminate thickness.

An empirical relationship to estimate elastic strain energy, E absorbed at the point of delamination failure under transverse impact is given by[3, 19]

$$E = \frac{2\tau^2 w L^3}{9E_f t} \tag{10}$$

where t = thickness, τ = interlaminar shear strength (ILSS), w = width, L = unsupported length and E_f = flexural modulus. The elastic strain energy absorbed reduces with increase in thickness and flexural modulus (high stiffness).

2.4.3 Delamination initiation and interaction with matrix cracking. Delamination caused by transverse impact occurs after a threshold energy has been reached. Delamination develops in the presence of a matrix crack. Delaminations do not always progress precisely in the interface region, but can propagate adjacent to the interface.

Delamination is initiated predominantly as a mode I fracture (although mixed mode fracture can also occur) due to high out-of-plane normal stresses caused by the presence of the matrix cracks and high interlaminar shear stresses along the interface. The matrix crack initiated delamination is typically due to the development of the interlaminar normal and shear stresses at the interfaces. Bending cracks and shear cracks can initiate delamination, but the delamination induced by shear cracks is unstable and that bending crack induced delaminations grow in a stable manner and proportional to the applied load.

2.4.4 Delamination growth. If 1 and 2 represent the in-plane directions of the laminate and 3 the out-of-plane direction, then delamination growth is governed by interlaminar longitudinal shear stress σ_{13} and transverse in-plane stress σ_{22} in the layer below the delaminated interface and by the interlaminar transverse shear stress σ_{23} in the layer above the interface. The energy absorbed per unit area of delamination growth is found to be constant[4]. The interlaminar fracture toughness is independent of delamination size and the delamination area can be predicted from peak impact force generated. The peak force and delamination area follows a linear relationship. Mode II fracture is the dominant failure mode of propagation[20-21].

2.4.5 Fiber failure. Fiber failure occurs later than matrix cracking and delamination in the fracture process. Fiber failure occurs just below the striker due to locally high stresses, and indentation effects (mainly governed by shear forces), and on the non-impacted face due to high bending stresses. Fiber failure is a precursor to catastrophic penetration mode. The energy required for fiber failure due to back surface flexure is given by[19]:

$$E = \frac{\sigma^2 wtL}{18E_f} \tag{11}$$

where σ= flexural strength, E_f = flexural modulus, w = width, L = unsupported length and t = specimen thickness.

2.4.6 Penetration. Penetration is a macroscopic mode of failure and occurs when the fiber failure reaches a critical extent, enabling the impactor to completely penetrate the material. The impact energy penetration threshold rises rapidly with specimen thickness. The major forms of energy absorption during laminate penetration are; shear-out (shear plug), delamination and elastic flexure. Of these mechanisms, 'shear-out' accounts for 50-60% depending on the plate thickness. Various factors including tow size, fiber sizing, orientation, weave architecture, matrix type and interface have an influence on the penetration process[1,22].

A suggested analytical model of penetration to give the energy absorbed as:

$$E = \pi\gamma^2 td \tag{12}$$

where γ = fracture energy, d = diameter of impactor, and t = plate thickness.

2.4.7 Damage in randomly oriented fiber laminates. When the fibers are oriented randomly, the crack patterns are less easy to establish. Sheet molding compound (SMC), glass mat thermoplastics (GMT), long fiber thermoplastics (LFT) and continuous filament mats (CFM) are common examples of randomly oriented short and long fiber composites[23]. The fiber length and fiber aspect ratio plays an important role in the impact response of randomly oriented composites[23-24]. As seen in Figure 8, the impact resistance increases with increasing fiber length. For fiber length less than the critical fiber length, the short fibers enhance stiffness of the composite, but are not effective in resisting impact (Figure 9a).

With increasing fiber length and fiber aspect ratio, the reinforcement engages more effectively with the striker by mechanisms of fiber pull-out, increase fiber/matrix shear and fiber breakage, as observed in long fiber glass reinforced thermoplastics (Figure 9b and 9c). The impact damage resistance and energy absorbed is much higher due to mechanism of fiber pull-out and enhanced fiber/matrix interface as illustrated in Fig. 9c[25]. The progression of damage in such randomly oriented composites approximately follows: (1) indentation (crushing of matrix under the impactor), (2) fiber/matrix pull-out, (3) bending fracture, and (3) perforation (i.e. damage resulting from penetration and associated fracture).

At low impact energy, in randomly oriented composites, matrix cracks are formed on the impacted surface in the form of short cracks and series of rings away from the point of contact[3]. This type of failure is caused by the tensile strain wave moving out from the center of impact.

2.5 IMPACTOR GEOMETRY AND MASS

The impactor size, shape, mass, material, and angle of incidence all have a strong influence on the response of the specimen [1,9,10,26, 27]. Sharp impactors create more surface and fiber damage, while blunt impactors tend to cause more internal delamination[22] (Figure 10). A blunt object (a large diameter projectile) emulates the effects of debris, and produce larger damage area dominated by delamination. A sharp conical impactor produces more fiber breakage which results in localized damage and high indentation/penetration depth. The momentum transfer to graphite/epoxy targets by flat impactors is reported to be about four times greater than that of a sharp or conically tipped impactor[28].

The impactor mass has a significant effect on energy absorption in composites while maintaining impactor size and shape. A heavier impactor promotes a global panel response causing significant bending, and energy absorption in the panel. A lighter mass concentrates the energy locally resulting in lower penetration energy.

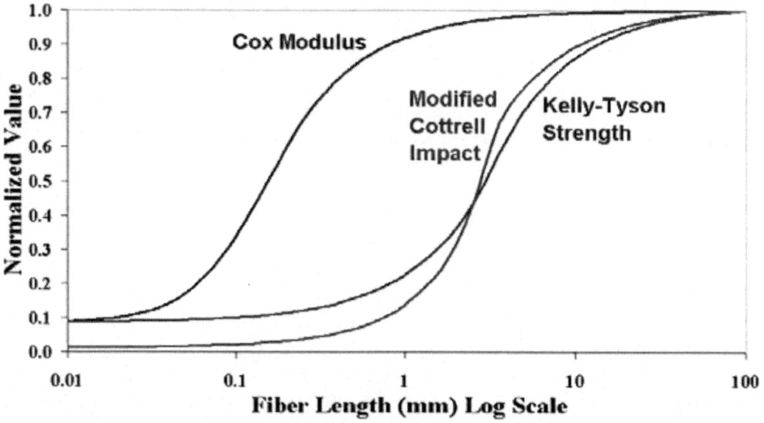

Figure 8. Normalized mechanical properties as a function of fiber length scale[50]. The details of the Cox model, the Modified Cottrell Impact and Kelly-Tyson provided in reference [50].

2.6 OTHER PARAMETRIC TRENDS

Delamination area reduces at low energy levels when fibers with a small strain to failure are used, because some of the impact energy is absorbed through fiber breakage as opposed to predominantly through delamination formation and growth[29]. Larger changes in fiber direction between adjacent plies increase the bending stiffness mismatch and therefore produce more delamination area[7]. Increased resin toughness results in smaller damage area as shown in Figure 11.[29, 30] Thin laminates bend when impacted creating a reverse pine tree damage pattern. The specimen is subjected to flexural stresses (less shear) which cause failure in the form of matrix microcracking and debonding which occur progressively from the impact side. Thick laminates deform less when impacted, producing high contact stresses, high transverse shear stresses causing a pine tree shape damage pattern, as shown in Figure 12.[31]

Figure 9. Impact resistance as a function of fiber length for glass/polypropylene. Failure of (a) 3 mm length short fibers; (b) 12 mm length long fibers, and (c) 25 mm length long fibers. Fiber pull-out and enhanced fiber/matrix interaction are effective energy absorbing mechanisms for long fibers[25].

2.7 THERMOSET AND THERMOPLASTIC MATRIX COMPOSITES

There is increasing emphasis on use of thermoplastic matrix composites in recent years over thermoset composites. While a bulk of the LVI work has been published on thermoset matrices (epoxy, vinyl ester, phenolics), there is limited data on the LVI of thermoplastic composites. Advanced thermoplastic matrix composites are currently being used in structural applications as replacement to aluminum parts such as aircraft wing leading edges, flaps or ailerons, keel beams, bumper beams, golf shafts etc. The advantages of using thermoplastic composites in structural components are; ease of fabrication, rapid cycle times, unlimited shelf life, high toughness, and recyclability[32].

Carbon reinforced polyphenylene sulphide (C/PPS) is an attractive thermoplastic composite structural material combination since it can provide high stiffness, excellent flexural rigidity, and outstanding high temperature and creep behavior (Heat Deflection Temperature up to 300°C)[33]. Figure 13a shows impact damage at increasing energy levels of 30, 50 and 70J respectively, with corresponding exponential increase in damage area for C/PPS [34] (Figure 13b). At low energy levels the failure of the plate is dominated by fiber fracture. Once there is sufficient energy to propagate damage and perforate the panel, the load bearing capacity of the panel falls abruptly to 50% of the maximum sustained load and gradually fails until it is unable to carry any load (Figure 14). The damaged area grows exponentially with the increase in impact energy for this type of impact. The failure of carbon/PPS panels subjected to LVI occurs by the interaction of several failure mechanisms; fiber longitudinal failure, transverse failure, longitudinal shear, buckling, and resin tensile failure different stages of the damage progression.

LVI response studies of carbon/epoxy, glass/epoxy and hybrid glass/carbon epoxy at different energy levels provide the designer information about the peak force and energy absorbed[35]. Figure 15a and 15b represent the force-time histories for these composites. At 10 J impact energy, the peak load attained by the carbon/epoxy (CE) is higher than the other configurations, however at 40 J impact energy, the CE perforates, and exhibits the lowest peak load and hence lowest impact damage resistance .

Oftentimes there is need to choose between a thermoset and thermoplastic composite for an application. Figure 16 provides a comparison of a carbon/toughened epoxy composite with C/PPS and C/nylon[36]. The peak loads attained by the thermoset and thermoplastic systems are comparable, with the differences attributed to moduli and fracture strain of the polymer. In the study in [36], the peak load attained by C/PPS composite was the highest, due to the crystalline nature of PPS. This results in a asymmetrical force-time response and a sudden load drop upon perforation. The C/nylon and C/epoxy composites exhibit a more gradual unloading response due to more number of damage modes – pullout, tensile side fiber breakage and fiber/matrix interfacial friction. When plotted on a normalized load (peak load/thickness) basis, for identical damage states, the C/nylon thermoplastic absorbs 20% more impact energy than the corresponding C/epoxy or the C/PPS composite.

2.8 LOW VELOCITY IMPACT OF SANDWICH COMPOSITES

A sandwich composite consists of high strength, high modulus faces separated by a low density core. The core is generally lightweight and its thickness is greater than that of the faces. The two faces are generally identical in material and thickness although faces may differ in thickness, material or fiber orientation. The faces resist in-plane and lateral (bending) loads. The primary function of the core is to resist shear, and transverse compression loading. The faces are adhesively bonded to the core to provide load transfer between the constituents. The core of a sandwich structure can be of almost any material or architecture, but in general, the core falls into four types - (a) foam or solid core, (b) honey comb core, (c) truss core, and (d) corrugated core[37].

The typical LVI damage in a sandwich composite occurs as follows: (a) the impact side facesheet is subjected to a transverse shear force. If the facesheet resists penetration, there is extensive damage limited to the impact side facesheet. Some debonding can occur between the facesheet and the core adjacent to the impact point due to strain mismatch between them. At higher impact energy, the impactor penetrates the top facesheet and progresses into the core; (b) The core damage is in the form of cell crushing, shear failure and debonding of the facesheet to the core; (c) The tensile side facesheet gets loaded by the impactor, causing tensile bending on the back face. As a result the debonding area between the core to the back facesheet is extensive. Significant delamination of the back facesheet can occur before complete penetration of the impactor[39].

The mode of impact damage to a sandwich composite depends upon the panel support condition, projectile shape, geometric and material properties of facesheet and core. When a facesheet is thin (<10 times the core thickness), the deflections tend to be large (many times facesheet thickness) and high in-plane tensile forces cause Mode I tensile cracking in the core. When the facesheet is thick (<2-5 times the core thickness), the deflection is small (less than face sheet thickness). Transverse shear forces in regions surrounding the projectile are large high and cause Mode II transverse shear cracking in the core[38].

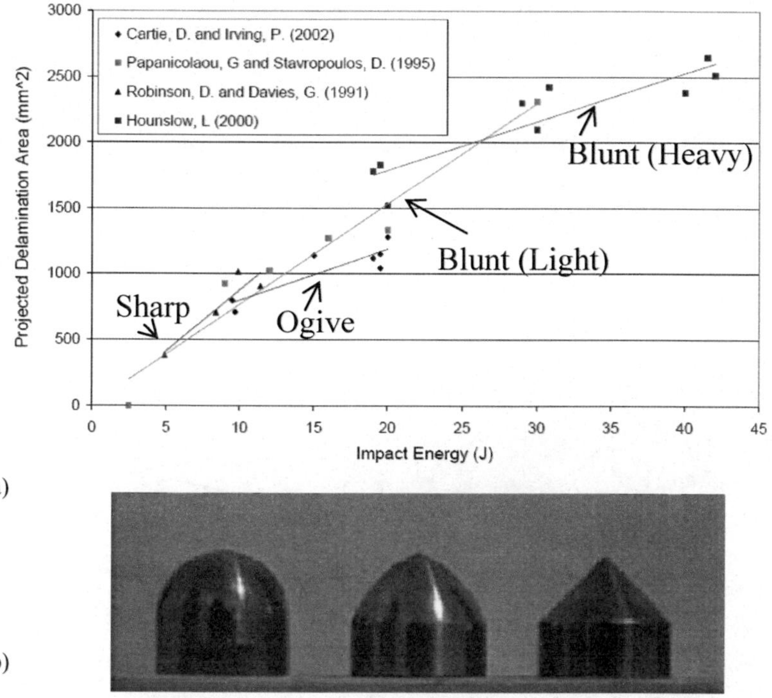

(a)

(b)

Figure 10. LVI response. (a) Projected delamination area as a function of impact energy for different shape impactors; and (b) Different shape projectiles – hemispherical, ogive and pointed[10].

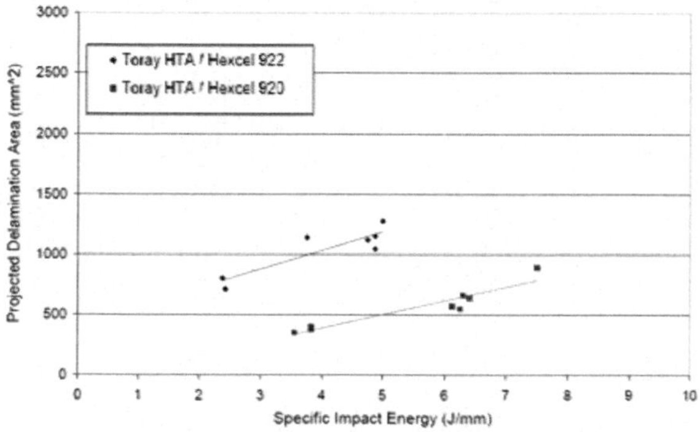

Figure 11. Effect of resin toughness on impact response of carbon fiber reinforced composites. The HTA/920 system has higher toughness than the HTA/922[10]

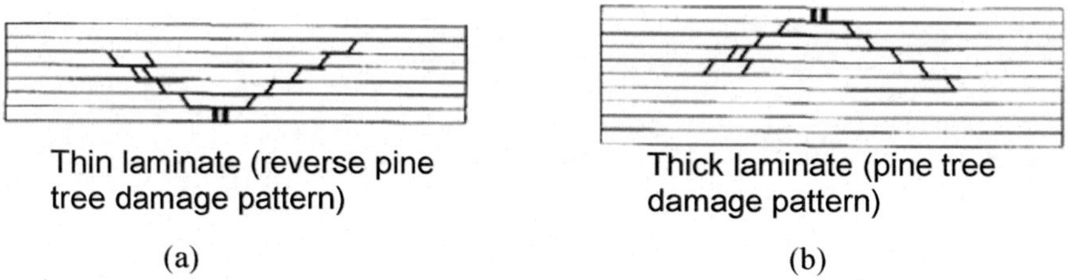

Figure 12. Low velocity impact response of thin versus thick laminates[31]

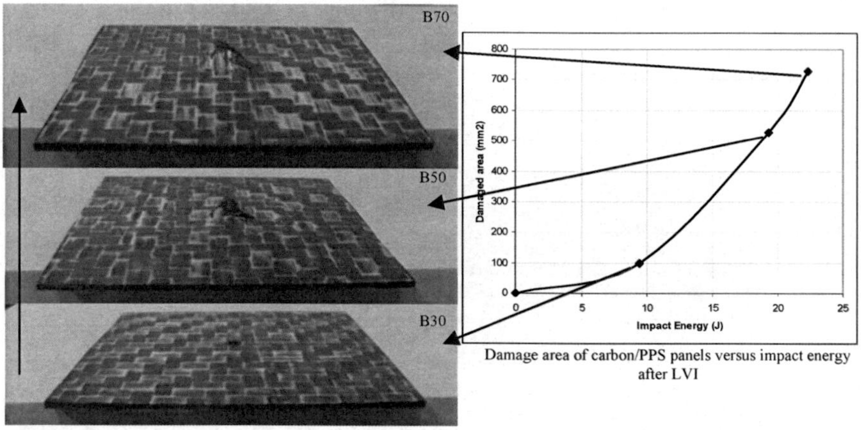

Figure 13. Low velocity impact testing of carbon/poly phenylene sulfide (C/PPS). Increasing damage area with impact energy[34].

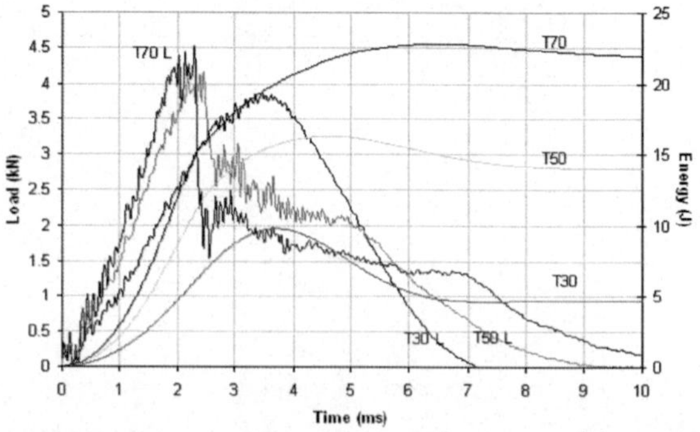

Figure 14. Low velocity impact testing of 3 mm thick carbon/poly phenylene sulfide (C/PPS) composites. The specimens exhibit different failure extent - rebound (T30), onset of back face penetration (T50), and perforation (T70)[34].

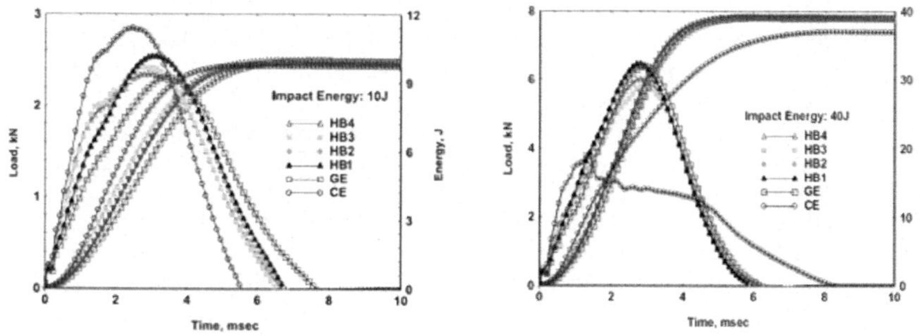

(a (b

Figure 15. LVI testing of different materials at different energy level for selecting suitable material. CE – carbon epoxy; GE-Glass epoxy; HB-Hybrid (glass plus carbon). (a) Impact energy = 10 J, (b) Impact energy = 40 J[35.]

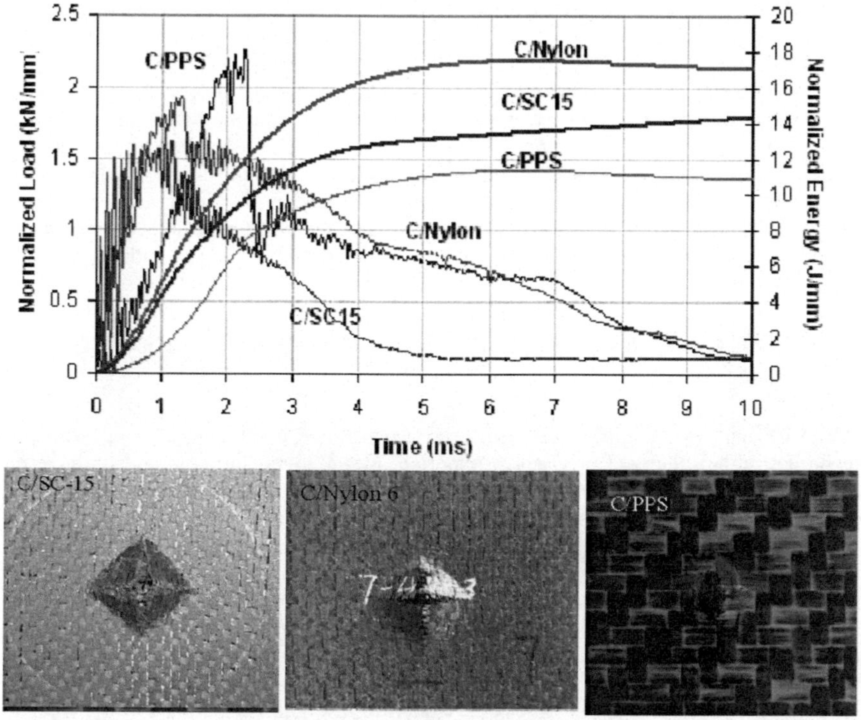

Figure 16. Low velocity impact testing of carbon/nylon, carbon/polyphenylene sulfide and carbon/epoxy composites tested for equivalent impact conditions[36].

A typical force-time and energy-time history of LVI failure of a E-glass/balsa wood core sandwich composite is shown in Fig. 17. The figure illustrates distinct stages of failure progression - namely front face penetration (I), failure within the core (II), and back face penetration (III) can be noted from the force-time history and corresponding energy absorption[40]. Similar penetration response has been observed for honeycomb core sandwich composites by Mines et al. with carbon/epoxy facesheets and aluminum honeycomb cells[41].

New buildings are featuring structural insulated panels (SIPs) comprising sandwich designs of plywood faces sandwiching a foam core. These are marketed as hurricane resistant structures to resist high velocity wind debris. However, in low velocity, the SIPs panels can exhibit perforation as illustrated in Figure 18a. Here the plywood facesheets are vulnerable to flying wood debris which can penetrate the faces readily. The polystyrene foam undergoes progressive crushing under impact. A newer development in place of SIPs is composite structural insulated panels (CSIPs) which feature a E-glass/polypropylene (E-glass/PP) facesheet with a polystyrene foam core[43, 44]. The E-glass/PP faces exhibit much higher resistance to penetration compared to the plywood used in SIPs. At identical LVI loads, the CSIPs designs resist impact more effectively (Figure 18b).

Figures 19a-d illustrates the LVI response of aluminum foam core sandwich composites with Kevlar, S2-glass, carbon and E-glass facesheets[42]. Aluminum foam cells 0.1 to 1 mm in diameter provide lightweight characteristics and can progressively crush under impact. When used as a core along with fiber reinforcement, the aluminum foam has potential to provide effective impact damage tolerance. The facesheets have a strong influence on the impact characteristics of the sandwich composites. In the study conducted in [42], the Kevlar, carbon and E-glass facesheets perforate readily on the impact side. The metal foam core cells fail in a highly localized area below the impact location. The S2-glass resists penetration and exhibits higher impact side facesheet damage than the other reinforcements.

2.9 IMPACT PERFORMANCE OF COMPLEX GEOMETRY SPECIMENS

The vast majority of impact testing has been performed on flat coupons in beam or plate format, either clamped or simply supported. Many composite components have a complex geometry and the response of stiffened panels is a particularly important area[3, 27]. The energy to cause barely visible impact damage (BVID) drops significantly near the stiffeners, where the structure is less compliant, and that the stiffeners caused damage to spread asymmetrically, as would be expected over an area of non-uniform stiffness. Impact forces tend to be higher in stiffened regions, but the reduced deflections may lead to smaller strains and therefore less strain-induced failure. At the edge of the stiffeners, delaminations are formed, while an impact directly over the stiffener causes debonding between the plate and stiffener. The damage tends to extend down the stiffener because the induced forces follow the stiffest path. Cratering also can occur due to high forces induced in the stiffened regions[3, 27].

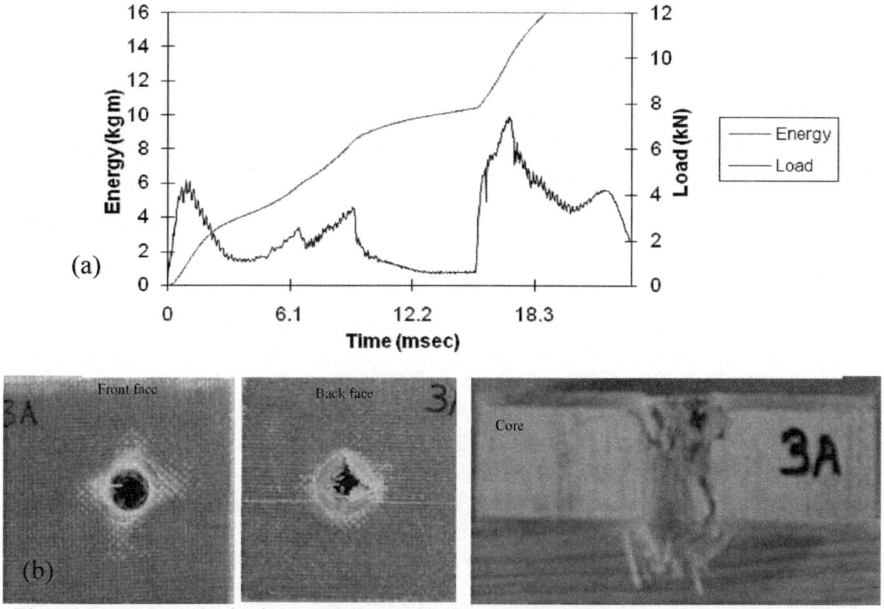

Figure 17. Low velocity impact failure of balsa wood core/E-glass vinyl ester sandwich composite. (a) Force –time, and energy-time curve; (b) Perforation of top facesheet, core and bottom facesheet[40].

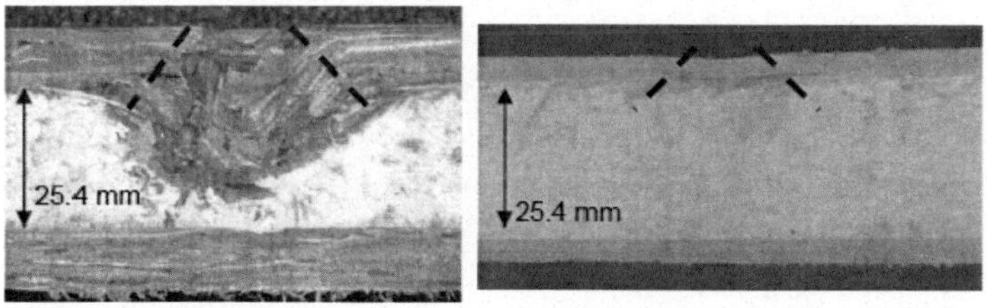

Figure 18. Low velocity impact failure of (left) building structural insulated panel (SIP) comprising plywood face and styrofoam core; and (right) glass/polypropylene face with styrofoam core[44].

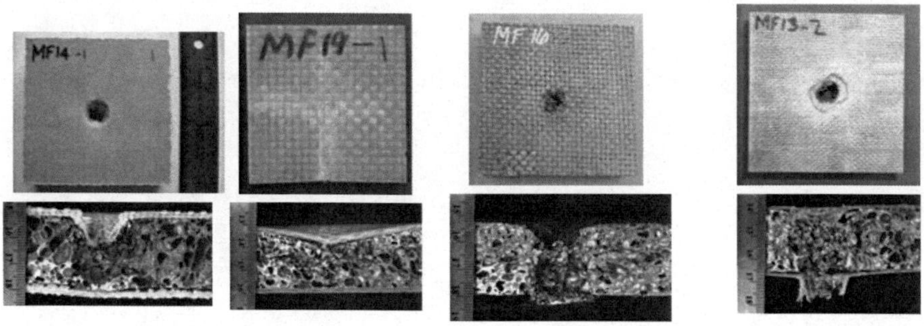

(a (b (c (d

Figure 19. Low velocity impact failure of aluminum foam core with (a) Kevlar facesheet, (b) S2-glass facesheet, (c) carbon facesheet and (d) E-glass facesheet. The S2-glass resists the impact and energy is concentrated on the impact side facesheet[42].

2.10 POST-IMPACT RESIDUAL STRENGTH

Due to susceptibility of composite materials to impact damage, dramatic loss in residual strength and structural integrity can result. The term *damage tolerance* refers to a composite materials' ability to perform post-impact. Even BVID can cause strength reductions of up to 50%. Residual strengths in tension, compression, bending and fatigue will be reduced to varying degrees depending on the dominant damage mode.

2.10.1 Residual tensile strength. Residual tensile strength normally follows a trend illustrated in Figure 20. In region I, no damage occurs as the impact energy is below the threshold value for damage initiation. Once the threshold has been reached, the residual tensile strength reduces quickly to a minimum in region II as the extent of damage increases. Region III sees a constant value of residual strength because the impact velocity has reached a point where clean perforation occurs, leaving a neat hole. In this region the tensile residual strength can be estimated by considering the damage to be equivalent to a hole the size of the impactor. The minimum in region II is less than the constant value in region III because the damage spreads over a larger area than is produced at a higher velocity when the damage is more localized (resulting in a cleaner hole). As the fibers carry the majority of tensile load in the longitudinal direction, fiber damage is the critical damage mode[1, 19].

2.10.2 Residual compressive strength. CAI is one of the most common means of assessing residual properties of the composite laminates. This is mainly due to local instability resulting from delamination causing large reductions in compressive strength. As delamination can be produced by low-energy impacts, large strength reductions in compression can occur for BVID. Delamination divides the laminate into sub-laminates which have a lower bending stiffness than the original laminate and are less resistant to buckling loads. Under a compressive load, a delamination can cause buckling in one of three modes: global instability/buckling of the laminate, local instability (buckling of the

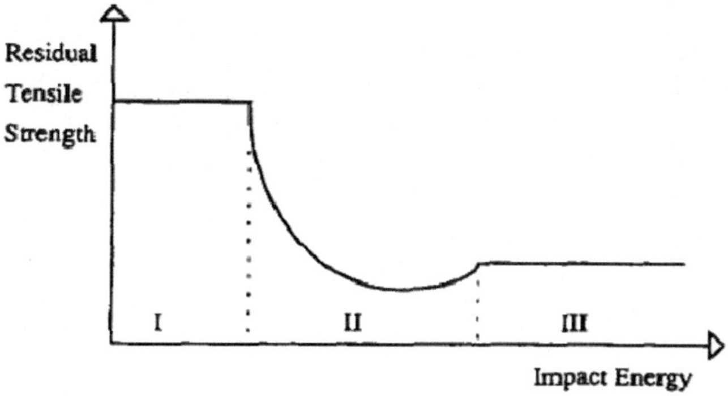

Figure 20. Residual strength curve for a composite laminate[19]

thinner sub-laminate), or a combination of the above. The mode of failure generally changes from global, to local, to mixed mode as the delamination length increases. In some cases, anti buckling guides are used to support the specimen to prevent global

$$E_{absorbed} = E_{damage} + E_{SL} \qquad (13)$$

buckling, but at the same time must not prevent local instability.

2.11 ENHANCING IMPACT DAMAGE TOLERANCE

Various approaches have been used to improve the damage tolerance of composite materials. These include control of fiber-matrix interfacial adhesion[45], matrix modifications such as rubber toughening[29], lamination design (selection of laminate stacking sequence[46], introduction of through-the-thickness reinforcements by braiding, three-dimensional weaving and stitching [47], insertion of interlaminar layers[48], fiber hybridization [35]and use of high strain fibers[49].

3.0 INTERMEDIATE VELOCITY IMPACT

Intermediate velocity impact (IVI) is considered to fall between the low and high velocity regimes. IVI occurs in the range of 20-100 m/s usually with a blunt impactor[50]. Depending on the projectile mass, large deformation may occur in the intermediate velocity range, particularly in the case of massive projectiles, but may differ from LVI with regard to loading rate and momentum.

Typically IVI occur from events such as road debris impact to an automobile, a lower end velocity of a projectile (bullet), hail impact or a baseball bat striking a ball. In addition, the importance of this test methodology lies not only in the characterization of debris hits, but also energy dissipation and failure mechanisms under high loading rate.

The energy absorbed in the system, $E_{absorbed}$ can be described as having two components: E_{damage} and E_{SL} and given by[50]:

where E_{damage} is the energy absorbed in creating damage in the specimen. The E_{damage} term includes the energies associated with the specimen indentation, matrix damage, fiber breakage, fiber debonding, and fiber pull-out; and E_{SL} is the energy absorbed by the system through vibration, heat, elastic response of the specimen, and neglecting the elastic behavior of the supports. The energy absorbed in the system can be equated to the energy imparted to the system, e.g. the kinetic energy (KE) of the impactor neglecting, the elastic behavior of the supports, E_{SL}, the absorbed energy is given by:

$$E_{absorbed} = KE = \frac{1}{2} m V_o^2 - \frac{1}{2} m V_f^2 \quad (14)$$

where m is the impactor mass, and V_o and V_f are the initial and residual projectile velocities.

The gas gun apparatus used in testing ballistic impact (Section 4.0) is also used for intermediate velocity testing. The impact energy absorption is the highest at the ballistic limit of the composite material, i.e. the velocity at which the projectile has zero exit velocity.

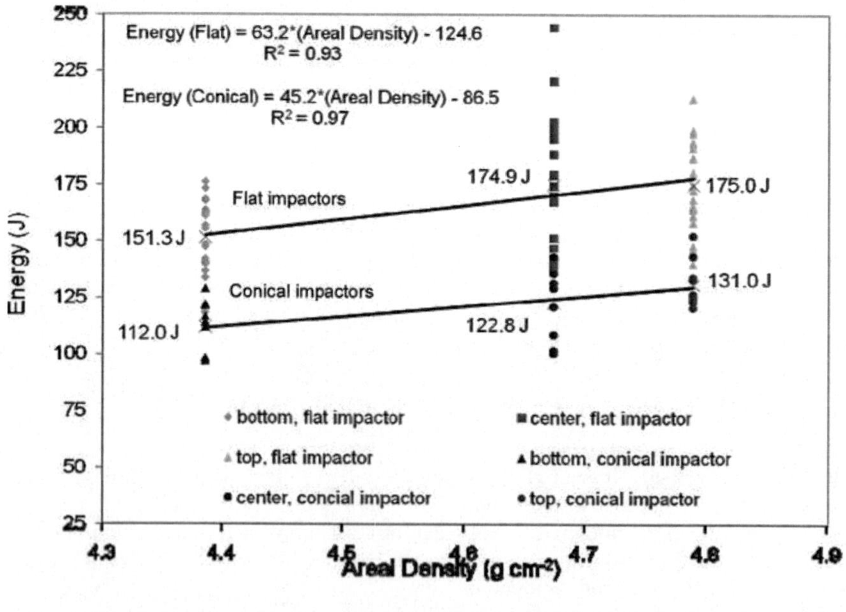

Figure 21. Intermediate velocity impact to long glass fiber/polypropylene composite by flat and conical impactor. Panels of typical size 18" x 8" of differential thickness were machined to obtain the test specimens. The locations *top, center* and *bottom* indicate different locations on the panels and their corresponding areal densities[50].

The V_f term is zero, unless the projectile perforates the specimen and has a detectible exit velocity. In thin specimens, there is considerable specimen deflection and a spall effect, resulting in V_f=0 (no penetration). Since a standard test method does not exist for IVI, the critical velocity is considered on the basis of a standard deviation in velocity no greater than 10 m/s, in which 50% of the projectiles do not perforate the specimen, also referred to as V_{50} in ballistics terminology. In a study by Bartus and Vaidya[51] composite plates of 25 mm long glass thermoplastic, E-glass/PP composites, 40% fiber weight fraction were investigated for IVI. A flat and conical shaped impactor of mass 140 g and 38 mm diameter, 50 mm length traveling at an average speed of 180 m/s was used to impact long fiber thermoplastic plates. The average impact energy dissipation at the critical velocity (ballistic limit) was 167 J and 121 J for a 4.61 g cm^{-2} specimen impacted by flat and conically shaped projectiles, respectively (Figure 21). The impact energy dissipation increased linearly with increasing areal density. In the long fiber thermoplastic composite, the impact damage propagates preferentially along the fiber orientation direction as illustrated in Figure 22.

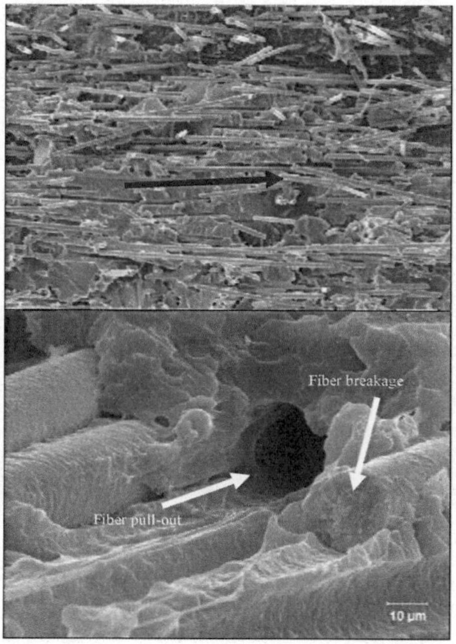

(a)

(b)

Figure 22. Intermediate velocity impact. (a) Impact damage progression along preferential fiber orientation. The arrow indicates the direction of crack propagation and also that this coincides with the direction of preferential fiber orientation. and (b) resistance to pull-out of glass/polypropylene long fiber thermoplastics[50].

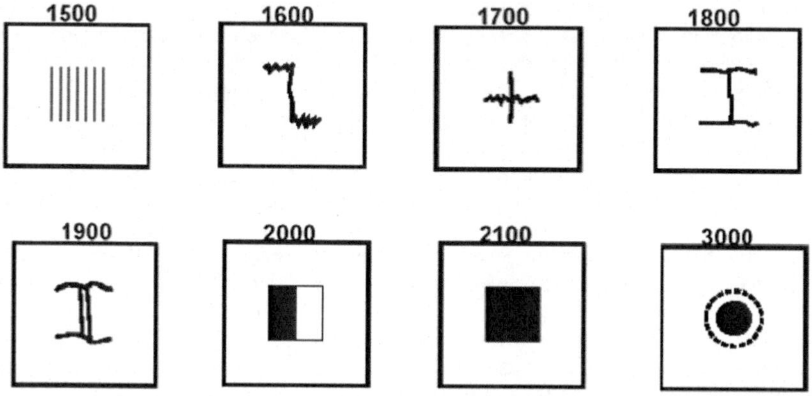

Figure 23. Typical failure modes of unidirectional carbon/epoxy composites preloaded in tension impacted by a soft projectile. Each illustration represents a laminate, and simplified failure modes in the laminate. The values of impact velocity are in ft/s [50].

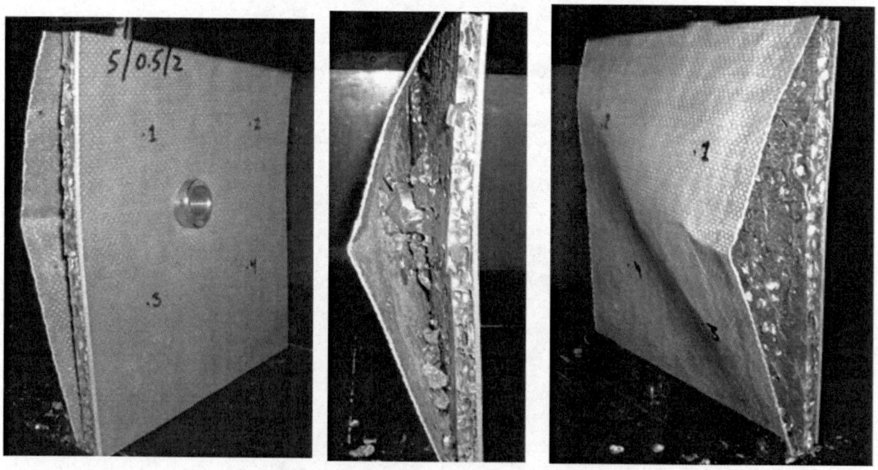

Figure 24. Penetration through a seven-ply Kevlar/vinyl ester with metal foam core sandwich composite impacted at the ballistic limit with a blunt object, 38 mm diameter, impact energy 756 J. (a) front face penetration; (b) core penetration, and (c) projectile pushing the back face[42].

Blunt objects inflict planar cracking of composites upon impact. Typical sequence of damage modes for a unidirectional C/epoxy laminate impacted with a soft projectile are shown in Figure 23. The damage models progress from matrix splitting, interplay debonding, transverse and longitudinal matrix cracking and fiber bridging at velocities ranging from 1500 to 1900 ft/s and subsequently localized perforation and penetration at higher velocities (2000 to 3000 ft/s). Damage from flat projectiles initiated along the periphery of the impact area and propagated radially along two or three planes, away from area of initial damage.

In another study on sandwich composites with Kevlar/vinyl ester facesheets with an aluminum foam core, the blunt object penetration at 145 J impact energy results in a localized front face perforation followed by penetration through the core, and significant tensile side deformation of the Kevlar facesheet[52] (Figure 24).

4.0 BALLISTIC IMPACT

Fiber reinforced composites are being extensively used in ballistic armor applications, and understanding of the failure and perforation mechanisms is important. The reader is referred to Chapter No. 7 for additional details on terminology on ballistic impact terminology and complementing information. A brief description on the experimental aspects of ballistic impact is provided here.

Composite structures are oftentimes subjected to damage from high velocity projectiles and small fire arms. In a composite laminate subjected to ballistic impact, the kinetic energy of the projectile is dissipated through several mechanisms.
Abrate[1] defines the high velocity impact regime as the ratio between the impactor velocity and the transverse compressive wave velocity being greater than the maximum strain to failure in that direction. The response is governed by wave propagation and not by the impactor velocity. High velocity is thus characterized by penetration induced fiber breakage, and low velocity by delamination and matrix cracking.

Small mass, high velocity impact can lead to a higher degree of local loading with a corresponding increase in damage for equivalent impact energy in contrast to the loading condition with a quasi-static impact response[5]. Cantwell and Morton[6] found small mass, high velocity impact to be more detrimental to carbon fiber reinforced laminates than low velocity drop tower impact.

4.1 GAS GUN

A gas gun or a powder gun is generally used to propel a projectile towards the target at velocities ranging from 100 m/s to 1000 m/s. A number of variations have been devised and configured for ballistic testing of materials [50, 53]. A single-stage light gas gun consisting of a pressure chamber, a barrel and a nitrogen/helium tank is used to launch the projectile (Figure 25 and 26).

The projectile dimension and shape varies based on the range of velocity and energy desired. Spherical steel balls, 0.22 (5.58 mm), 0.30 (7.92 mm) and 0.50 (12.70 mm) caliber bullets or fragment simulating projectile (FSP) or blunt edge cylinder of dimension 12 mm to 38 mm diameter and 25 mm to 50 mm on length are common (Figure 27). The FSPs are made of alloyed tool steel. A mass of 2.67 g for a 0.30 caliber and 13.40 g for a 0.50 caliber is consistent with the mass of the corresponding NATO Fragment Simulating Projectiles.

A laboratory scale gas gun is briefly described below. The limiting factors on achievable velocities for a gas gun or powder gun are the speed of sound in the working fluid (e.g. air, nitrogen), launch package mass, barrel length, and firing pressure.

4.1.1 Firing Mechanism. A pressure vessel is constructed to hold the working fluid. It is typically a flange bolted cylinder to eliminate welds. Typically steel tensioners are used to bolt the aluminum end caps. The force in the tensioners is equal to the axial force exerted on the end caps as a result of the internal pressure, divided by the number of tensioners.

The firing valve is a critical component. It must have a very fast response in order to release the firing pressure uniformly. In addition, in order to provide the maximum precision, it must be repeatedly open at the same rate. Otherwise inconsistencies in the launch package velocity will arise. Diaphragm valves are commonly employed as the firing valve in which the diaphragm is clamped between two mounting blocks at the breach between the pressure vessel and launch package[54]. They are typically made from a metal or polymer plate that is either scored such that it will fail at a predetermined pressure or utilize a conductive wire across the polymer face[55]. In the case of the later, an electrical current heats the conductive wire, causing the diaphragm to fail.

A curve of velocity of pressure must be generated to calibrate the gas gun. An example velocity versus pressure curve is shown in Figure 28. Based on this curve, the firing pressure is adjusted to achieve a desired projectile velocity.

4.1.2 Firing Control and Valve. A soft seat, fast acting valve can be used for releasing firing pressure because of good repeatability and ease of operation. Butterfly valves are commonly used because they are durable, need little maintenance, and require a minimal amount of torque for actuation. A digital timer is used for fire control.

4.1.3 Working Fluid. Work is required to accelerate the launch package and the firing fluid itself. Lighter gas requires less work to accelerate. Helium, nitrogen and argon can be used as a working fluid. The parameters measured during the test are the firing pressure, and incident and residual projectile velocity. Firing pressure is monitored using a pressure gage.

4.1.4 Barrel. A large caliber barrel for a gas gun allows a wide array of projectile threats to be evaluated. Large caliber threats (>20mm) are secondary blast debris, tornado and hurricane debris, runway kick-up and road debris[50]. Also, high velocities can be achieved using sabots for smaller caliber projectiles since the pressure is acting over a greater area (for e.g. a 38.1 mm diameter barrel has a factor of 24.5 times greater cross sectional area versus a .30 caliber projectile). The barrel is to be supported on rings and the assembly is typically mounted to a stand.

4.1.5 Capture chamber. The sample is housed in a capture chamber to house the test specimen of required dimensions (typical surface dimension of 100 mm x 100 mm to 300 mm x 300 mm) equipped with optical clear windows for incident and residual velocity acquisition. The chamber can be accessed via a hinged door which opens vertically, running the full length of the chamber (Figure 29). The capture chamber can be constructed out of mild steel.

A kinetic energy deflector (projectile trap) is placed opposite the barrel in order to catch penetrating projectiles. Its purpose is to provide an extra margin of safety against penetrating projectiles and to serve as a mechanism to keep them from ricocheting around in the capture chamber. A soft recovery trap can also be placed within the projectile trap allowing the post impact deformation of projectiles to be examined and save projectiles from deformation so they can be reused.

4.1.6 Specimen Fixture. The specimen fixture can be configured to house different specimen sizes (for e.g. 100 mm x 100 mm (4" x 4") to 300 mm x 300 mm (12" x 12"). The boundary conditions are typically simply supported or fully clamped. Toggle clamps can be used to apply a clamping force.

4.1.7 Instrumentation. There are several means of measuring projectile velocity, and all methods utilize some form of detection (e.g. photoelectric, break screen, magnetic) in which two or more detectors are placed a known distance apart. The time between detections is recorded, and the velocity is the known distance divided by time. The most common form of velocity measurement is the photoelectric chronograph. The only requirement is light from the sun, infrared LED, or incandescent light bulb. Fluorescent lights cannot be used because they pulse at 60 Hz (U.S) causing errors in readings. Operation begins with the photodetector sensing a break in the light. At that point, a chronograph which uses a crystal oscillator is used to measure the time until the second photodetector senses a break. Most of the chronographs automatically calculate the velocity based on the distance between the sensors.

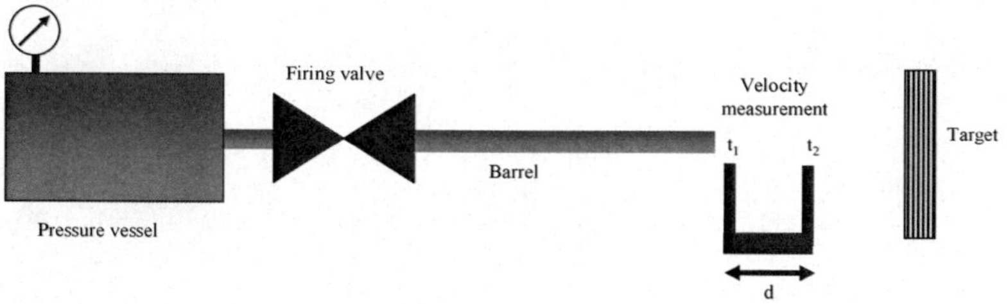

Figure 25. Schematic of a ballistic gas gun. Velocity (v) = distance/time = d/(t2-t1)

4.2 UNIVERSAL RECEIVER

For velocities ranging from 200 to 1000 m/s, a universal receiver powder gun can be used to evaluate ballistic behavior of composites, Fig. 30a. Different caliber barrels can be interchanged in the universal receiver facilitating testing for different threat levels. Different caliber barrels such as the 2 ¾" 12-gauge, .300 Winchester Magnum, and .50 caliber Browning Machine Gun (BMG) are typically used[53]. The velocity range is approximately 200 $m{\cdot}s^{-1}$ (12-gauge, 4 fragments) to 1000 $m{\cdot}s^{-1}$ (.50 BMG, .50 caliber FSP). Custom fixture can be built to allow the universal receiver powder gun to articulate to adjust for windage and elevation. The articulating fixture allows the user to aim the gun at the desired impact location.

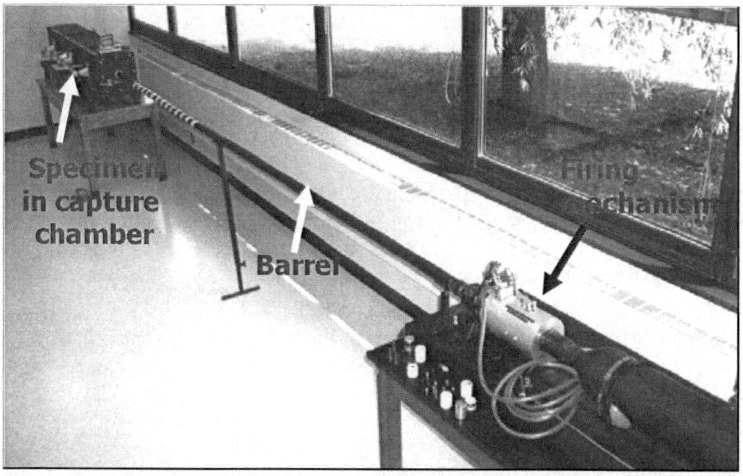

Figure 26. A typical gas gun. The firing mechanism, the barrel and the capture chamber are illustrated. The barrel dimensions are 38 mm diameter, 5 m length and velocity range is 50 to 400 m/s[51].

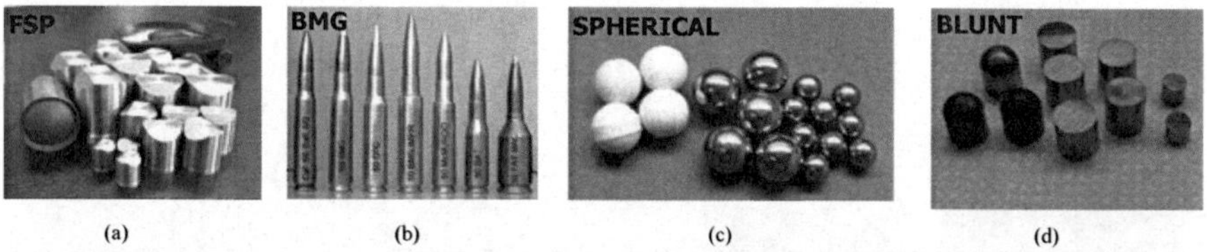

Figure 27. Typical projectiles used in ballistic impact; (a) Fragment simulating projectile; (b) Browning machine gun ammunition; (c) spherical hard ceramic or metal balls, and (d) blunt, hemispherical and flat head. Typical diameter range from 0.22 caliber to 0.50 caliber.

A clamping fixture holds the specimen and aligns it with the universal receiver and velocity screens (Fig. 30b). Clamping is considered to provide a boundary condition between simply supported and fully clamped.[56] The assembly comprising the test specimen in the clamping fixture and the velocity screens (one pair in front of the target, and the second behind the target) are shown in Fig. 30c.

The fixture must also support an aluminum witness plate with normal orientation, placed 0.15 to 0.30 m behind the specimen. A witness plate is used as a proof of penetration, a 'go/no go' impact test. If the witness plate is held up to a light bulb and no light passes through after fully penetrating impact, the projectile is considered defeated. A chronograph can also be placed behind the specimen on a standard camera tripod to measure residual velocity.

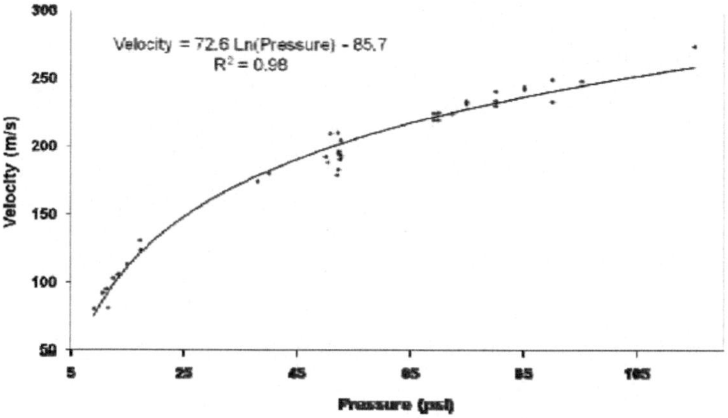

Figure 28. Velocity versus pressure calibration curve for Fragment Simulating Projectile (FSP)[51].

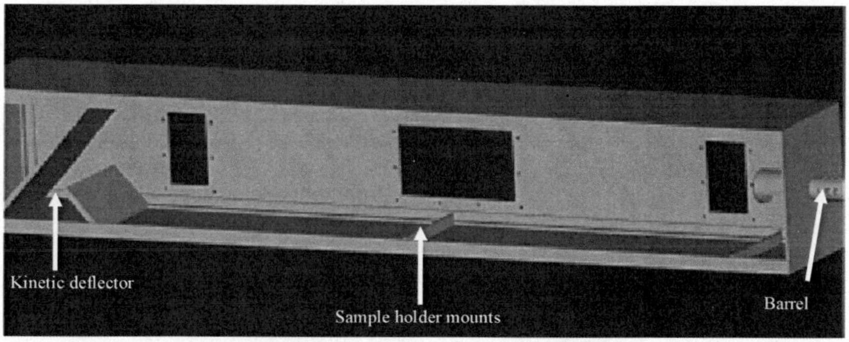

Figure 29. Capture chamber for a gas gun[50].

A modern center fire ammunition, illustrated in Figure 31a (a) and (b) consists of a hull, primer, wad, and projectile. The primer provides the spark to ignite the propellant. When the propellant ignites, it produces a subsonic deflagration wave that rapidly expands producing a pressure gradient across the sabot/projectiles, accelerating them down the barrel. Smokeless propellant consists of nitrocellulose often combined with up to 20% nitroglycerin [57]. Since smokeless propellants only burn at the surfaces of the granules, the shape and size of the granules affects the burn rate. In addition, flame-deterrent coating are often used to retard the burn rate such that a more or less constant pressure is exerted on the projectile but yet burn at a sufficient rate to fully combust while the launch package is in the barrel. Figure 31b provides a comparison of typical cases and

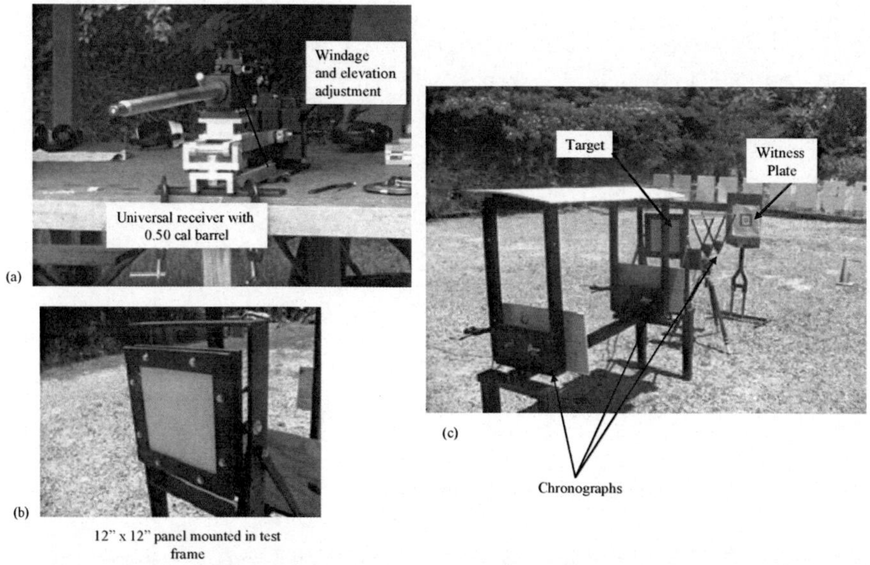

Figure 30. Universal receiver setup for ballistic impact. (a) Assembly showing incident and exit chronographs pairs and the target; (b) closer view of the target and mounting; and (c) universal receiver barrel and adjustments[52].

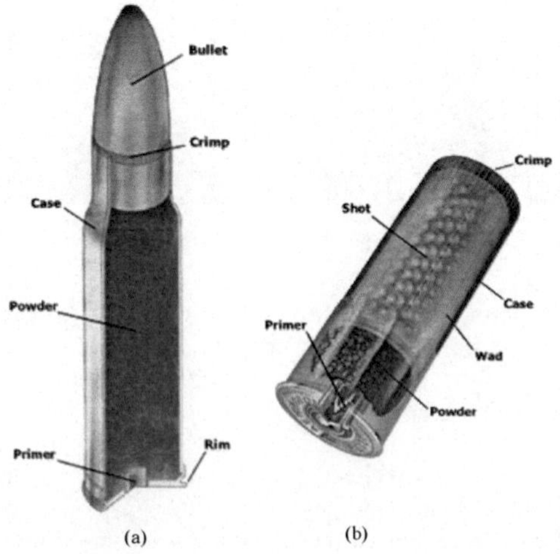

Figure 31a. Modern centerfire ammunition. The typical components are - (a) bullet, primer, propellant, case; and (b) shot shell cartridge and parts.

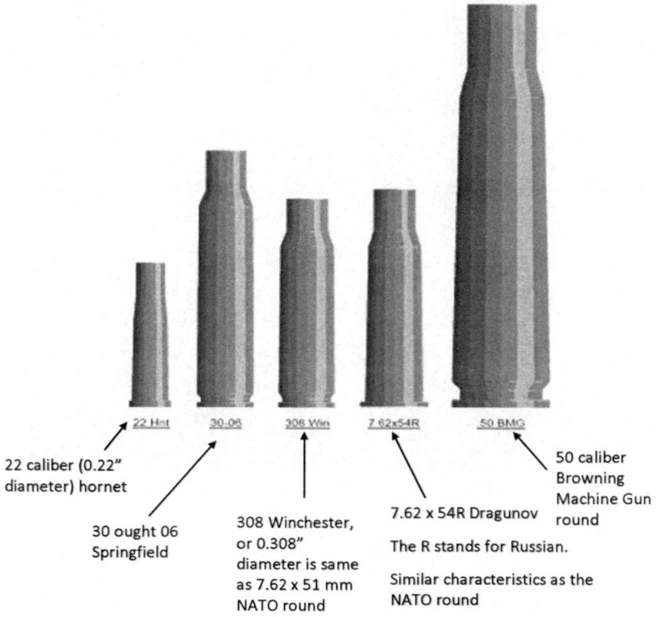

Figure 31b. Typical cartridges and their classification.

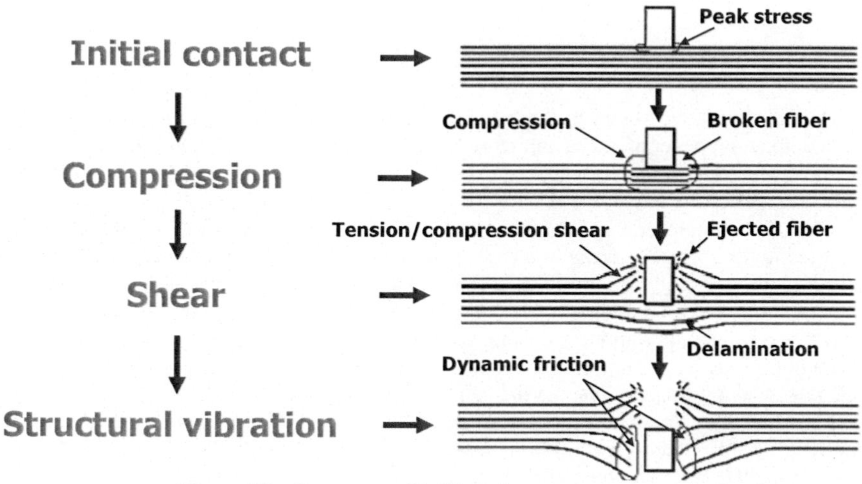

Figure 32. Sequence of ballistic impact to a composite laminate.

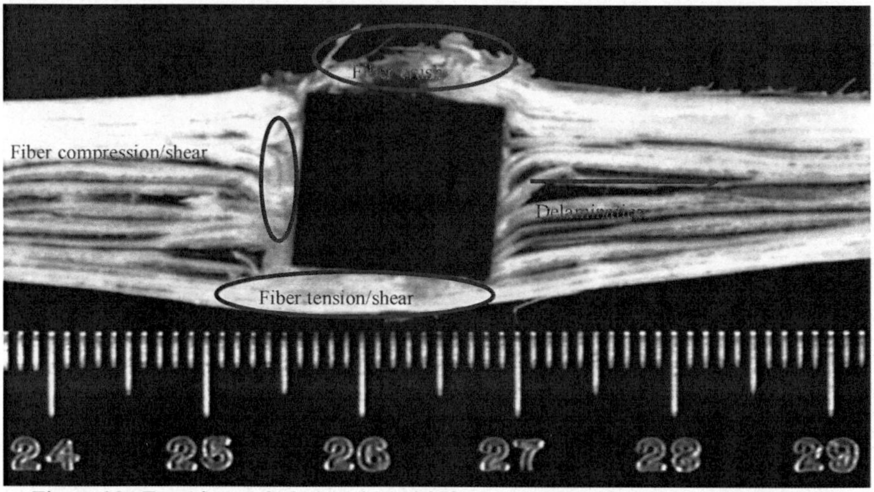

Figure 33. Experimental observation of failure modes for laminate at ballistic limit.

specifications used in ballistic testing; 22 caliber, 30 ought 6, 308 Winchester, 7.62 x 54R Dragunov and 50 caliber BMG.

The desired velocity and loads can be produced by hand loading the shells using a reloading press. A velocity versus propellant weight must be established. Deviating from published loading data can be extremely dangerous so extra care was taken to insure excess chamber pressures are not encountered. Flattened and pierced primers indicate excessive chamber pressure but excessive pressures can be encountered without these indications present.[58]

4.3 BALLISTIC IMPACT PARAMETERS

Ballistic limit, V_B refers to the minimum velocity at which a projectile consistently and completely penetrates the target of given thickness and physical properties at a specified angle of obliquity, whereas V_{50} is the velocity at which penetration is likely to occur for 50% of the impacts[50, 53].

The predominant energy absorption mechanisms are in the form of kinetic energy imparted to the specimen, namely cone formation on the distal side of the laminate and/or spall formation, energy absorption as a result of shear plugging, tensile fiber failure of the primary yarns, fiber debonding, fiber pull-out, elastic deformation of the secondary yarns, matrix cracking, interlaminar delamination, and frictional energy absorbed during interaction of the penetrator and laminate[59]. These damage mechanisms are illustrated in Fig. 32 and supporting evidence of the damage modes is provided in Fig. 33.

The ballistic impact response of thermoset and thermoplastic composites is significantly different. Figure 34a-f illustrates multisite ballistic damage to a thermoset multidirectional carbon/epoxy composite. The back face matrix crack splitting along the fiber directions is observed in Figs. 34 a and b, and the impact side clean perforations in

Figs. 34c-f. The global deformation through the thickness is very limited for a thermoset

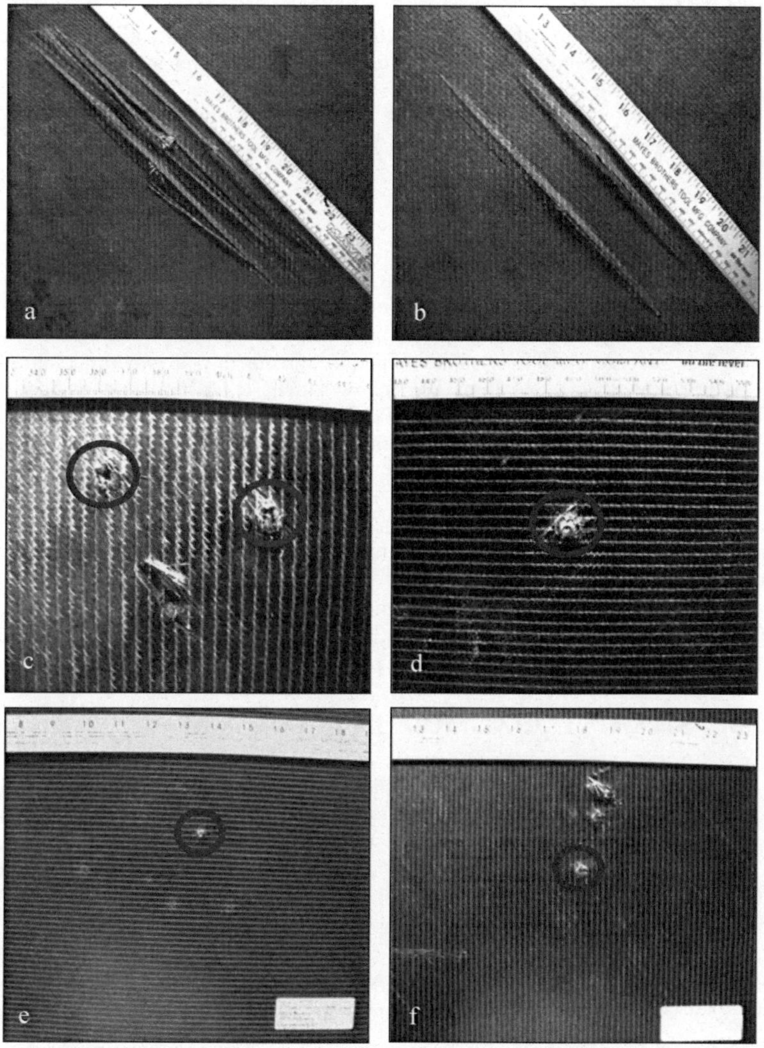

Figure 34. Ballistic impact failure modes for carbon/epoxy laminate; (a,b) Typical back-face damage for perforating and non-perforating fragment cluster; (c), (d), (e), & (f) show impact-face damage, circled region indicates an embedded projectile[52].

composite. For the case of thermoplastic composites, significant global thickness deformation is observed, as illustrated in the case of ballistic impact on a polyethylene reinforced composite (Fig. 35). The large deformation in thermoplastic composites can be attributed to the high strain (elongation) of the thermoplastic polymer.

The reader is referred to Chapter 7 for experimental studies on ballistic behavior of composites. The next section of this chapter will focus on the modeling aspects of ballistic impact of composites.

5.0 MATERIAL MODELS AND SIMULATION

Many phenomenological theories have been proposed to predict failure in an unidirectional lamina under plane stress condition. They are classified into three groups; limit or non-interactive theories (maximum stress, maximum strain); interactive theories, Azzi-Tsai-Hill[60], Tsai-Wu[61]; and partially interactive or failure mode based theories Hashin[62], Cheng-Chang[63] and Puck [64, 65]. The validity and applicability of a given theory depends on the convenience of its application and its agreement with experimental results.

The basic theoretical foundations of finite element (FE) modeling of laminated composites using a non-linear FE code LS-DYNA is discussed in this section. Although LS-DYNA based studies are presented, the approach is generic to the use of any other comparable FE code or user-defined programs.

LS-DYNA is a general purpose finite element code for analyzing impact problems and dynamic response of structures. The solution methodology is based on explicit time integration. Spatial discretization is achieved by the use of four node tetrahedron and eight node solid elements, two node beam elements, three and four node shell elements, eight node solid shell elements, truss elements, membrane elements, discrete elements, and rigid bodies. LS-DYNA currently contains over one-hundred constitutive models and ten equations of state to cover a wide range of material behavior[66].

5.1 COMPOSITE MATERIAL MODELS IN LS-DYNA

LS-DYNA is used for modeling the elastic and inelastic behavior response of composite structures. There are some specific material models which can be used for modeling polymer matrix composites[66]. In LS-DYNA these are identified by MAT 2, 22, 54, 55, 58, 59, 116, 117, 118, 161 and 162. These material models use different constitutive equations to simulate the behavior of a polymer matrix composite. The most relevant constitutive equations and progressive failure criteria are discussed here.

The finite displacement and rotations for small strains in a laminate are predicted assuming an orthotropic laminate. The material law that relates second Piola-Kirchhoff stress, σ to strain, ε in geometric coordinate system is given by Equation (14) where the transformation matrix, T and the stiffness matrix, C_l is defined in terms of the material symmetric axes as shown in Equation (15).

$$\sigma = C.\varepsilon = TC_l T'\varepsilon \Rightarrow \begin{bmatrix} \sigma_x \\ \sigma_y \\ \sigma_z \\ \sigma_{xy} \\ \sigma_{yz} \\ \sigma_{zz} \end{bmatrix} = TC_l T' \begin{bmatrix} \varepsilon_x \\ \varepsilon_y \\ \varepsilon_z \\ \varepsilon_{xy} \\ \varepsilon_{yz} \\ \varepsilon_{zz} \end{bmatrix} \quad (15)$$

$$C_l^{-1} = \begin{bmatrix} \frac{1}{E_1} & -\frac{\nu_{21}}{E_2} & -\frac{\nu_{31}}{E_3} & 0 & 0 & 0 \\ -\frac{\nu_{12}}{E_1} & \frac{1}{E_2} & -\frac{\nu_{32}}{E_2} & 0 & 0 & 0 \\ -\frac{\nu_{13}}{E_1} & -\frac{\nu_{23}}{E_2} & \frac{1}{E_3} & 0 & 0 & 0 \\ 0 & 0 & 0 & \frac{1}{G_{12}} & 0 & 0 \\ 0 & 0 & 0 & 0 & \frac{1}{G_{23}} & 0 \\ 0 & 0 & 0 & 0 & 0 & \frac{1}{G_{31}} \end{bmatrix} \quad (16)$$

Because of the symmetric properties, Poisson's ratio, ν_{ji} has the following form

$$\nu_{ji} = \nu_{ij} E_{jj} / E_{ii}$$

During high velocity impact testing on a composite laminate, material at the impact region undergoes large deformation or erosion due to plug formation or spalling. To model failure at the impact region an element erosion criterion is adopted in conjunction with the orthotropic model. In LS-DYNA, the erosion criterion is referred to as MAT_ADD_EROSION[66]. The erosion criterion defines failure according to the following; (a) Pressure at failure in compression; (b) Maximum principal stress criterion; (c) Equivalent stress at failure; (d) Maximum principal strain at failure, and (e) Maximum shear strain at failure.

5.1.1 Progressive Failure of Composite Laminate (Material Model 161). The composite failure occurs in a progressive manner and the constitutive equations must account for the damage progression. The determination of three-dimensional (3D) stress field as the laminate fails progressively is determined using the models MAT161 and MAT162 in LS-DYNA. Different failure modes include - fiber failure, matrix damage, and delamination under all conditions – opening, closing, and sliding failure of a laminate. MAT 161 applies Hashin's theory[62] for composite failure.

The discussion following is for a woven fabric composite, however the modeling approach is generic for unidirectional and multidirectional composites. The mechanical properties for a woven composite with plain weave architecture are assumed equal in both in-plane directions. Therefore fiber breakage in the fill and warp direction can be caused by tensile and shear stresses leading to the following failure criteria[62, 68]:

$$f_{tensile/shearfill} = \left(\frac{\langle \sigma_1 \rangle}{S_{lT}}\right)^2 + \left(\frac{\tau_{12}^2 + \tau_{31}^2}{S_{XFS}^2}\right) - 1 = 0 \quad if \quad \sigma_1 > 0 \tag{17}$$

$$f_{tensilewpa} = \left(\frac{\langle \sigma_2 \rangle}{S_{tT}}\right)^2 + \left(\frac{\tau_{12}^2 + \tau_{23}^2}{S_{YFS}^2}\right) - 1 = 0 \tag{18}$$

where S_{lT}, S_{tT} are the axial tensile strengths in the fill and warp directions, respectively and S_{XFS}, S_{YFS} are the fiber shear strengths.

During transverse impact, the impact region of the composite laminate is compressed by the projectile leading to high in-plane compressive stress generation in fill and warp direction expressed by the maximum stress criterion[62, 68]:

$$f_{compressio\ nwarp} = \left(\frac{\langle \sigma_2' \rangle}{S_{tC}}\right)^2 - 1 = 0, \quad \sigma_2' = -\sigma_2 + \langle -\sigma_3 \rangle \tag{19}$$

$$f_{compressionfill} = \left(\frac{\langle \sigma_1' \rangle}{S_{lC}}\right)^2 - 1 = 0, \quad \sigma_1' = -\sigma_1 + \langle -\sigma_3 \rangle \tag{20}$$

where S_{lC}, S_{tC} are the compression strengths in the fill and warp directions, respectively.

The penetration failure mechanism caused by fiber crush under compressive pressure is modeled using the following criterion:

$$f_{crush} = \left(\frac{\langle p \rangle}{S_{FC}}\right)^2 - 1, \qquad p = -\frac{\sigma_1 + \sigma_2 + \sigma_3}{3} \tag{21}$$

and S_{FC} is the fiber crush strength and P is the hydrostatic compressive pressure.

Matrix mode failure occurs in a plane where the normal and two-shear stress components reach a maximum value. Under loading conditions transverse matrix cracks propagate through the layer thickness and extend along fiber-matrix interface leading to subsequent delaminations. Matrix cracking and interface delamination in a plain weave layer is given by Equations 22 and 23 respectively:

$$f_{delamination} = S^2\left\{\left(\frac{\langle\sigma_3\rangle}{S_{zT}}\right)^2 + \left(\frac{\tau_{23}}{S_{23}}\right)^2 + \left(\frac{\tau_{31}}{S_{31}}\right)^2\right\} - 1 = 0 \tag{22}$$

$$f_{matrixcrack} = \left(\frac{\tau_{12}}{S_{IS}}\right)^2 - 1 = 0 \tag{23}$$

where S_{zT} is out of plane tensile strength and S_{23} and S_{12} are the failure shear strength properties and $\sigma_2, \sigma_3, \tau_{23}$ and τ_{31} are corresponding stress state. A scale factor S in Equation 23 is introduced to achieve better correlation of simulated delamination area with experimental values.

Higher S value provides more delaminated area and vis-versa. The layer shear strengths, S_{23} and S_{31}, of the plain weave layer under loading conditions depends on the compressive normal stress state, $-\sigma_3$, i.e.,

$$\begin{Bmatrix} S_{23} \\ S_{31} \end{Bmatrix} = \begin{Bmatrix} S_{23}^0 \\ S_{31}^0 \end{Bmatrix} + \tan(\varphi)\langle -\sigma_3 \rangle \tag{24}$$

where S_{23}^0 and S_{31}^0 are the experimental shear strength values, and $\tan(\varphi)$ is the coefficient of friction. The friction angle (φ) can be determined by the Coulomb-Mohr theory.

The failure equations (17-24) provide a particular failure mode in a composite lamina according to a given stress state in that particular lamina. When fiber in tension/shear mode is predicted in a layer, all the stress components are reduced to zero instantaneously and the load bearing capacity of the layer is completely eliminated. During compressive failure the lamina carries a residual strength, S_{RC} (= $SFFC * S_{IC}$) in axial direction, where SFFC denotes scale factor for residual compressive strength. When the fiber crush occurs, the lamina behaves elastically for compressive pressure, p > 0, and does not carry any load for tensile pressure, p < 0.

During matrix failure, transverse shear strength values S_{23}^{0} and S_{31}^{0} are reduced to zero. For tensile mode, $\sigma_3 > 0$, the through thickness stress components are reduced to zero. For compressive mode, $\sigma_3 < 0$, σ_3 is assumed to be elastic, while τ_{23} and τ_{31} act as frictional shear stress on the fractured surface. The detailed description of Mohr-Coulomb failure criterion for composite laminates under interlaminar shear (ILS) and compression is given by Xiao et al.[69]. They introduced a transition parameter, β for a complete failure of a laminate under ILS and through thickness axial stress:

$$f = \left\{ \left(\frac{\langle \sigma_3 \rangle}{S_{zT}} \right)^2 + \left(\frac{\tau_{23}}{S_{23}} \right)^2 + \left(\frac{\tau_{31}}{S_{31}} \right)^2 + \left(\frac{\left\langle \left(\frac{\sigma_{3C}}{S_{zC}} \right) - \beta \right\rangle}{1 - \beta} \right)^2 \right\} - 1 = 0 \qquad (25)$$

where S_{zT}, S_{zC} are out of plane tensile and compressive strength of the lamina and σ_3, σ_{3C} are the corresponding tensile and compressive stress state.

It can be noted that an extra term, $\left(\frac{\left\langle \left(\frac{\sigma_{3C}}{S_{zC}} \right) - \beta \right\rangle}{1 - \beta} \right)^2$ is added in Equation 25. The parameter β is assumed to separate the medium and high pressure regimes. Under tensile loading on a laminate the first three terms in Equation 24 are taken into account without the frictional term, $\tan(\varphi)\langle -\sigma_3 \rangle$. In the medium compressive region, the failure envelop is governed by middle two terms in Equation 24 only. At higher compressive regime, quadratic interaction between compression and shear is considered by adding the last term in Equation 24.

5.1.2 Progressive Failure of a Composite - Material model 162. MAT 162 is known as 'Damage model', extension of MAT 161 by adopting continuous damage mechanics approach (CDM) proposed by Matzenmiller et al.[70]. The CDM formulation takes into consideration the post failure mechanisms in a composite plate as characterized by a reduction in material stiffness (E_{red}, G_{red}). A set of damage variables, ϖ_i with

$i = 1,....6,$ were introduced to indicate the state of anisotropic damage maintaining the orthotropic damage nature of the material throughout the damaging process:

$$E_{red} = (1 - \varpi_i)E_i, \; G_{red} = (1 - \varpi_i)G_i \qquad (26)$$

$$\varpi_i = 1 - e^{\frac{1}{m_i}\left(1 - r_j^{m_i}\right)} \qquad i = 1,.....,6 \qquad (27)$$

where ϖ_i = damage variable, m_i = strain softening parameter and r_j = damage threshold. The damage variable ϖ_i varies from 0 to 1.0 as r_j varies from 1 to ∝, respectively. There are four strain softening input parameters (m_1-fiber damage in x direction, m_2-fiber damage in y direction, m_3-fiber crush and punch shear damage and m_4-delamination damage) required for MAT 162. Xiao et al.[69] proposed certain values (fiber damage, m_1/m_2 =2, fiber crush and shear damage, m_3=0.50, delamination, m_4= 0.2) for the above mentioned strain softening parameters for S2-glass/epoxy composites using quasi-static punch shear technique. Detailed description on "m" parameters is given in different studies[69-72]. The damage function is formulated to account for the overall nonlinear elastic response of a lamina including the initial 'hardening' and the subsequent softening beyond the ultimate strengths.

The effect of the layer strength values of the composite failure modes is modeled by the strain rate dependent functions for the strength values $\{S_{RT}\}$ as:

$$\{S_{eff}\} = \{S_{RT}\}\left(1 + C_{rate} \ln \frac{\{\dot{\varepsilon}\}}{\dot{\varepsilon}_0}\right) \qquad (28)$$

where

$$\{S_{RT}\} = \begin{Bmatrix} S_{lT} \\ S_{tT} \\ S_{lC} \\ S_{tC} \\ S_{FC} \\ S_{FS} \end{Bmatrix} \qquad \{\dot{\varepsilon}\} = \begin{Bmatrix} |\dot{\varepsilon}_a| \\ |\dot{\varepsilon}_a| \\ |\dot{\varepsilon}_b| \\ |\dot{\varepsilon}_b| \\ |\dot{\varepsilon}_c| \\ \left(\dot{\varepsilon}_{ca}^2 + \dot{\varepsilon}_{bc}^2\right)^{1/2} \end{Bmatrix}$$

and C_{rate} are the strain rate constants and $\{S_{eff}\}$ are effective the strength values of the laminate at strain rate $\dot{\varepsilon}$. S_{RT} are the strength values of the laminate under quasi-static strain rate $\dot{\varepsilon}_0$.

In the damage model (MAT 162), the effect of strain rate on the stiffness of a composite layer is modeled by:

$$\{E_{eff}\} = \{E_i\}\left(1 + \{C_{rate}\}\ln\frac{\{\dot{\varepsilon}\}}{\dot{\varepsilon}_0}\right) \quad (29)$$

where

$$\{E_i\} = \begin{Bmatrix} E_a \\ E_b \\ E_c \\ G_{ab} \\ G_{bc} \\ G_{ca} \end{Bmatrix} \quad \{\dot{\varepsilon}\} = \begin{Bmatrix} |\dot{\varepsilon}_a| \\ |\dot{\varepsilon}_b| \\ |\dot{\varepsilon}_c| \\ |\dot{\varepsilon}_{ab}| \\ |\dot{\varepsilon}_{bc}| \\ |\dot{\varepsilon}_{ca}| \end{Bmatrix} \quad \text{and} \quad \{C_{rate}\} = \begin{Bmatrix} C_{rate1} \\ C_{rate2} \\ C_{rate2} \\ C_{rate4} \\ C_{rate3} \\ C_{rate3} \\ C \end{Bmatrix}$$

and $\{C_{rate}\}$the strain rate constants are $\{E_{eff}\}$ are the modulus values of $\{E_i\}$ at the strain rate $\dot{\varepsilon}$.

Under high strain rate and high pressure loading conditions, a composite lamina or laminate undergoes large deformation around the loading area. When simulated under those conditions, large distortion in the elements around the loading region occurs and this creates numerical instability. To avoid numerical distortion element erosion is necessary. Sections 6 and 7 provide various simulations using material model MAT162.

5.2 MATERIAL PROPERTY INPUTS TO MAT161/162

Appropriate material input parameters are required to incorporate into the material model to simulate the impact response accurately. To optimize and design for ballistic performance, the dynamic material properties such as strength and strain at high strain rates, failure mechanisms, strain rate regime (e.g. low, intermediate, or high velocity) and laminate architectures must be known. Since material models MAT161/162 account for fiber failure in tension, compression and crush as well as matrix mode crushing and delamination, the elastic properties of the composite (i.e. in-plane tensile, compressive and shear modulus and strength respectively) have to be obtained by standard ASTM tests (for e.g. D 3039, D 695 and D 5379).

The material models MAT 161/162 require 32 input parameters to fully evaluate damage in a composite laminate. Typical input parameters needed for MAT162 are listed in Table 1. Currently there is no standard ASTM (or equivalent) test to measure the fiber crush, fiber shear, damage parameters and strain rate co-efficient parameters. Fiber crush and fiber shear data are estimated by comparing the experimental results with the simulation[72]. As will be discussed in Section 6.0, the different strain softening parameters (m_i) are experimentally determined using the quasi-static punch shear (QS-PST) loading approach. The load versus displacement curve obtained from a QS-PST experimental test represents the combined effect of fiber damage, matrix damage and delamination of the laminate[69, 72].

Table 1. Input parameters for MAT 162 and corresponding technique

Density, r	ASTM C-20
Tensile modulus, E_A, E_B, E_C	AST D 3039
Poisson's ratio v_{21}, v_{31}, v_{32}	Calculated from tensile and shear modulus
Shear modulus, G_{AB}, G_{BC}, G_{CA}	ASTM D 5379/D 5379M
In-plane Tensile Strength, S_{AT}, S_{BT}	ASTM D3039
Out of plane tensile strength, S_{CT}	ASTM D 6415
Compressive strength, S_{AB}, S_{BC}, S_{CA}	ASTM D 3410/D, 3410 M-03, ASTM D 695
Fiber crush, S_{FC}	N.A.
Fiber shear, S_{FS}	N.A.
Matrix mode shear strength, S_{AB}, S_{BC}, S_{CA}	ASTM D 5379/D 5379M or ASTM D 3846
Residual compressive scale factor, S_{FFC}	N.A.
Friction angle, P_{HIC}	N.A.
Damage parameter, AM1, AM2, AM3, AM4	N.A.
Strain rate parameter, C_1, C_2, C_3, C_4	Split Hopkinson Pressure Bar
Delamination, S_DELM	N.A.
Eroding Strain, E-LIMIT	N.A.

In the table, A, B and C correspond to the 3-orthotropic directions; density, tensile modulus, shear modulus, in-plane tensile strength, out-of-plane tensile strength and compression strength, matrix mode shear strength are standard parameters; Fiber crush occurs at the point of the contact of the projectile with the plate; Fiber shear occurs during penetration; residual scale factor corresponds to partial strength of a failed region in compression; friction angle corresponds to the Coulomb-Mohr friction between the fiber and matrix; damage parameters represent the progressive strain softening of the laminate; strain rate parameter corresponds to strain rate dependency of the plate; delamination criteria guides the in-plane matrix failure mode, and eroding strain corresponds to eroding of the failed element(s) when the strain limit is exceeded.

6.0 QUASI-STATIC PUNCH SHEAR (QS-PST) and BALLISTIC RESPONSE

Several researchers have concluded that the failure modes under impacts exceeding perforation, and those obtained from static punch through tests are identical and can be used for modeling and prediction of high-speed ballistic impact and penetration[73-76].

Goldsmith et al.[73] conducted experimental and analytical investigation of quasi-static and ballistic perforation of woven T-300/934 epoxy laminates by a cylindro-conical projectile. Major damage mechanisms were crack propagation, fiber breakage, petaling, delamination and hole enlargement. Sun et al.[74-76] evaluated a punch curve from quasi-static punch experiments and used the punch curve as a 'structural constitutive model' to capture nonlinearity during high speed and penetration of graphite epoxy laminates of various thicknesses and size. In addition, dynamic work/energy of perforation was found to be higher than static for the same specimen thickness, even though the modes of failure are similar for the two cases. Lee and Sun[77] characterized the penetration process of graphite epoxy laminates using a model based on a static punch curve. They predicted the ballistic limit of the laminate and compared the results from the dynamic impact test. Vlot[78] investigated the quasi-static, low velocity and ballistic penetration behavior of fiber metal laminates using hemispherical nose projectiles. He reported that the absorbed energy in low velocity impact and ballistic perforation is higher than quasi-static punch shear experiments. Ursenbach et al.[79] studied the quasi-static response of CFRP laminated plates using a combined experimental and analytical method. They identified two major damage modes caused by delamination and plugging during penetration by cylindrical indenters.

Quasi-static punch shear (QS-PS) damage mechanisms are a function of support span and are - (a) transverse matrix damage, (b) initiation of shear plug and delamination, (c) delamination progression and completion of shear plug formation, (d) push-out of the shear plug, and (e) tensile fiber fracture and push out of the plug and punch.

6.1 QUASI-STATIC PUNCH SHEAR TESTING

The test fixture used in quasi-static punch is illustrated in Figure 35. The test fixture comprises a thin cover plate (12.7 mm thick) with a circular hole at the center, a thick support plate (70 mm thick) with a central hole similar to the cover plate, and a cylindrical punch. The test specimen is clamped between the plates with an exposed area equal to the circular hole.

The punch geometry represents the projectile diameter and shape of interest for high velocity impact. Generally, an indentor diameter 7.62" (0.3", 0.30 caliber) or 12.7 mm (0.5", 50 caliber) is used in quasi-static test to represent a projectile of high velocity impact[72].

During QS-PST, the load increases with punch displacement until damage initiation occurs at 'A', Figure 36a. From points O to A, the laminate elastically deforms under bending. After the first peak load drops to 'B' where delamination starts in the laminate. From 'B' to 'C' delamination progresses until the plug is formed at C. The drop in load at point C is caused by fiber breakage on the rear surface. The load drop from points C to D is insignificant compared to that at point D which results in a complete loss of load carrying capacity. The load drop DE corresponds to the complete formation of the plug.

After the plug formation, the plug is being pushed by the indenter and a frictional force is generated between the plug and the surrounding material. This frictional force corresponds to the long tail of the punch curve, E to F. Unlike a brittle epoxy matrix composite, the load drop at point A does not occur in case of a rubber toughened epoxy matrix composite during QS-PST, Figure 36b[72].

Figure 35. Test set up for quasi-static punch shear.

6.2 QUASI-STATIC PUNCH COMPARISON TO BALLISTIC IMPACT

The QS-PST testing was conducted for two span-to-punch radius (SPR) conditions – SPR=2 and SPR=4 respectively. The maximum load decreases as the SPR increases from 2 to 4 in the 8, 12 and 16 layer laminates. This is because specimen compliance increases as the SPR increases (longer span). Figure 37 illustrates the load-displacement curve for the laminates at SPR= 2 and 4. During QS-PST, load increases with punch displacement until damage initiation occurs at point A, B and C for 8, 12 and 16 layers respectively. Upto the points, A, B and C the laminate deforms elastically under bending load. The load-displacement curve deviates from these points A, B and C and increases non-linearly up to maximum load at points P, Q and R. In the displacement range A-B, B-Q and C-R, progressive accumulation of matrix damage, fiber-matrix debonding, delamination growth and initiation of a shear plug is observed. The total energy under the punch curve is the energy dissipated during quasi-static penetration. The total energy E_{total} absorbed by the laminate during QS-PST is given by:

$$E_{total} = E_{se} + E_d + E_f \tag{30}$$

where E_{se} is the strain energy, E_d is the energy dissipated in damage creation, and E_f is the friction energy.

A series of numerical simulations were carried out to predict the ballistic limit of C/PPS laminates struck by the projectile by incorporating the calibrated strain softening parameters from the quasi-static test. Figure 38 illustrates a typical quasi-static punch shear of C/PPS and the corresponding simulation. The matrix cracking, interply shearing, fiber shear and plug formation is captured in the simulation as is in agreement to the experiment.

From the punch-shear test, the load versus displacement curve was obtained. The results from the experimental QS-PS were reproduced in the simulation for different strain softening parameters (the *m* parameters). The selection of the *m* parameters to match the experiment was based on trial and error. These *m* parameters were then incorporated in the ballistic simulation to predict ballistic response.

The ballistic impact response of a C/PPS laminate subjected to impact velocities of 195 m/s to 210 m/s is shown in Figs 40 a and b respectively[75]. These simulations were generated using the *m* parameters calibrated from the QS-PST. The experimental mean ballistic limit of the 16 layer C/PPS laminate is measured to be 197 m/s which is in agreement with the simulation. The cross-section of the laminate and corresponding simulation is shown in Figure 39a and 39b. At 194 m/s there is partial penetration of the laminate. Complete penetration occurs at 210 m/s, Figure 39b. Fiber-matrix damage long the primary yarns and fiber fracture is much more extensive than the partial penetration case. Due to compression-shear at impact velocity of 210 ms^{-1}, the projectile starts penetrating through the laminate and complete formation of shear plug is observed due to high tension shear in the area beneath the projectile, Figure 40b. The failure load or the penetration energy of the composite laminate for a dynamic event is higher than that of the QS-PST, due to strain rate effects[75].

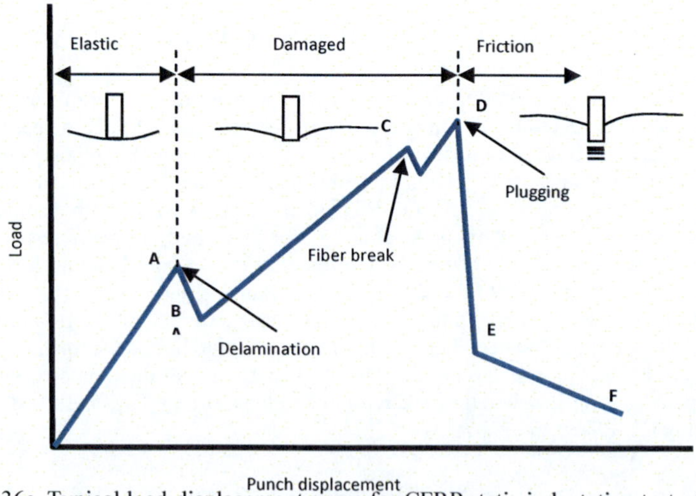

Figure 36a. Typical load-displacement curve for CFRP static indentation test - CFRP with brittle matrix [72]

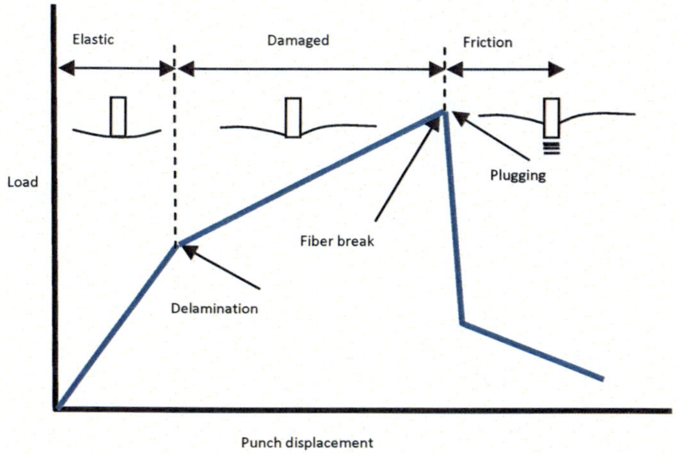

Figure 36b. Typical load-displacement curve for CFRP static indentation test - CFRP with toughened matrix[72].

Studies by Gama et al[80] also demonstrate that the ballistic impact and quasi-static response are similar in S2-glass/epoxy composites. The typical failure modes indicated in Fig. 40 are matrix cracking, delamination, fiber-matrix debonding, fiber breakage and fiber plugging, where only the quasi-static indentation response is shown. The ballistic damage is identical to the quasi-static indentation response.

7.0 BALLISTIC IMPACT SIMULATION

In a composite laminate subjected to ballistic impact, the kinetic energy of the projectile is dissipated in several ways. The predominant energy absorption mechanisms are in the form of kinetic energy imparted to the specimen, namely cone formation on the distal side of the laminate and/or spall formation, energy absorption as a result of shear plugging, tensile fiber failure of the primary yarns, fiber debonding, fiber pull-out, elastic deformation of the secondary yarns, matrix cracking, interlaminar delamination, and frictional energy absorbed during interaction of the penetrator and laminate[81, 82].

7.1 NUMERICAL APPROACH

7.1.1 Simulation Tools. A mesh generation tool such as Altair Hypermesh™ can be used for pre-processing[30]. The solver is LS-DYNA which is used to analyze perforation mechanisms, failure modes, and damage evaluation during high velocity projectile impact.

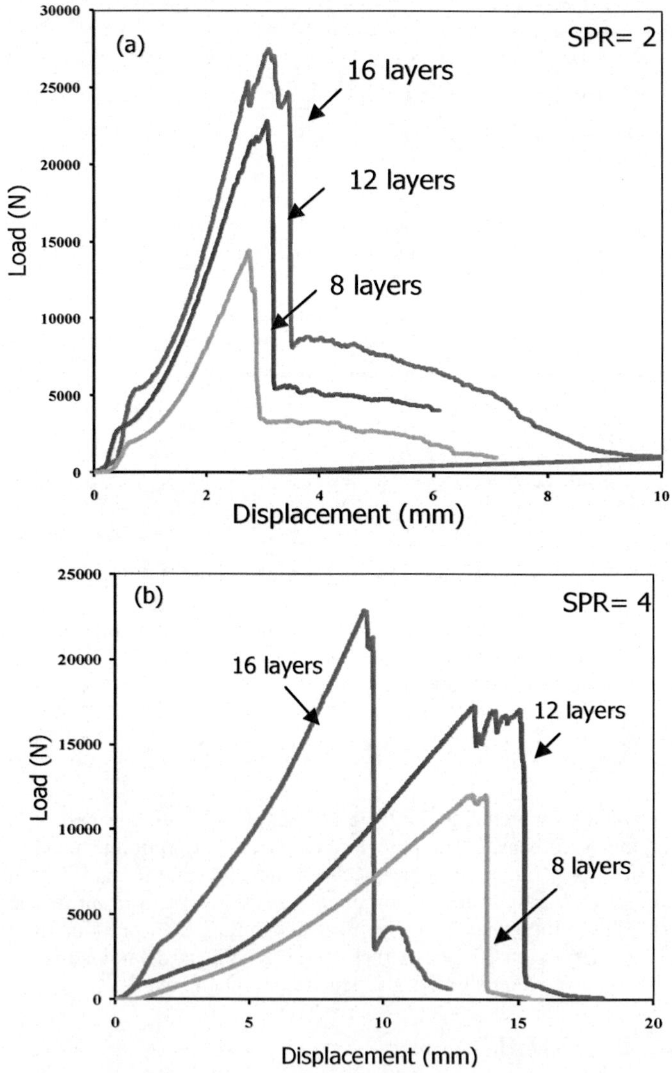

Figure 37. Load displacement results from quasi-static shear punch; (a) SPR=2, (b) SPR=4[72].

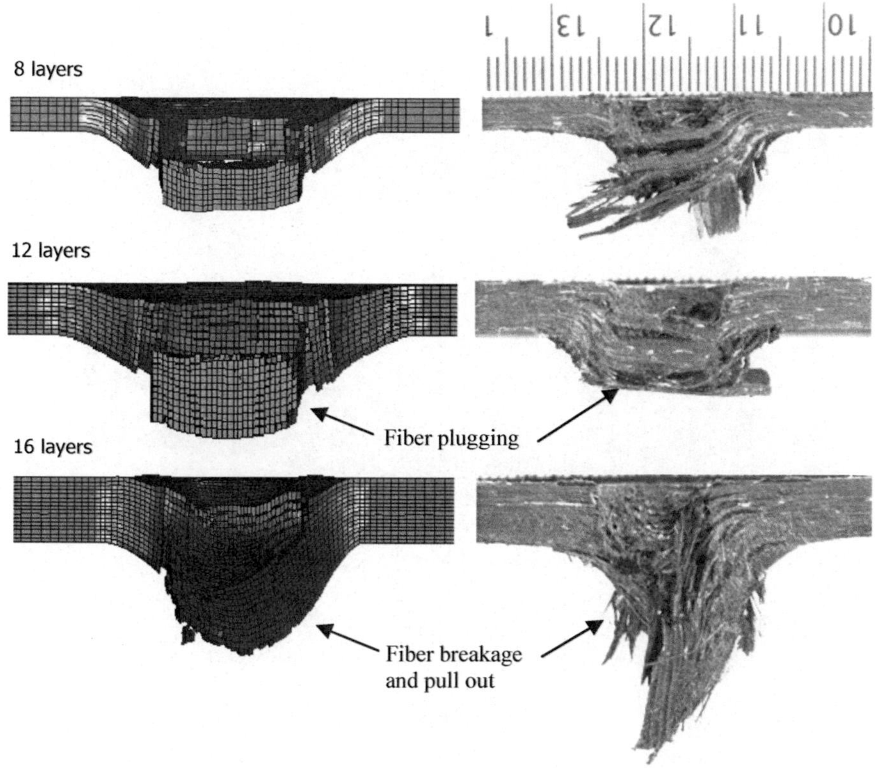

Figure 38. Quasi-static shear punch damage correlation[72].

7.1.2 Numerical model. In this example, the composite plate and the projectile is meshed using brick elements. The projectile is assigned an initial velocity. A typical meshed geometry of the layered composite plate is shown in Figure 41. A fine grid relative to the boundary is used in the impact region of the target to obtain a smooth stress gradient. The mesh size is increased gradually toward the outer edges to maximize computational efficiency. A quarter symmetry model is adopted to reduce computational time.

Representative material properties for the composite plate and the projectile used in this simulation are shown in Tables 2 and 3 respectively, and Table 4 provides the model dimensions. Since the damage mechanisms in high velocity impact are similar to those in quasi-static loading, the strength properties of the composite laminate are taken from quasi-static tests.

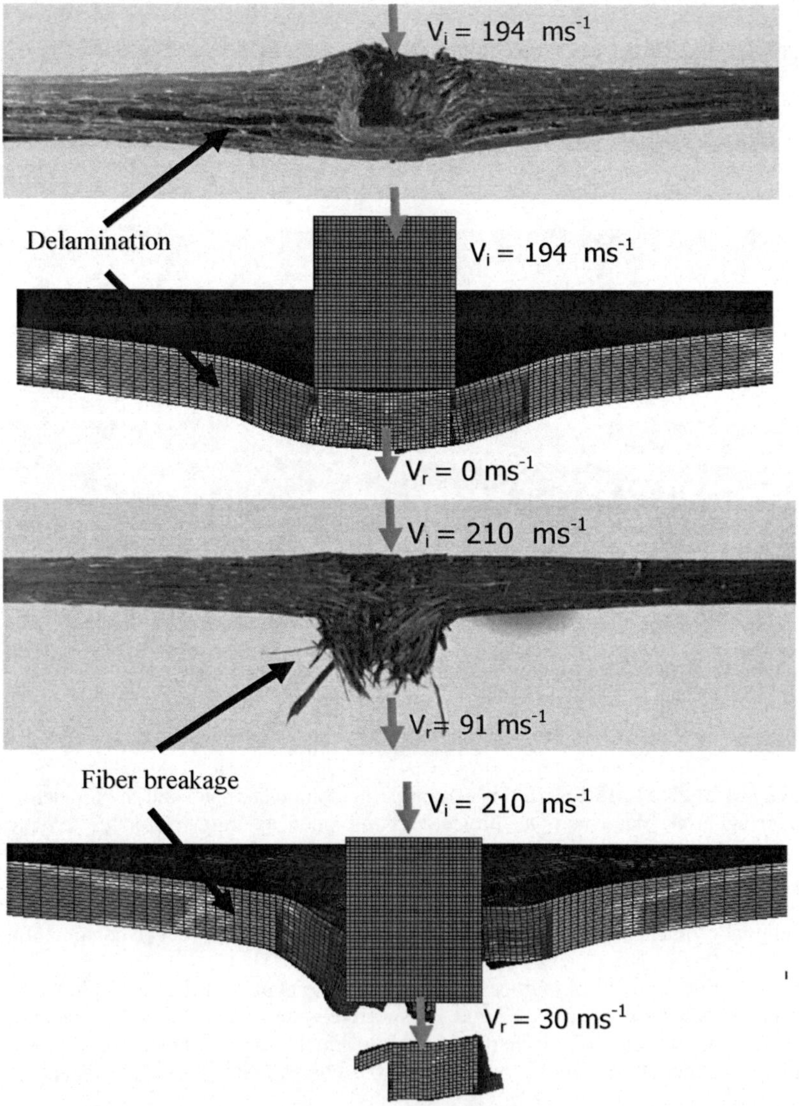

Figure 39. Quasi-static shear punch damage correlation to ballistic impact. (a) Experimental test and simulation for impact velocity = 194 m/s.

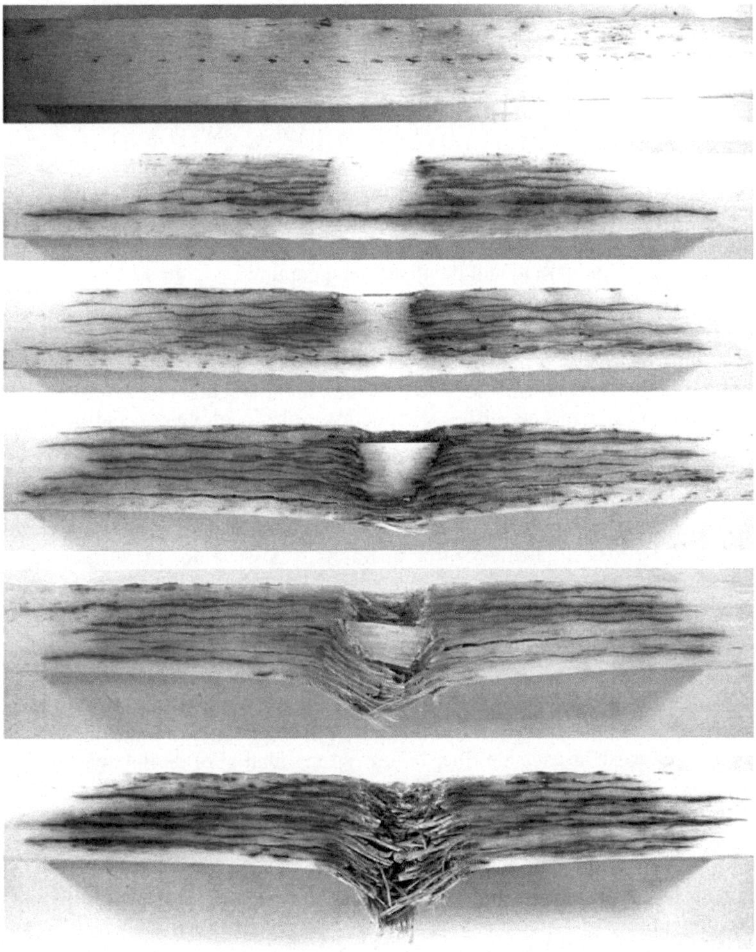

Figure 40. Quasi-static shear punch damage progression in S2-glass/epoxy composite[69].

7.1.3 Material model. As explained in Section 5.0, Material model 162 (MAT_COMPOSITE_ DMG_MSC) is used to evaluate progressive damage in the composite laminate[66]. MAT 162 is based on the Hashin's failure criteria[62], it considers five failure modes; tensile and compressive fiber failure, fiber crush, through-the-thickness matrix failure and delamination. During impact simulation, some elements in the impact region undergo large distortions, which may lead to numerical instabilities; hence an element erosion criterion is incorporated in Material model 162.

The projectile is modeled using Material model 3 (MAT_PLASTIC_KINEMATIC)[66]. MAT 3 is a bi-linear elastic-plastic model that contains formulations combining isotropic and kinematic hardening. In the experimental study negligible amount of deformation of the impacted face of the cylindrical projectile was observed. The other option for modeling a non-deformable projectile is Material model 20 (MAT_RIGID).

7.1.4 Contact Type. Proper contact definition between projectile and the laminate is required to model high velocity transverse impact. Three different types of contacts are adopted between the impactor (slave) and the target (master), namely: kinematic constraint method, the penalty method and the distributed parameter method[66].

7.1.5 Progressive Damage and Damage Parameters. A CDM formulation (also refer to Section 4.0) is incorporated in MAT 162 by adopting the MLT damage mechanics approach[70]. Post failure mechanisms in a composite laminate are characterized by reduction in material stiffness. The stiffness reduction (elastic modulus) and damage variables were provided in equations (25) and (26).

The effect of the strain softening parameter m on the stress-strain response of the element is shown in Figure 43. High values of m (for e.g. $m = 50$) results in brittle failure of the material. Once the tensile stress reaches the maximum value, it becomes zero which means there is no loss in stiffness prior to failure. Low values of m indicate a ductile failure response resulting in more energy absorption prior to complete damage with a gradual loss of stiffness after failure.

7.2 IMPACT SIMULATIONS AND COMPARISON TO EXPERIMENTS

Damage growth can be attributed to the combination of fiber failure modes including punch, shear plug, fiber crush and tensile fiber failure. A simulation of partial penetration and full penetration of E-glass/PP composite is illustrated in Figure 44a and b respectively. During the penetration process, a peak stress is generated in the contact region and propagates along the primary yarns which undergo tensile failure, while the secondary yarns undergo elastic deformation.

The tensile fiber failure in a plain weave layer were expressed by Equations (16) and (17) respectively. In high strain rate loading, the material experiences higher stiffness because of strain rate sensitivity. The effect of strain rate on the layer strength values and the effective modulus were provided earlier in Equations 27 and 28.

The strain softening parameter has a significant influence on the ballistic simulation. Figure 43 illustrates the projectile penetration at 181 m/s with a residual velocity of 131 m/s for a strain softening parameter m value of 0.57. By trial and error, a m value of 0.03 provided a residual velocity of zero which was in agreement with the experiment. The strain softening parameter has to be calibrated for each material type and subsequently various velocities and material geometries can be investigated.

The Kinetic Energy (KE) plots with respect to time are shown in Figure 45 for different damage modes and projectile penetration on a woven fabric laminate at different time steps. Damage growth can be attributed to the combination of fiber failure modes including punch, shear plugging, fiber crush and tensile fiber failure. During the penetration process, a peak stress is generated in the contact region and propagates along

Table 2. Material properties of a plain weave E-glass/PP composite layer		
Density (kg/m^3)	ρ	1500
Tensile modulus (GPa)	E_{11}	14
	E_{22}	14
	E_{33}	5.3
Shear modulus (GPa)	G_{21}	1.79
	G_{31}	1.52
	G_{32}	1.52
Poisson's ratio	V_{21}	0.08
	V_{31}	0.14
	V_{32}	0.15
Tensile Strength (GPa)	X_T	0.43
	Y_T	0.43
	Z_T	0.15
Compressive strength (GPa)	X_C	0.23
	Y_C	0.23
Matrix mode shear strength, (GPa)	S_{12}	0.032
	S_{23}	0.032
	S_{31}	0.03
Fiber shear strength (GPa)	SFS	0.25
Fiber crush strength (GPa)	SFC	0.65
E-limit	^{0}	2
Delamination factor	S	0.3
Friction angle		20
Strength properties strain rate coefficient	C_1	0.024
Longitudinal moduli strain rate coefficient	C_2	0.0066
Shear moduli strain rate coefficient	C_3	-0.07
Transverse moduli strain rate coefficient	C_4	0.0066

Table 3. Material properties for the tool steel projectile		
Density (kg/m^3)	ρ	7860
Young's modulus (GPa)	E	210
Poisson's ratio	ν	0.28
Yield strength (GPa)	σ_y	1.08

the primary yarns which undergo tensile failure, while the secondary yarns undergo elastic deformation.

The steeper slope in projectile KE vs. Time curves in Figure 44 from approximately 0 to 0.01 ms is due to fiber crush which provides maximum resistance to penetration. The decrease in resistance to the penetration from approximately 0.01 to 0.03 ms is believed to be due to cone formation in the vicinity of the impact zone. This resulting deformation leads to reduced slope in the kinetic energy curve via kinetic energy transfer to the target. The increase in slope in the next time interval of 0.03 to 0.05 ms can be attributed to the bending stiffness of the laminate which offers resistance to the projectile penetration. At ballistic limit, the projectile came to rest at about the same time i.e. 12 ms for 8, 12 and 16 layer plates.

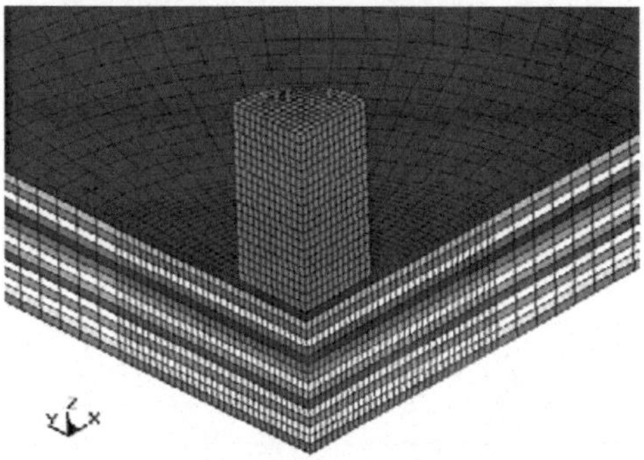

Figure 41: Quarter symmetry geometry of the projectile and 16 layer laminate showing the mesh refinement along the periphery of the laminate[72].

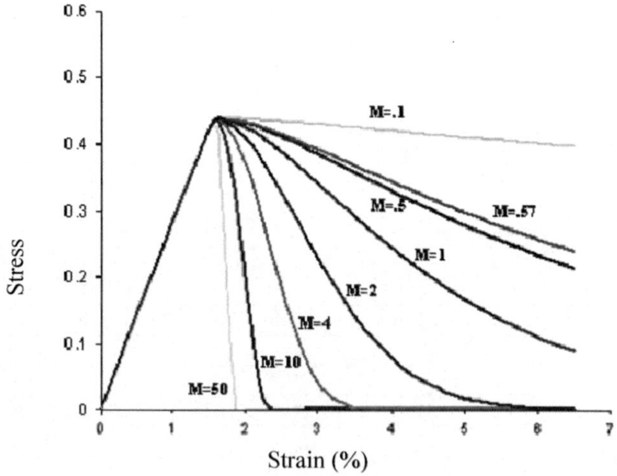

Figure 42: Stress-strain response of the single element loaded in-plane tension with different values of m parameter. (Note: M in the above figure is the same as *m* in the text for strain softening parameter)[72].

Table 4. Dimensions of quarter symmetry model		
¼ plate		
Length (mm)	50	
Width (mm)	50	
Layer thickness (mm)	0.68	
¼ projectile		
	0.30 caliber	0.50 caliber
Radium (mm)	3.98	6.35
Length (mm)	6.96	13.65

Table 5. Expt. impact result and numerical prediction for 0.50 caliber impact on the 8, 12, and 16 layer E-glass/PP plates

Specimen	Incident velocity ($m\ s^{-1}$)	Incident kinetic energy (J)	Residual velocity ($m\ s^{-1}$)	Residual kinetic energy (J)	Energy absorbed (J)	Mean experimental ballistic limit ($m\ s^{-1}$)	Standard deviation for ballistic limit ($m\ s^{-1}$)	Numerical prediction of ballistic limit ($m\ s^{-1}$)
1 (8 layers)	237.1	376.2	NR	NR	376.2	181.3	18.7	179.5
2 (8 layers)	226.2	342.2	NR	NR	342.2			
3 (8 layers)	207.9	289.1	NR	NR	289.1			
4 (8 layers)	187.1	234.3	63.7	27.1	207.2			
5 (8 layers)	186.8	233.5	0.0†	0.0†	233.5			
1 (12 layers)	260.3	455.3	NR	NR	455.3	272.5	26.2	250
2 (12 layers)	330.4	733.6	179.8	217.3	516.3			
3 (12 layers)	321.9	695.7	179.5	216.4	479.2			
4 (12 layers)	265.2	472.2	NR	NR	472.2			
1 (16 layers)	332.5	742.5	153.9	159.1	583.5	288.8	31	286.1
2 (16 layers)	319.1	683.9	121.3	98.8	585			
3 (16 layers)	279.5	524.6	0.0‡	0.0‡	524.6			
4 (16 layers)	285.3	543.3	0.0‡	0.0‡	543.3			

‡ = Ballistic limit Velocity, † = Partial Penetration, NR = No Reading (data point not valid)

Note: Penetration was not achieved for the 12- and 16-layer laminates using the 0.30 caliber projectile for the velocity range 150 to 250 m s^{-1}; hence this study was limited to 8-layer laminates only.

7.2.1 Damage Simulation. Delamination at the interface is one of the major failure mechanisms at the matrix mode. It is caused by the interlaminar stresses $(\sigma_z, \tau_{yz}, \tau_{zx})$ which initiate matrix microcracks which span the fiber-matrix interface and propagate along the fiber. The delamination of the composite plate was given by equation (22).

The delamination scale factor, S (Refer to Equation 22) has a significant effect on ballistic limit velocity or energy absorption of composite plates. S is introduced to achieve better correlation with experimental values of delamination area by a scaling

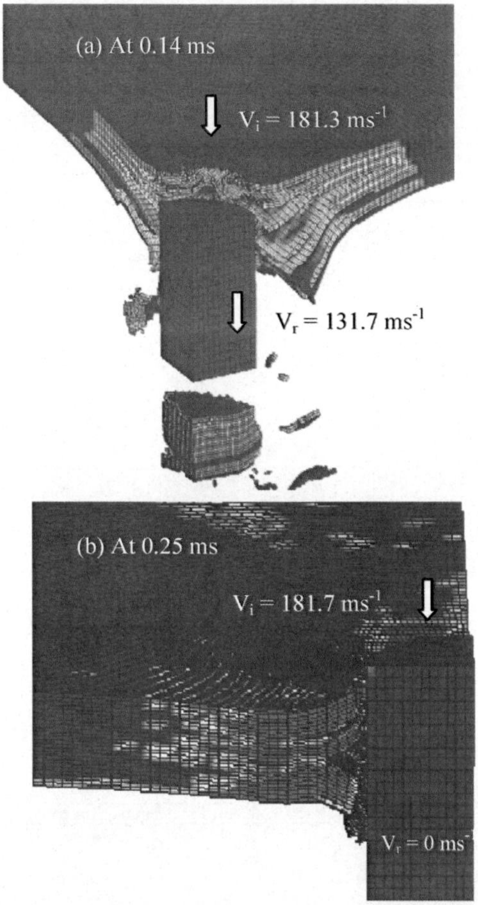

Figure 43: (a) Simulated damage showing the projectile at an incident velocity of 181.3 ms^{-1} perforating a 8 layer laminate (a) with a residual velocity of 131.7 ms^{-1} for a strain softening parameter m value of 0.57 (b) zero residual velocity; projectile arrested by the plate for a m value of 0.03. The different gray and white regions represent various stress levels in the laminate as the projectile penetrates through the thickness[72].

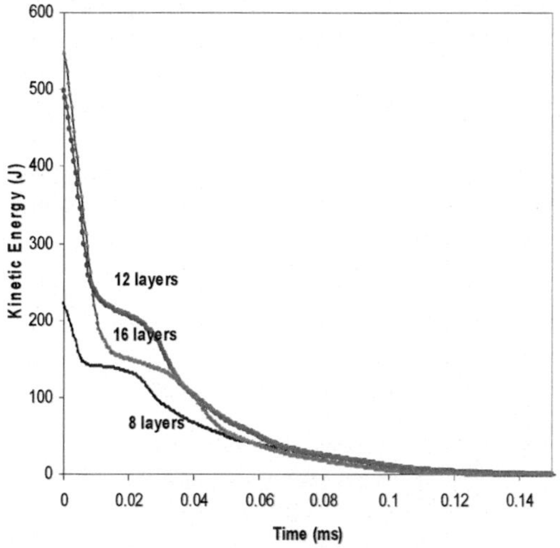

Figure 44: Kinetic energy lost by the projectile vs. time for the 8, 12, and 16 layer laminates at ballistic limit at damage parameter, m=0.57. 0.325 and 0.42 respectively[72].

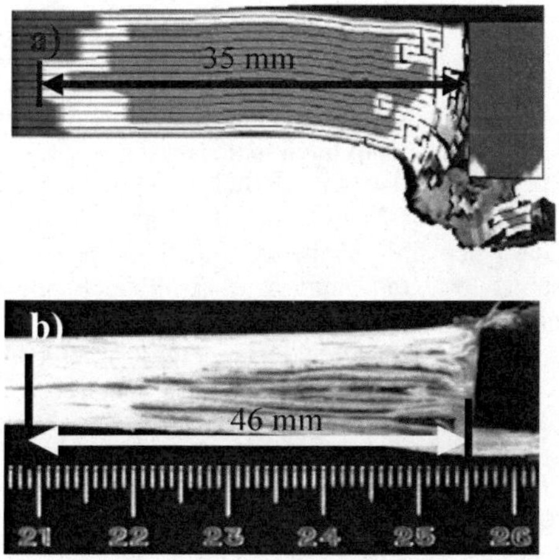

Figure 45. Delaminated area of 16 layer composite at ballistic limit (a) simulation (b) experimental (cross section)[72].

factor. Figure 45 illustrates debonding of the fibers from the surrounding matrix and represents delamination failure of the laminates for both numerical and experimental observation for typical a 16-layer E-glass/PP composite laminate at its ballistic limit.

7.3 BALLISTIC IMPACT OF A SANDWICH COMPOSITE

Under transverse impact loading of a sandwich composite, the projectile penetrates the impact side facesheet by fiber breakage, matrix cracking and delamination, followed by penetration through the core, progressing to the back facesheet. At higher impact velocities a critical condition is reached when the local contact stress exceeds the local strength, which may be the laminate bending strength, core compression strength or interface delamination strength. This stress leads to partial or complete penetration of the projectile into the sandwich composite.

As described in the section on LVI of sandwich composites, foam, balsa wood and honeycomb cells are some of the common core materials in sandwich composites. Balsa wood is considered in this discussion, as a commonly used core material.

The cellular microstructure of balsa wood core is anisotropic, possesses excellent specific strength, stiffness and specific energy dissipation capacity. Wood has been used as a protective material for high velocity impact events for several centuries [81, 82]. There have been some studies on the behavior of wood that have investigated high rates of loading from impact events[83-87].

7.3.1 Modeling of a Sandwich Plate. The grid geometry of the sandwich plate is modeled by brick elements; in this case 3 layers, with one element through the thickness per layer, and ten elements through the thickness of the core. Each facesheet lamina and the balsa core is modeled with brick elements. 0.30 and 0.50 caliber steel spherical projectiles are modeled with brick elements as well. The impact velocity considered here ranges from 220-280 m/s. A gradient in mesh density is applied with high mesh density in the impact area and coarser elements towards the edges, thereby providing computational efficiency with smooth stress gradient from the impact position to the edge of the plate .

The contact between the projectile and the sandwich composite plate is defined using CONTACT_ERODING _SINGLE_SURFACE which is a penalty method[66]. Eroding contact type is recommended when solid elements in the contact definition are subjected to erosion (element deletion) to avoid numerical disturbance due to large element distortion at the impact region [59, 69, 71, 88, 89].

Penetration through the facesheets is handled using eroding elements with strain based failure criterion, whereas penetration through the balsa wood core is modeled with damage induced stress reduction criteria. If an element undergoes tensile failure and exceeds the axial tensile strain, then it is automatically eroded. The material properties used for the simulation of the laminate and the balsa wood core are summarized in Table 6 and Table 7, and for the projectile(s) in Table 8 respectively.

7.3.2 Wood material model. Wood is a porous, fibrous, complex anisotropic material. Wood behaves linearly in longitudinal and transverse tension, while significant non-linearity is observed in the stress-strain relationship in compression and shear. For analytical purposes, wood can be assumed to be an orthotropic material because it possesses different properties in three directions; longitudinal, tangential, and radial directions[89]. It exhibits different properties with time, temperature, moisture content and loading rate[87, 89]. Under static conditions or low strain rate, wood can be treated as a linear elastic material. With increasing loading rate the cell walls start to buckle locally and the response of the wood is non-linear. The localized deformation of different wood subjected to quasi-static compression and dynamic loading progresses as compressive (crushing) wave fronts.

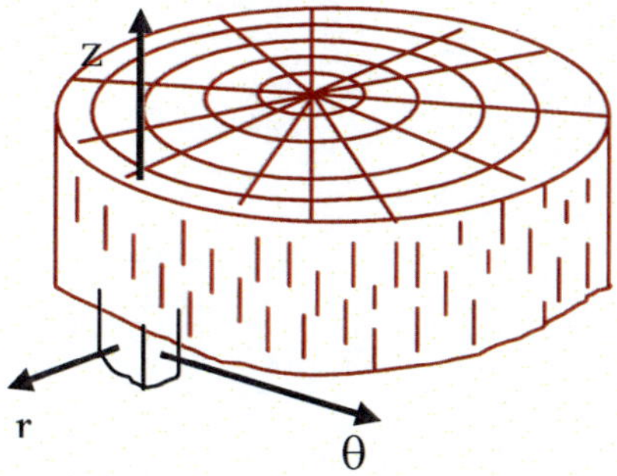

Figure 46. The three principal directions in wood

For most practical purposes, balsa wood can be categorized as transversely isotropic with an isotropic plane being perpendicular to the axis of the tree[86, 89]. The material model MAT 143 is generally used for modeling balsa wood. MAT 143 is based on transverse isotropy and accounts for compression hardening. It has peak-softening characteristics under tension. In material model MAT 143, separate damage parameters are incorporated to account for parallel and perpendicular grains. With progression of damage, the stiffness of the wood reduces along these directions.

In balsa wood, the planes perpendicular to the longitudinal (z), radial (r) and tangential (θ) directions shown in Figure 46 are considered as planes of elastic symmetry (orthotropic). Hence, the element erosion criteria is based on the fact that the elements automatically erode when six parallel stress components reach 99% damage via a parallel (longitudinal) damage parameter.

Energy absorption of the composite facesheets and the balsa wood core provides an indication of ballistic efficiency of the constituents. The composite facesheets for this study have an average thickness of 2 mm and fiber volume fraction of 40%. Commercial balsa wood (trade name Baltek® SB100 structural end-grain balsa) with a nominal density of 150 kg m^{-3} is used. Detailed information on the composite facesheet can be found in [72].

7.3.3 0.30 caliber impact. The cross-section of the sandwich plate subjected to impact energies between 50 and 67 J subjected to .30 cal. single projectile impact are shown in Figures 47 and 48. These energy levels are less than the ballistic limit of the specimen for the .30 cal. impact.

For the 50 J impact the top composite facesheet undergoes complete perforation exhibiting fiber fracture and delamination at the facesheet-core interface. The balsa core exhibited localized crushing directly below the point of impact (Figure 47a) up to a distance of 12 mm through the core thickness. The projectile remains embedded within

Table 6. Material properties of a plain weave S2-glass/epoxy laminates[72]	
Density, kg/mm^3	1.85E-06
Tensile Modulus, E_A, E_B, E_C, GPa	27.1, 27.1, 12.0
Poisson's Ratio, ν_{21}, ν_{31}, ν_{32}	0.11, 0.18, 0.18
Shear Modulus, G_{AB}, G_{BC}, G_{CA}, GPa	2.9, 2.14, 2.14
In-plane tensile strength, S_{AT}, S_{BT}, GPa	0.604
Out of plane tensile strength, S_{CT}, GPa	0.058
Compressive strength, S_{AC}, S_{BC}, GPa	0.291
Fiber crush, S_{FC}, GPa	0.85
Fiber shear, S_{FS}, GPa	0.30
Matrix mode shear strength, S_{AB}, S_{BC}, S_{CA}, GPa	0.075, 0.058, 0.058
Residual compressive scale factor, S_{FFC}	0.30
Friction angle, P_{HIC}	10
Damage parameter, AM1, AM2, AM3 and AM4	0.6, 0.6, 0.5, 0.2
Strain rate parameter, C_1	0.10
Delamination, S_DELM	1, 5
Eroding strain, E_LIMIT	1.20

Table 7: Material properties of end-grain balsa wood core[72]	
Density, kg /mm^3	1.55E-07
Moisture content, %	1.20E+01
Stiffness	
Parallel normal modulus, E_L, GPa	5.30
Perpendicular normal modulus, E_T, GPa	0.20
Parallel shear modulus, G_{LT}, GPa	0.166
Parpendicular shear modulus, G_{LR}, GPa	0.085
Parallel major poisson's ratio	0.25
Strength	
Parallel tensile strength, X_T, GPa	0.0135
Perpendicular tensile strength, Y_T, GPa	0.0004
Parallel compressive strength, X_C, GPa	0.0127
Perpendicular compressive strength, Y_C, GPa	0.0023
Parallel shear strength, S_{XY}, GPa	0.003
Parpendicular shear strength, S_{YZ}, GPa	0.004

the balsa core by pushing the cells in the vicinity of the projectile. Small amounts of delamination at the distal side can be attributed to the energy imparted by the wood cells that are pushed down by the projectile towards the bottom facesheet shown in Figure 47a.

Table 8. Material properties for the tool steel spherical projectile	
Density, kg/mm^3	7.86E-06
Young's Modulus, E, GPa	210
Poisson's Ratio	0.28
Yield Strength, GPa	1.08

The simulated damage of the sandwich composite plate is illustrated in Figure 47. For the impact energy of 67 J the damage through the top facesheet and core progresses identically to the 50 J impact, however the back facesheet is engaged in the impact event. The back facesheets exhibits small amounts of debonding at the core to facesheet interface as shown in Figure 48a,b.

7.3.4 0.50 caliber impact. Fiber breakage is observed on the front facesheet along with a local compression of the wood core. The balsa wood core crushes directly under the point of impact (Figures 49a,b).

The .50 caliber projectile was arrested within the sample for the 233.18 J impact; while complete projectile penetration was observed for the higher impact energy, 301.4 J. The composite facesheets and wood core material absorbs a significant amount of energy through indentation, delamination/splitting and progressive collapse for the 233.18 J impact. The bottom facesheet suffers negligible fiber breakage in the bottom most ply with extensive interlaminar delamination and core-facesheet splitting as shown in Figure 49a. The specimen subjected to impact energy 301.4 J exhibits significant amount of fiber breakage on the top and bottom facesheet, Figure 49b.

The projectile penetrates the top facesheet by localized shear. Due to pressure exerted by the projectile to the sandwich plate the entire plate goes into flexure. The balsa core cells in the vicinity of the projectile crush. Due to the combined effect of core crushing and flexure, the balsa core debonds at the facesheet to core interface. Due to the action of flexure (bottom face is in tension) the bottom facesheet has 178% higher damage than the top.

7.3.5 Energy dissipation in .30 and .50 caliber impact – Laminate versus Sandwich. Energy is dissipated during penetration of each constituent i. e. top facesheet, balsa core and bottom facesheet of the sandwich plate. The projectile looses velocity as it penetrates the top facesheet, core and the bottom facesheet. This loss in energy for different specimens is shown in Figure 50.

The front facesheet absorbs lesser energy for a smaller projectile (0.30 caliber), while it absorbs higher energy for a larger projectile (0.50 caliber). In this case, the front composite facesheet absorbs an average energy of 36.16 J for the 0.30 caliber projectile, while 121.3 J for the front facesheet for the 0.50 projectile impact, Table 10. The energy difference is higher for the constituent laminate (i.e. laminate without a sandwich core).

In a sandwich construction, the core restricts the flexural motion of the facesheet. In the study conducted by Deka et al[72] It has been shown that the energy absorption of the

Table 9. S2-glass/epoxy balsa core sandwich plate, single projectile impact

Spec - imen	Projec tile (cali -ber)	Incident Velocity (m/s)	Resi- dual Veloci ty (m/s)	Impact Ene- rgy (J)	Resi- dual Ene- rgy (J)	Ene- rgy Absor- ption (J)	New Surface creation (cm^2)	
							Top face	Bottom face
1	0.30	220.37	0	49.50	0	49.50^	13.45	15.22
2	0.30	254.80	0	66.18	0	66.18^	16	39.64
3	0.30	256.34	0	66.99	0	66.99^	17.64	38.53
4	0.30	266.10	0	72.23	0	72.23&	12.50	35.40
5	0.30	307.24	113.69	96.28	13.18	83.10^	14.60	78.65
6	0.50	220.37	0	203.48	0	203.48 ^	20.52	60.20
7	0.50	235.91	0	233.18	0	233.18 ^	21.48	165.20
8	0.50	268.10	118.20	301.16	58.54	242.62 &	19.12	154.40
9	0.50	267.23	103.20	299.22	44.62	254.59 &	20.42	157.80
10	0.50	268.20	108.70	301.39	49.05	252.34 &	21.44	159.80

^Bottom facesheet delamination with no penetration; &full penetration

Table 10. Numerical prediction of single project impact of S2-glass/epoxy balsa core sandwich plate

Spec - imen	Inci- dent velo- city (m/s)	Resi- dual Velocity (m/s)	Energy Absorption (J)				New Surface creation (cm^2)	
			Top face (J)	Bot- tom face	Wood core	Total energy absorp tion (J)	Top face (cm^2)	Bottom face (cm^2)
1	220.37	0.00	32.58	0.00	16.92	49.50	14.20	0.00^
2	256.34	0.00	33.59	11.90	21.50	66.99	15.60	43.50^
3	235.91	0.00	115.43	75.74	42.01	233.18	20.40	154.00^
4	268.20	110.20	119.67	86.77	43.09	249.53	19.40	163.00&

1,3 specimens impacted by 0.30 caliber projectile
8,11 specimens impacted by 0.50 caliber projectile
^Bottom facesheet delamination with no penetration; &full penetration

top facesheet is ~17% *lower* for a smaller (.30 caliber) projectile due to the restriction to bending offered by the core. However, for a larger projectile, i.e. 0.50 caliber impact projectile the energy absorption of the top facesheet is ~16% *higher* than that of the constituent laminate. This can be attributed to localized crushing of increased number of wood cells (area ~300%) in the impact region which contributes to the resistance against the penetration process.

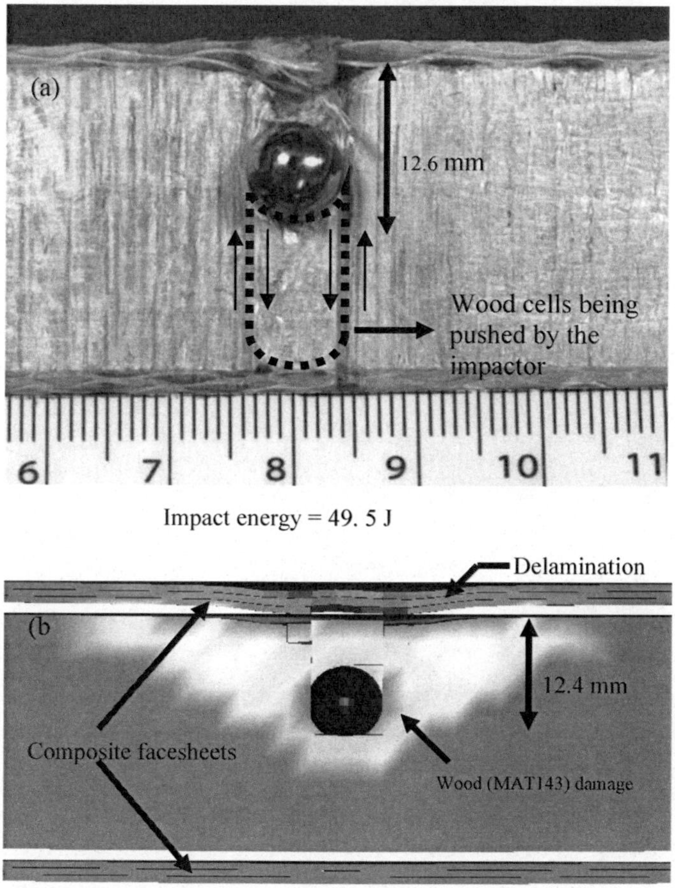

Figure 47. (a) 0.30 caliber impact to S2-glass/epoxy balsa core sandwich panel subjected to impact energy of 49.5 J (b) simulation showing wood damage and delamination[72].

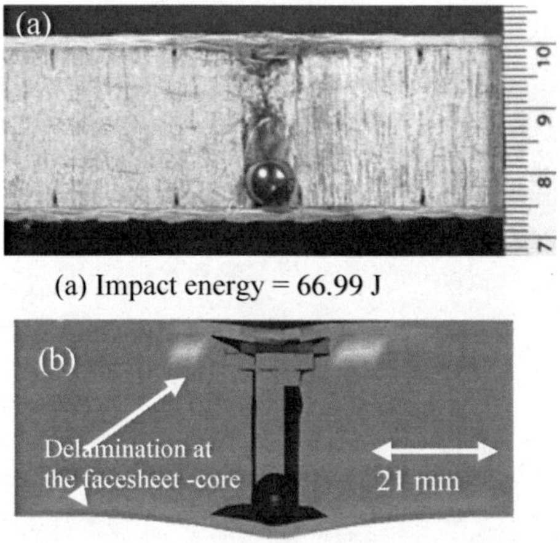

Figure 48. 0.30 caliber projectile impact at 66.99 J (a) Experimental damage (b) Simulation showing delamination (red area) at the top and the bottom facesheet[72].

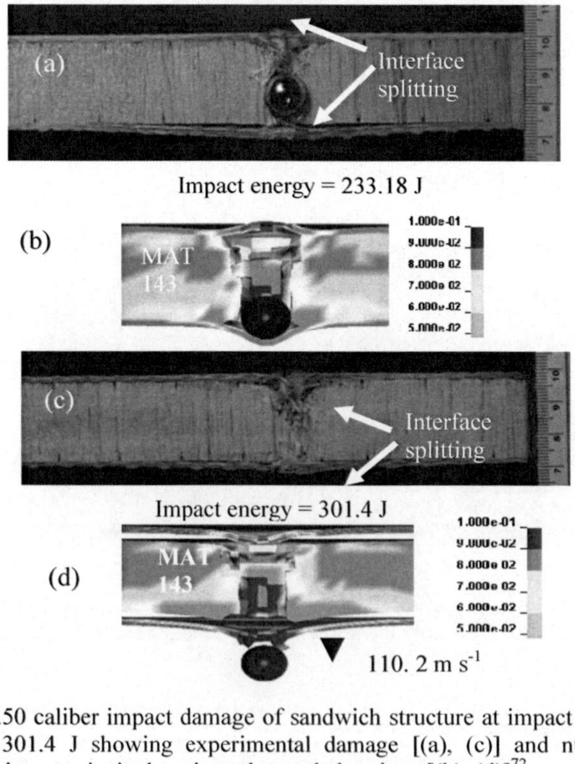

Figure 49. 0.50 caliber impact damage of sandwich structure at impact energies between 233.18 and 301.4 J showing experimental damage [(a), (c)] and numerical analysis showing maximum principal strain at the eroded regions [(b), (d)][72].

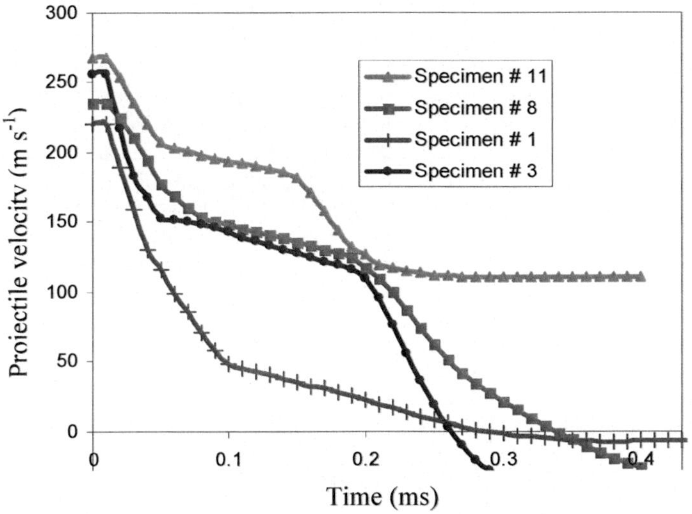

Figure 50. Predicted projectile velocity lost vs. time history plot during single projectile impact on the sandwich specimens 1, 3, 8 and 11[72].

8.0 HIGH STRAIN RATE TESTING

High strain rate testing is important for many engineering structural applications and metalworking operations [90, 91]. Events of practical importance for composite materials include foreign object damage, blast loading, structural impacts and terminal ballistics. In structural composite applications, various components must be design to function over a broad range of strain rates and temperatures. When a composite material is subjected to impact, the material undergoes large amounts of strains depending upon the magnitude of the impact, temperature range and strain rates. The composite material must, therefore, be characterized at the strain rates and temperatures of the intended application. Conventional servohydraulic machines are generally used for testing at quasi-static strain rates of 1/s or less (Figure 51). For higher strain rates, special test methods are required. Table 11 summarizes various methods in terms of the strain rates that can be achieved.

High strain rate measurement and test methods have been developed by Hopkinson[92], Davies[93] and Kolsky[94] over number of years. Considerable data has been generated on the high strain rate behavior of many materials including steels, aluminum, copper alloys, beryllium, titanium, magnesium and zinc. However, high strain rate behavior of composite materials is less studied and data is limited.

The common method of measuring high strain rate is using a split-Hopkinson pressure bar [91, 94, 95] (also, referred to as a Kolsky bar), which allows the deformation of a sample of a ductile material at a high strain rate, while maintaining a uniform state of stress within the sample (Figure 52). The split-Hopkinson pressure bar can be configured for compression, tension or shear mode of testing; compression being the more common one. The compression Hopkinson pressure bar is described in detail later. Other methods for compression tests at high strain rates include Cam Plastometer, Drop Tower

Compression and Taylor Rod Impact. Some additional methods for high strain rate tension testing are; Expanding Ring Test, Flyer Plate and Rotating Wheel tests. In case of high strain rate shear tests, High-speed Hydraulic Torsion machine, Torsional Impact, Double-Notch Shear, Punch Loading, Shear Testing with Hat-Shaped Specimen, Drop-Weight Compression Shear Testing and Pressure Shear Plate Testing[91, 96, 97].

In the split-Hopkinson pressure bar, which is more commonly used for composite materials, the test sample is sandwiched between an input and output bar. This technique provides the capability to measure the stress-strain response of ductile materials and high strain rate, usually between 50/s to 10^4/s. For composite materials, strain rates of 50/s to 3000/s are of interest. Strains exceeding 100% can be achieved with the Hopkinson bar method. The maximum strain rate that can be attained in a Hopkinson bar varies inversely with the length of the test specimen. The maximum strain rate is also limited by the elastic limit of the Hopkinson bars that are used to transmit the stress pulse to the test sample.

The bars used in a split-Hopkinson bar setup are traditionally constructed from a high-strength structural metal, AISI-SAE 4330 steel/maraging steel, or a nickel alloy such as inconel. Such construction is used because the yield strength of the selected pressure bar material determines the maximum stress attainable within the deforming specimen given that the pressure bars must remain elastic.

Inconel bars have been used previously for elevated-temperature Hopkinson bar testing because this alloy's elastic properties are essentially invariant up to 800°C. Because a lower-modulus material increases the signal-to-noise level, the selection of a bar material with lower strength and lower elastic modulus material for the bars is sometimes desirable to facilitate high-resolution dynamic testing of low-strength materials such as polymers or foams.

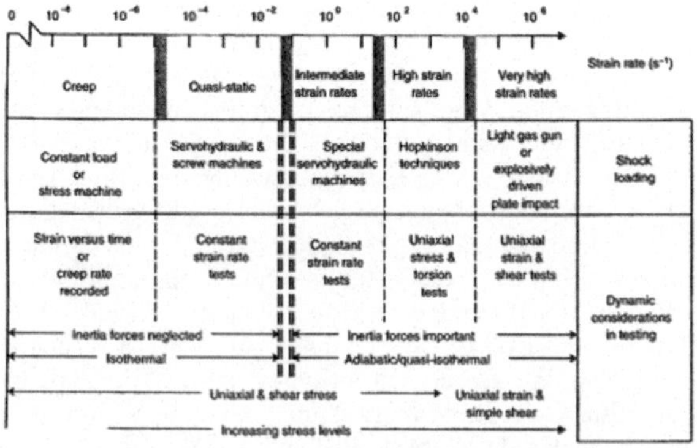

Figure 51. Strain rate regimes and associated instruments and experimental conditions[91].

Table 11. Experimental methods for high strain rate impact testing[91]

Applicable strain rate, s^{-1}	Testing technique
Compression tests	
<0.1	Conventional load frames
0.1–100	Special servohydraulic frames
0.1–500	Cam plastometer and drop test
200–10^4	Hopkinson (Kolsky) bar in compression
10^3–10^5	Taylor impact test
Tension tests	
<0.1	Conventional load frames
0.1–100	Special servohydraulic frames
100–10^3	Hopkinson (Kolsky) bar in tension
10^4	Expanding ring
>10^5	Flyer plate
Shear and multiaxial tests	
<0.1	Conventional shear tests
0.1–100	Special servohydraulic frames
10–10^3	Torsional impact
100–10^4	Hopkinson (Kolsky) bar in torsion
10^3–10^4	Double-notch shear and punch
10^4–10^7	Pressure-shear plate impact

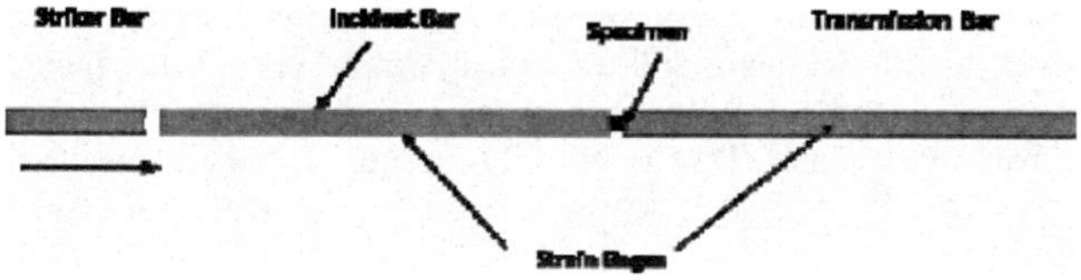

Figure 52. Conventional split Hopkinson pressure bar for high strain rate compression testing[91]

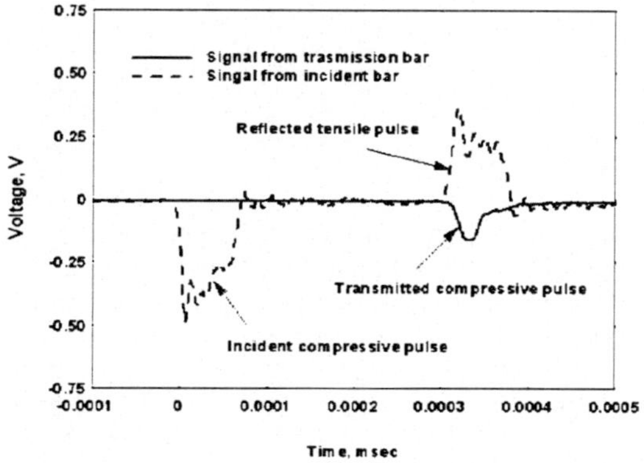

Figure 53. Typical voltage versus time response recorded from the strain gages of the Hopkinson bar[100]

Researchers have selected bar materials possessing a range of elastic stiffness from maraging steel (210 GPa) to titanium (110 GPa) to aluminum (90 GPa) to magnesium (45 GPa) and polymer bars (<20 GPa). Alternately, the signal-to-noise ratio of a Hopkinson bar used to test polymeric materials can be increased using a hollow tubular transmitted pressure bar. While this technique can yield increased transmitted wave measurement sensitivity, the absolute resolution of the sample stress-strain data for polymeric materials must still address the elastic wave dispersion in the tubular-transmitted bar.

The length l and diameter, d, of the pressure bars are chosen to meet a number of criteria for test validity as well as the maximum strain rate and strain level desired in the sample. The strain rate is defined as the change in strain with respect to time. The length of the pressure bars must first ensure one-dimensional wave propagation for a given pulse length; for experimental measurements on most engineering materials, this propagation requires approximately 10 bar diameters.

To readily allow separation of the incident and reflected waves for data reduction, each bar should exceed a length-to-diameter (L/D) ratio of ~20. In addition, the maximum strain rate desired will influence the selection of the bar diameter because the highest strain-rate tests require the smallest diameter pressure bars.

The third consideration affecting the selection of the bar length is the amount of total strain desired to be imparted to the specimen; the absolute magnitude of strain is related to the length of the incident wave. The pressure bar must be at least twice the length of the incident wave if the incident and reflected waves are to be recorded without interference. Depending on the sample size, for strains >30% it may be necessary for the split-Hopkinson bars to have an L/D ratio of 100 or more. There are similar requirements for bar L/D ratios to allow wave separation for compression, tensile and torsion Hopkinson bars.

The most common method of generating an incident wave in the input bar is to propel a striker bar to impact the end of the incident bar. The striker bar is normally fabricated from the same material and is of the same diameter as the pressure bars. The length and velocity of the striker bar are chosen to produce the desired total strain and strain rate within the test specimen. While elastic waves can also be generated in an incident bar through the adjacent detonation of explosives at the free end of the incident bar, it is more difficult to ensure a one-dimensional excitation within the incident bar by direct explosive loading.

8.1 DYNAMIC EQUILIBRIUM

A fundamental difference between a high strain rate and a quasi-static test is that inertia and wave propagation effects become more pronounced at higher strain rates[91]. Table 12 shows the various strain rate regimes in compression testing. At high strain rates, uniform deformation within the test piece becomes a factor; hence the test sample is of relatively small dimensions to minimize wave propagation times. Inertia initially opposes uniform deformation, and a stress wave is transmitted and reflected within the test piece at the speed of sound. If the deformation is purely elastic, then the longitudinal sound velocity C_{EL}, is simply given by

$$C_{EL} = \sqrt{\frac{E}{\rho}} \quad (31)$$

where E is the elastic modulus, and r is the density. If the material has deformed into the plastic region, the plastic-wave velocity is more appropriate and generally can be an order of magnitude smaller than the elastic wave velocity. One-dimensional strain-rate-independent theory predicts that the plastic wave propagates at a velocity, C_{PL}, determined by:

$$C_{PL} = \sqrt{\frac{d\sigma / d\varepsilon}{\rho}} \quad (32)$$

where $d\sigma / d\varepsilon$ is the slope of the true stress/true strain curve. For many materials, the initial work-hardening rate $d\sigma / d\varepsilon$ is approximately 1% of E; thus for these materials, the plastic wave velocity is approximately 10% of the elastic-wave velocity.

The most important characteristic of Kolsky's split-Hopkinson compression apparatus is that it allows high strain rate deformation while the sample is, in fact, in dynamic equilibrium, that is, the stress gradient is essentially zero along the sample. It is thus possible to develop the uniaxial stress-strain response of many materials at a variety of strain rates.

Because the response of most materials depends on both the strain rate and the temperature, the technique allows developing constitutive relations that express the uniaxial stress to the corresponding strain rate and temperature. From such results, one is able to produce experimentally based, three-dimensional constitutive models for numerous materials.

8.2 THEORY

Strain rate, ε is the rate of change of strain ε with time t

$$\dot{\varepsilon} = \frac{d\varepsilon}{dt} \quad (33)$$

where ε can be either the engineering or the true strain. From a constant strain rate experiment, the strain rate is the total strain divided by the test duration.

$$\dot{\varepsilon} = \frac{\varepsilon}{t} \quad (34)$$

when ε in Equation 34 is the engineering strain, then

$$\frac{d\varepsilon}{dt} = \frac{1}{L_0}\frac{dL}{dt} = \frac{V}{L_0} \tag{35}$$

where L is the length of the specimen of original length L_0, and V is the velocity at which the specimen is being deformed. A constant crosshead speed in a mechanical testing machine yields a constant engineering strain rate.

In a Hopkinson bar the pressure bar remains elastic, the displacements in the pressure bar are directly related to the stresses and the length of the wave in the bar is related to the duration of the impact through the velocity of the sound in the bar. Two elastic pressure bars sandwich the specimen between them. Upon impact of a striker bar, and the time-dependent strain in the pressure bar is measured at the strain gage A at the midpoint of the incident bar (Figure 52). At the incident bar/specimen interface, the wave is partially reflected and partially transmitted into the specimen. A portion of the incident wave is reflected back along the incident bar as a tensile wave. This reflected strain is measured by a strain gage A. Strain measurements are also taken on the output bar with strain gage B. A typical pulse recorded from the Hopkinson bar is shown in Figure 53.

These strain measurements on the pressure bars are used to determine the stress-train behavior of the specimen if two basic conditions are met. First, the wave propagation within the pressure bars must be one-dimensional. Secondly, the specimen must deform uniformly. Under these two conditions, stress-strain behavior can be related.

The strain rate in the deforming specimen is:

$$\frac{d\varepsilon}{dt} = \frac{V_1 - V_2}{L} \tag{36}$$

where V_1 and V_2 are the velocities at the incident bars/specimen and specimen/output bar interfaces, respectively. The velocity V_1 is the product of the longitudinal sound velocity, C_o and ε_L. In the pressure bar, the total strain at the incident bar/specimen interface is ε_L- . ε_R. Similarly, the velocity V_2 is equal to $C_o\varepsilon_T$. In this development the incident and transmitted strains ($\varepsilon_{L,}$ ε_T) and ε are all compressive strains, but are considered positive, whereas the reflected strain (ε_R) represents a tensile strain and is negative. Replacing V_1 and V_2 in Equation 36, with these expression yields:

$$\frac{d\varepsilon(t)}{dt} = \frac{C_o}{L}[\varepsilon_I(t) - \varepsilon_R(t) - \varepsilon_T(t)] \tag{37}$$

The average stress on the specimen is:

$$\sigma(t) = \frac{P_1(t) + P_2(t)}{2A} \tag{38}$$

where P_1 and P_2 are the forces at the incident bar/specimen and specimen/output bar interfaces, respectively, and A is the instantaneous cross-sectional area of the specimen. At the incident bar/specimen interface, the force is:

$$P_2(t) = E\varepsilon_T(t)A_0 \tag{39}$$

Combining the above equations, yields

$$\sigma(t) = \frac{E}{2}\frac{A_0}{A}[\varepsilon_I(t) + \varepsilon_R(t) + \varepsilon_T(t)] \tag{40}$$

When the specimen is deforming uniformly, the stress at the incident bar/specimen interface, equals that at the specimen/output bar interface and from Equations 39 and 40.

$$\varepsilon_I(t) + \varepsilon_R(t) = \varepsilon_T(t) \tag{41}$$

which may be simplified as:

$$\frac{d\varepsilon(t)}{dt} = \frac{-2C_0\varepsilon_R(t)}{L} \tag{42}$$

$$\sigma(t) = E\frac{A_0}{A}\varepsilon_T(t) \tag{43}$$

Thus, the stress-strain behavior of the specimen is determined simply by measurements made on the elastic pressure bars in a split Hopkinson pressure bar test.

Two important conditions are met in developing these equations. The first is that wave propagation within the pressure bars must be one-dimensional if surface displacements measurements are used to represent the axial displacement over the entire cross-sectional area of the pressure bar. The second is that the specimen must deform uniformly; this is opposed by both radial and longitudinal inertia and by frictional constraint at the specimen/bar interfaces.

At a constant strain rate, the maximum strain that can be achieved in the specimen is directly proportional to the length of the striker bar utilized, L. This strain is given by:

$$\varepsilon = 2\dot{\varepsilon}\frac{L}{c_b} \tag{44}$$

The nominal strain rate in the specimen may be conservatively approximated by considering momentum conservation between the striker bar and the incident bar. It can be similarly shown that the strain rate is a function of the striker bar length, l_s and the striker bar velocity, V.

$$\dot{\varepsilon} = \frac{V}{l_s} \tag{45}$$

The deformation is homogenous throughout the test only if frictional constraint is minimized at the interfaces between the platens and the specimen ends. The specimen typically starts out as a cylinder and maintains this cylindrical geometry throughout the test. This requires a suitable use of lubricant at the interfaces.

Minimizing the effects of friction in a compression test involves selection of a suitable length-to-diameter (l/d) ratio for the specimen and the selection of proper lubrication for the specimen/platen interfaces. A l/d ratio of 0.5 to 2 is generally considered optimum for composite materials. Ratios larger than this lead to buckling, which is particularly the case for unidirectional composites loaded along the fiber direction. Many lubricants such as viscous oils, powdered graphite, motor mica or grease based molybdenum disulfide have been used for room temperature testing.

Optimal data resolution requires careful design of the sample size for a given material as well as the selection of an appropriate striker bar length and velocity to achieve test goals. The determination of the optimal sample length requires consideration of the sample rise time t, required for uniform uniaxial stress state to be achieved within the sample. It has been estimated that this rise time is the time required for three (actually π) reverberations of the stress pulse within the specimen. For a plastically deforming solid that obeys the Taylor-von Karman theory[91], time follows the relationship:

$$t^2 \geq \frac{\pi^2 \rho_s l_s^2}{d\sigma / d\varepsilon} \tag{46}$$

where ρ_s is the density of the specimen, l_s is the specimen length, $\partial\sigma / \partial\varepsilon$ is the post-elastic regime work-hardening rate of the true stress/true strain curve for the material to be tested.

8.2.1 Pulse Shaping. Because the value of t from Equation 46 has a practical minimum, an alternate method to facilitate stress-state equilibrium at low strains is the increase the rise time of the incident wave. Because the highly dispersive short wavelength components arise from the leading and trailing edges in the incident wave, a longer rise-time pulse will contain fewer of these components that will a sharply rising pulse. Experimentally, the rise time of the incident wave can be increased by placing a soft, deformable metal shim between the striker and the incident bar during impact. Typically, the tip material is selected to have the same strength as the specimen and is 0.1 to 2 mm in thickness. This produces an uniform strain rate throughout the experiment.

8.2.2 Repeated Loading and Recovery Hopkinson Bar. One limitation of the classic split-Hopkinson bar technique has been the repeated loading of the sample by the stress pulses that travel back and forth along the bars. In composites, the repeated loading leads to progressive damage in the form of matrix cracking, debonding and delaminations. To measure the true response of the sample, techniques have been developed by Nemat-Nasser et al[98] to trap the reflected pulses at the free ends of the Hopkinson bars, making it possible to subject a sample to a single stress pulse and then recover the sample without it being subjected to any additional loading. Using this recovery technique, it is possible to develop isothermal stress-strain curves for many materials at a desired temperature and strain rate.

In case of a recovery mechanism for a compression split-Hopkinson pressure bar, the loading (striker) end of the incident bar includes a transfer flange, an incident tube, and a reaction mass as shown in Figure 54. The loading end of the incident bar begins with a transfer flange. In this setup, the incident tube, the striker bar, and the incident bar have the same impedance. The incident tube rests against the transfer flange at one end and against a reaction mass at the other. The reaction mass is a large steel cylinder through which the incident bar passes.

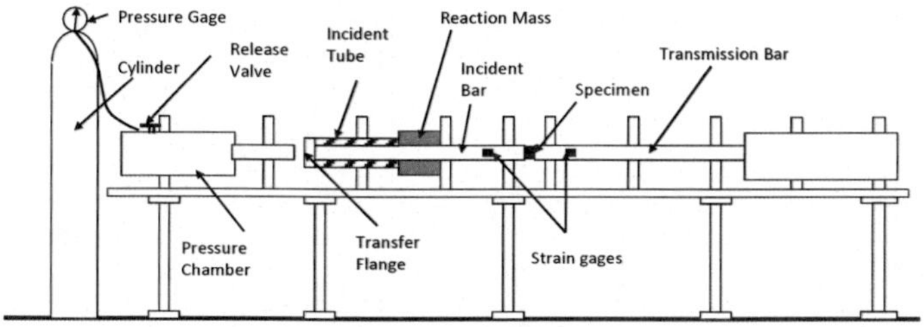

Figure 54. Stress reversal modification to the split Hopkinson bar to prevent multiple stress wave loading[100].

When the striker bar imparts the transfer flange at velocity V_o, it imparts a common axial strain to the incident tube and incident bar. The compression pulse in the incident bar travels along this bar toward the specimen. The compression pulse in the incident tube reflects from the reaction mass as a compression wave and reaches the transfer flange at the same instant that the tension release pulse – which is reflected from the free end of the striker – reaches the end of the striker bar in contact with the transfer flange (Figure 55). Since the combined cross section of the incident bar and the tube is twice that of the striker bar, having the same material properties, the striker bar beings to bounce back, away from the transfer flange, as the transfer flange is loaded by the compression pulse traveling along the incident tube. This compression pulse then imparts a tensile pulse to the incident bar, which follows the then-existing compression pulse in the incident bar

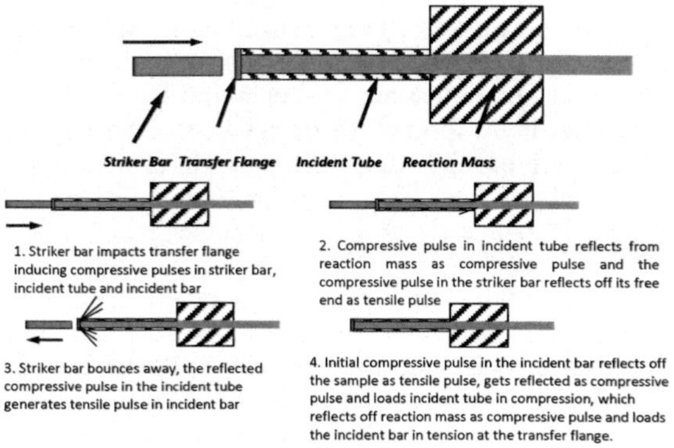

Figure 55. Mechanism of stress wave reversal in modified Hopkinson pressure bar[98].

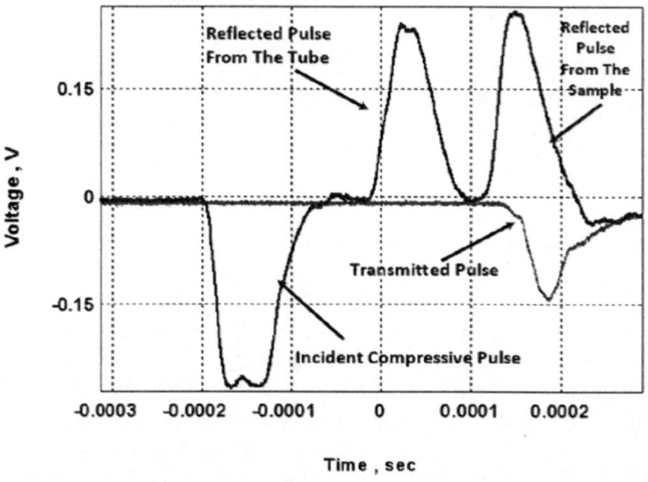

Figure 56. Voltage versus time signals recorded from a Hopkinson bar equipped with a stress reversal momentum trap[100].

both pulses traveling towards the sample. A typical pulse from the stress reversal Hopkinson bar is shown in Figure 56. Following the incident compressive pulses, the reflected pulse from the tube and from the sample respectively are illustrated in the figure.

8.3 HIGH STRAIN RATE OF LAMINATED AND SANDWICH COMPOSITES

The literature on high strain rate response of composites is relatively limited. The high strain rate behavior of composites is typically investigated for unidirectional, cross-ply, angle-ply, multidirectional and woven fabric architecture, generally under compressive loading [99-112].

Generally, the dynamic strength and modulus of composite materials increase compared to those at quasi-static loading. It is also observed that there is a significant scatter in failure strain. The specimen has considerable time for deformation and load redistribution under static loading, the strains are higher. Strain at peak stress was found to be 2–3 times higher in case of static loading as compared to dynamic loaded samples for all configurations.

8.3.1 Weave Architecture. The high strain rate impact response is sensitive to the weave architecture [99-103]. Studies have shown that satin weave samples exhibit higher peak stress and modulus as compared to plain weave samples. This is due to the lesser undulation (less crimp) in the satin weave fabric architecture of satin weave.

The compressive strength and failure strain are 30-40% higher along thickness direction compared with those along warp and fill directions for woven fabric composites, Figure 57 and 58. Compressive modulus is lower along thickness direction compared with those along warp and fill directions.

8.3.2 Temperature Effects. Temperature has a marked influence on high strain rate response of composites [104,105]. Hosur et a[104] reported high strain rate impact behavior of graphite/epoxy composite at room temperature up to 225^0F, at a strain rate range of 250/s to 950/s, Figure 59. The peak stress reduces with increase in temperature due to matrix softening. At a 250/s strain rate the damage initiates at 225F, while no damage is evident at room temperature (RT), 125 and 175F. With increasing strain rate, damage initiates at all temperatures (RT to 225F); however the mode of failure transitions to distinct single plane failure (Figure 60) to more shear dominated matrix failure. The failure occurs by multiple branched cracks originating from a weak location. Due to the increased ductility and degradation of fiber-matrix interface, the severity of failure decreased with increasing temperature.

In terms of cold temperature response, the high strain rate tensile response is shown to be dependent upon fiber orientation[103]. In studies by Navarro et al.,[105] the dynamic tensile strength was higher at -60^0C as compared to that at 20^0C for an unidirectional and cross-ply laminate. For quasi-isotropic laminate, the material was strain rate insensitive, Figure 61.

8.3.3 Off-axis plies. The high strain rate impact response is also sensitive to the orientation of the fibers to loading direction[106-108]. The samples loaded along off-axes angles exhibit a large nonlinear response which increases with increase in orientation up to 45^0. Experimental results from off-axis compression tests on the S2/8552 glass/epoxy composite indicate that fiber microbuckling failure occurs at small (5–15°) off-axis angles, Figure 62. Such angles can also result from misalignment of fibers during processing. Compressive strength of polymeric composites is rate-sensitive and that the presence of in-plane shear stress can appreciably lower the compressive longitudinal strength.

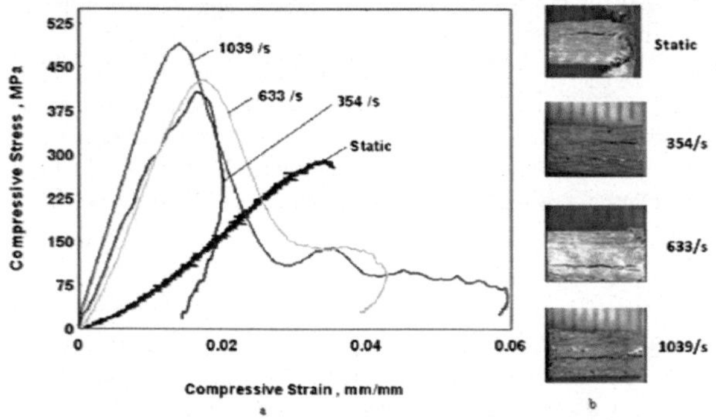

Figure 57. High strain rate stress versus strain response of a plain weave graphite/epoxy composite. Curves show quasi-static, 354/s, 633/s and 1039/s strain rates[102].

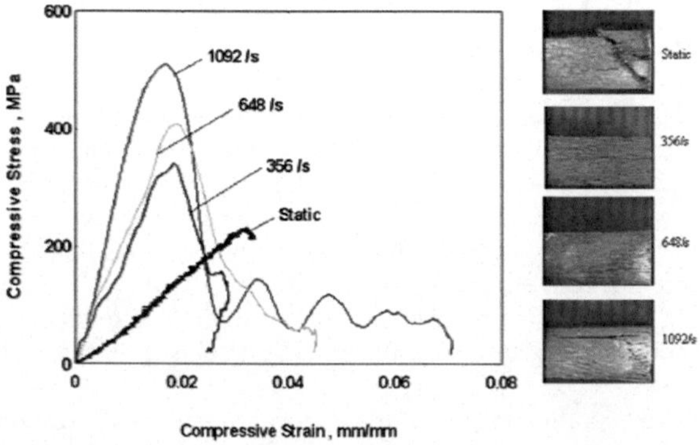

Figure 58. High strain rate stress versus strain response of a satin weave graphite/epoxy composite. Curves show quasi-static, 356/s, 684/s and 1092/s strain rates[103].

8.3.4 Sandwich Composites. Polymeric sandwich materials exhibit a strong strain rate sensitivity with the yield strength increasing by about 50–200% when the strain rate is increased from 10^{-4}/s to 10^{3}/s, due to their viscoelastic nature[109-111]. Figure 63 illustrates a typical dynamic stress-strain response of a polyvinyl chloride (PVC) foam core. The strain rate sensitivity of the PVC foam is evident for strain rates of 1000-1700/s in this study[109]. The corresponding quasi-static curve is shown for comparison. The initial loading corresponds to the elastic deformation of the core, followed by a plateau (core crushing) and finally, the densification stage. The dynamic curves follow the first two stages, i.e. elastic deformation, plateau, followed by partial recovery of the core with significant plastic deformation

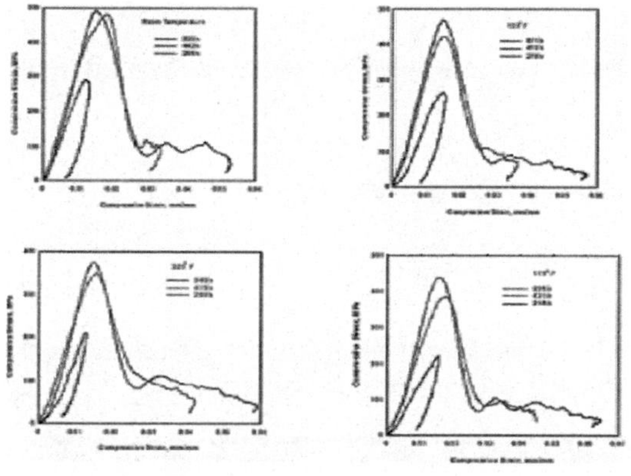

Figure 59. Effect of temperature on high strain rate response of graphite/epoxy composites. Curves are for room temperature, 125, 175 and 225 F temperatures, at an average strain rate of 260/s, 420/s and 900/s respectively[104].

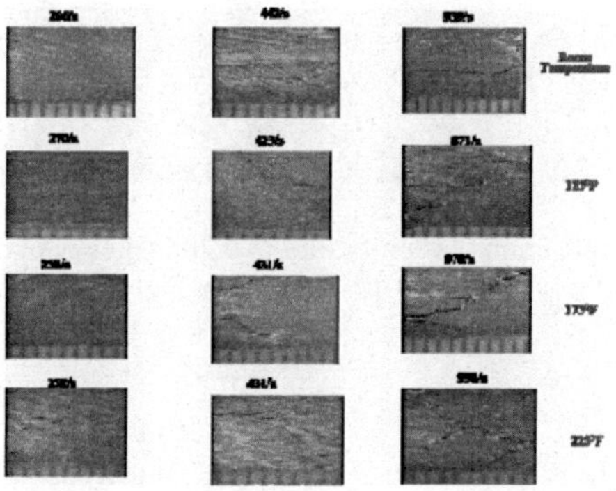

Figure 60. Effect of temperature on high strain rate response of graphite/epoxy composites. Failure modes for room temperature, 125, 175 and 225 F temperatures, at an average strain rate of 260/s, 420/s and 900/s respectively[104].

Cellular materials such as PVC foam and balsa wood are mildly strain rate sensitive in the range $(\dot{\varepsilon} = 10^{-2} - 4000s^{-1})$, whereas negligible strain rate sensitivity in the range $(\dot{\varepsilon} = 10^{-4} - 10^{-2}s^{-1})$. Studies on the dynamic properties of cellular materials such as foam, wood and honeycomb have attributed the strength increase under dynamic loading conditions to a series of concurrent factors; including strain sensitivity, micro inertial effects and shock wave propagation[111].

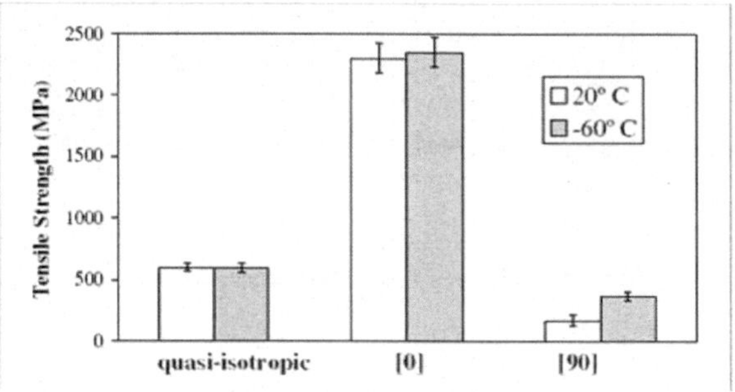

Figure 61. Effect of low temperature on high strain rate tensile strength of glass/epoxy composites of unidirectional, cross-ply and quasi-isotropic stacking sequence[105].

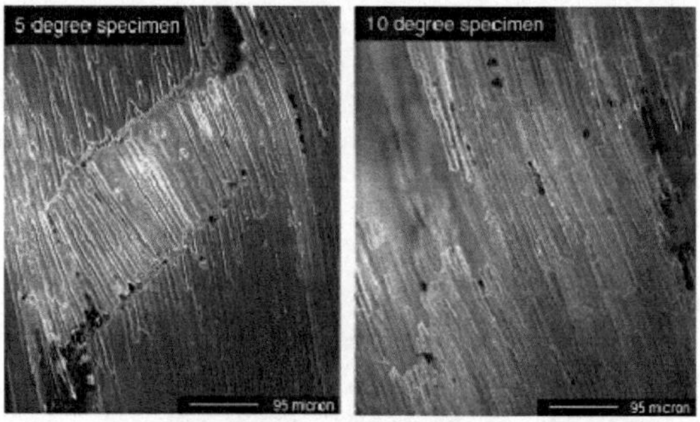

Figure 62. High strain rate compression failure of composites with off-axis plies. (a) 50 and (b) 100 fiber angles[108].

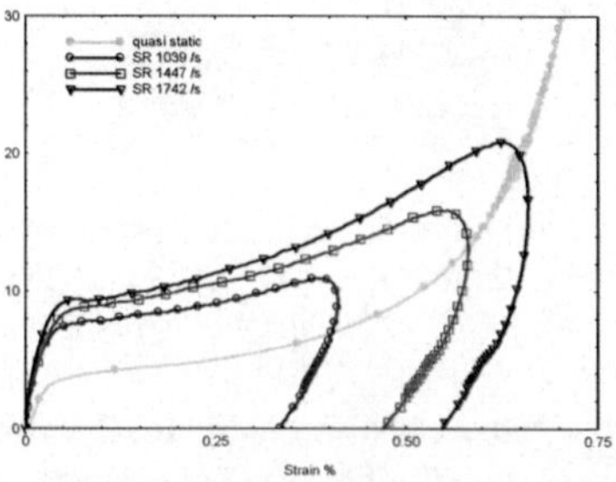

Figure 63. High strain rate compression failure of composites of PVC foam core at quasi-static loads and high strain rate of 1039 to 1742/s[109].

Isotropic, open-cell foams deform by plastic bending of cell walls. The macroscopic compressive plateau strength σ_{pl} of the foam is related to the yield strength of the parent material σ_{Ys}, by

$$\sigma_{pl} = C^{-3/2}\sigma_{Ys} \tag{47}$$

where C is the proportionality constant, of order unity, with its value depending upon the details of the foam microstructure.

Strain sensitivity of the cell-wall material: In foams, strain rate sensitivity is expected due to the inherent strain rate sensitivity of the parent material. The strain rate sensitivity of the cell-wall materials can be in the form of a power law relation[111]:

$$\sigma_{YS} = K_s \dot{\varepsilon}_s^{\,m} \tag{48}$$

where K_s is a proportionality constant and m_s the power law component. Thus, if other dynamic effects are not active, the strain rate sensitivity of the foam can be expected to have the same power law exponent as that of the solid parent material. The value of proportionality constant K depends on the foam relative density and topology.

Micro-inertial effects: The micro-inertial effects of individual cell walls can affect the deformation modes of cellular solids. Under dynamic loading, the collapse of a bending governed cellular material may transition from the quasi-static mode to a failure mode involving cell-wall stretching, which dissipates more energy and hence leads to an increase in the collapse strength.

PVC foams are nearly isotropic. They exhibit a flat-topped quasi-static compressive stress versus strain response under dynamic loading, micro-inertial effect plays little role and quasi-bending mode of collapse is maintained. In contrast, anisotropic materials like balsa wood exhibit strongly softening mode of collapse under quasi-static conditions.

When struck at high velocities, lateral inertia forces induce an initial phase of axial compression of the structure. A softening compressive stress versus strain response is observed. Consequently, the stresses and plastic work is enhanced before the bending mechanism is recovered and balsa exhibits an inertia sensitive response.

Shock-wave propagation: If a cellular material is impacted at high velocity, a plastic shock wave will propagate through the solid. This plastic wave propagation phenomenon leads to an increase of the collapse strength of the material, which can be quantified considering a simple one-dimensional shock model.

A comparison between the static ($\dot{\varepsilon} = 10^{-4}s^{-1}$) intermediate ($10^{-2}s^{-1} \leq \dot{\varepsilon} \leq 250s^{-1}$) and high strain rate ($500s^{-1} \leq \dot{\varepsilon} \leq 4000s^{-1}$) responses of the a low and high density foam, and balsa wood is presented in Figure 64[111].

An elevation of the plateau strength σ_{pl} with the applied strain rate is evident from comparing the stress versus strain curves. The PVC foams and balsa wood display negligible strain rate sensitivity for strain rates in the range $(10^{-4}s^{-1} \leq \dot{\varepsilon} \leq 10^{-2}s^{-1})$ but display a power law strain rate sensitivity at higher strain rates. The power law exponent *m* for the H250 PVC foam (a higher density foam) is approximately equal to that measured for solid PVC. By contrast, there is a decrease in strain rate sensitivity of PVC foams with decreasing density.

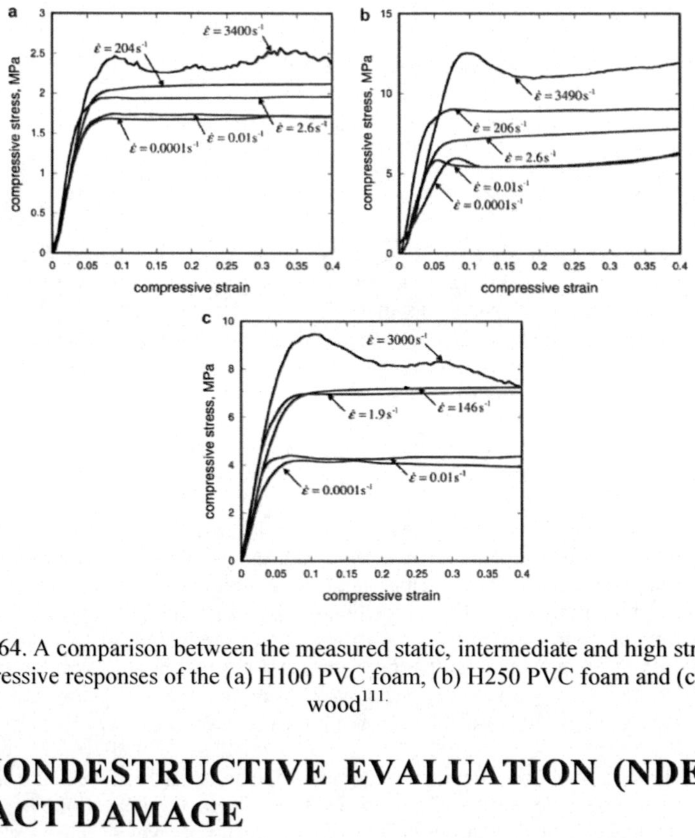

Figure 64. A comparison between the measured static, intermediate and high strain rate compressive responses of the (a) H100 PVC foam, (b) H250 PVC foam and (c) balsa wood[111].

9.0 NONDESTRUCTIVE EVALUATION (NDE) OF IMPACT DAMAGE

Impact damage to the composite laminate or sandwich is not always apparent on the surface. Complementary NDE tools are necessary to qualify and quantify defects due to impact damage. The various impact damage modes have been discussed in prior sections. The most common NDE techniques used for composites inspection are briefly discussed below.

9.1 VISUAL INSPECTION

Visual inspection can be conducted with the naked eye or with instrumented optics such as magnifying glass, borescopes or endoscopes[112]. It allows an assessment of low, intermediate and high velocity impact in composites. Surface characteristics such as the area of the visible delamination that appears as a discoloring at the target surface, microcracks or surface deformation such as bulging, fiber breakage can be readily quantified. Projected delaminations in translucent (or transparent) composites such as E-glass reinforced composites can be observed. Visual inspection has been used, for example, for assessment of the surface at the distal side of a target after impact in order to investigate interlaminar failure modes. The extent of 'whitening' can be used as a criterion of matrix damage in a composite. Measurements of the crack length or visual assessment of failure modes for composite targets are used in post-impact analysis.

9.2 WITNESS PLATE

As explained in the Ballistic Impact section, a witness plate is used to determine exit velocity and penetration of a composite[113]. Damage to a witness plate is a means of assessment of the primary target's impact response. Using the witness plate method, the extent of the spread of debris due to ballistic impact against a brittle target can be evaluated. The Depth of Penetration (DOP) of an impact threat in a witness plate is used as a criterion of the target response to the ballistic impact.

9.3 ULTRASONIC TESTING

Ultrasonic A, B and C-scan are historically used in metals inspection, but have been adopted extensively for inspecting composite materials[114-118]. Typically, ultrasonic methods are used in the pulse-echo and through-transmission (A-scan) or in a C-scan mode (Figure 65). Pulse-echo and through-transmission are used primarily for materials characterization. In pulse-echo, a single transducer or a pair of transducers interrogate a structure from one side only and rely on the pulsing ultrasonic signal and the corresponding echo. In through transmission testing, a pair of ultrasonic transducers is used, where the structure is sandwiched between the transducers. In general, pulse echo ultrasonics is applicable to structures where access is limited to a single side and for solid monolithic materials. Through transmission ultrasonics is useful to interrogate lossy foams and other porous structures. Porous materials are not effectively inspected in pulse-echo due to losses in the echo signal from pores and voids.

Ultrasonic modulus and density variations are readily detected based on the time of flight, t_f of the ultrasound. For a composite of a thickness d, the ultrasonic velocity, V is given by:

$$V = d/t_f \qquad (49)$$

The A-scan methods can also be used to determine local voids, pores, debond and discontinuities, (Figure 66). The frequency of the ultrasonic transducer is critical to establish the spatial resolution. For composites inspections, 1 MHz to 10 MHz frequencies is commonly used, with 2.25 to 5 MHz being common for laminates of thickness 2-5 mm.

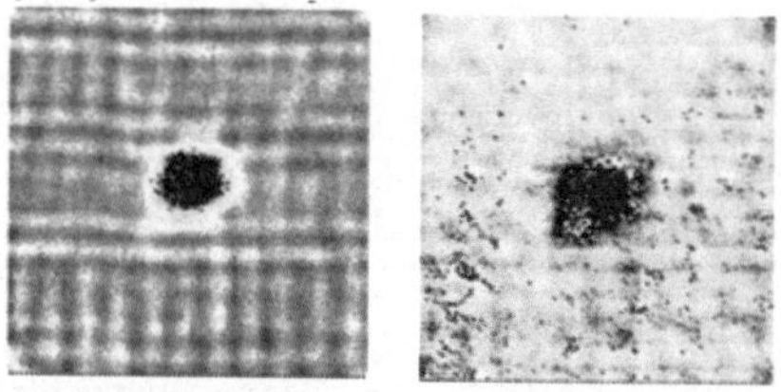

Figure 65. Ultrasonic images of low velocity impact to 6.4 mm thick glass/epoxy laminate. (a) Direct transmission impact, (b) First thickness reflection image[116].

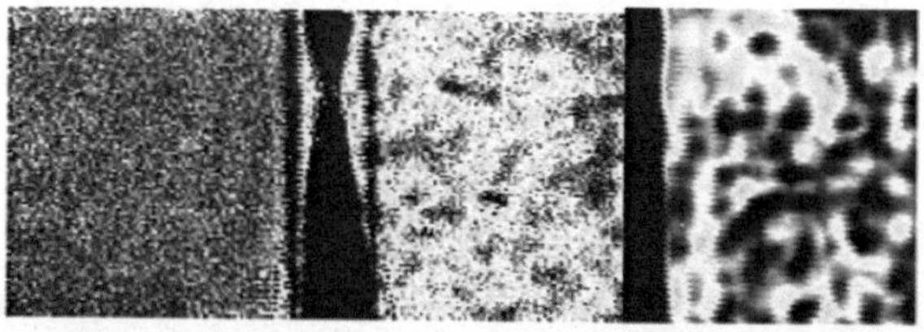

Figure 66. Porosity in CFRP composite. (a) zero porosity; (b) <1% porosity, (c) >4% porosity. Inspection done with 1 MHz transducer[114].

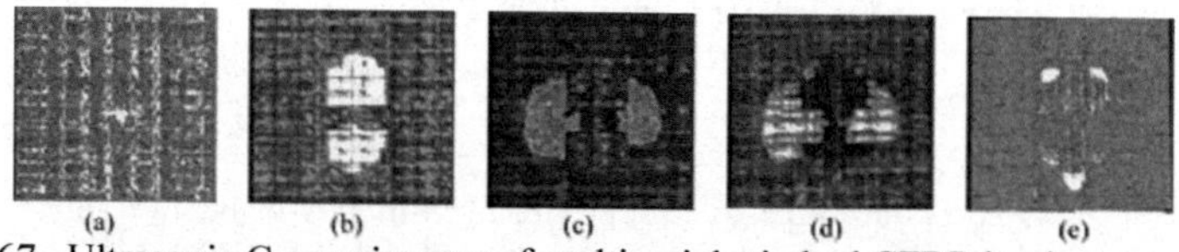

Figure 67. Ultrasonic C-scan images of multi-axial stitched CFRP laminates subject to low velocity impact. (a) 0.25 mm below impact face, (b) 0.75 mm, (c) 1.25 mm, (d) 1.75 mm and (e) 2.25 mm[117].

In a C-scan the plan view of the sample is mapped at each point in x-y scan mode. In advanced C-scans, layer wise scanning can be done by gating the signal to different through the thickness positions. The layer wise C-scan technique allows obtaining the two-dimensional distribution of flaws through the thickness of a sample[114].

Figure 67 illustrates an ultrasonic C-scan for an impact damaged multi-axial CFRP panel[117]. The damage state is distinctly different through the thickness which can be effectively captured using the layer wise C-scan approach. In a traditional C-scan, these damages would project on to each other, thereby masking information of their location in the laminate's thickness. Ultrasonic B-scan is used to determine through-the-thickness damage of a cross-sectional area.

Ultrasonic NDE is also conducted to damage in a sample with the change of speed based on Lamb waves[119, 120]. This technique employs a transmitter and receiver of ultrasound positioned at a distance from each other on a sample surface, so the acoustic waves (Lamb waves) propagating in the sample may change velocity and amplitude depending on a damage within the sample. The Lamb wave ultrasonic technique examines the material along the surface and sub-surface layers. Overt porosity, fiber misalignment, overall cure condition and surface cracks can be evaluated by the ultrasonic Lamb wave technique.

9.4 X-RAY PHOTOGRAPHY AND MICROTOMOGRAPHY

X-ray photography records a series of 2D-images; while microtomography generates 3D-picture of micro-structure or 2D-slices[27, 121]. Medium-resolution tomography allows one to assess samples with a relatively coarse structure of the order of tens of microns after a deformation event followed by unloading. High-resolution microtomography allows for in-situ analysis of materials with a finer structure, including metal matrix composites.

The X-ray-based techniques require a compromise between resolution focusing and exposure time[122-124]. Therefore, these techniques do not allow an in-situ study of processes which are faster than of the order of seconds, i.e., these techniques are not applicable to high-strain rate processes. Usually, composite samples are analyzed with this sort of techniques after the damaging event. X-ray inspection is used for an examination of matrix cracking and delamination and for an assessment of damage accumulation. X-ray inspection may be enhanced by computer tomography (CT) scanning which allows 3D-structure to be determined.

9.5 OPTICAL INTERFEROMETRY

Optical interferometry is used to measure deformation of a sample. Speckle and Moiré photography and interferometry utilize information from the specimen's surface [125, 126]. They can be used for measurements before and after impact, as well as for in-situ measurements. Digital speckle photography allows an evaluation of in-situ displacements and strains at the sample's surface during compression-after-impact tests. These methods cannot give detailed information about the stress state inside a sample and about changes in the sample's internal material structure.

9.6 ACOUSTIC IMPACT

Also referred to as the 'coin-tap' technique, acoustic impact is based on tapping a composite structure at different locations and listening to the sound that is emanated. Alternatively, an instrumented hammer can be used to lightly tap the structure[127, 128]. The technique is based on the principle of change if frequency in the audible range for a 'good' region to a 'defective' region. The defective region results in larger pulse width (increase in the force-time response) and reduction of frequency. The acoustic impact technique is most effective for sub surface defects in laminates such as voids, porosity, microcracks, debonds, and to core to facesheet debonds in sandwich composites.

9.7 VIBRATION BASED NDE

This is a global NDE technique in that the entire structure is evaluated through a single point measurement. The structure can be excited with an impulse hammer or an electromagnetic shaker setting resonances in the desired loading direction (flexure, axial, torsion) [129-132]. The frequency response function (FRF) of the structure is measured based on input force and output acceleration.

Random noise excitation is preferable for evaluating impact damage. The composite specimen can be mounted either on nylon cords to mimic a near free-free boundary condition, or attached to the impedance head mounted on the shaker, separated by a stinger rod (Figure 68). The stinger rod has high axial stiffness and low bending stiffness to readily transmit vibrations to the specimen from the shaker. The parameter for impact damage evaluation is the resonant frequency, damping ratio and the mode shapes.

The damping ratio values were measured are measured either by time domain logarithmic increment method, or frequency domain half power method[131]. The -3dB reduction of amplitude at resonance was determined. The damping ratio ξ is given by:

$$\xi = f_2 - f_1 / 2f_n \tag{50}$$

where f_1 and f_2 are the upper and lower frequencies, and f_n is the resonance frequency.

Vibration NDE conducted on a glass core/glass-kevlar hybrid facesheet laminate subjected to 5, 10, 20 and 30 J impacts show progressive reduction in flexural resonant frequencies with increase in impact damage (Figure 69)[133]. The corresponding damping ratio values exhibit an increase in damping ratio with increasing impact energy (Figure 70). The damping ratio increases with damage due to additional surfaces created that dissipate vibrational energy. This increase in damping ratio is oftentimes a measure of extent of damage in the structure.

9.8 ACOUSTIC EMISSION

Acoustic Emission (AE) methods are used for in-situ monitoring of composites subject to damage [129, 134-136]. The method is base of the measurement of amplitude and the number of acoustic signals generated during deformation of a composite material, when ultrasonic waves in the material result in pulses of the acoustic energy detectable by AE transducers. The development of a methodology for comprehensive interpretation of AE signals for different modes of fracture is still underway. However, when a database of signal patterns has been collected for probable fracture modes, the monitoring method may become a reasonable alternative to the more expensive methods which are currently used.

9.9 POLYVINYLIDENE FLUORIDE (PVDF), FIBER OPTICS

Different PVDF gages are used in order to measure acoustic emission using an artificial neural network approach as the analysis methodology to detect several failure mechanisms such as matrix cracking and local delamination during tensile loading of composite samples[137, 138].

Other novel methods of in-situ monitoring employ the principles of smart materials. For example, embedded sensors for monitoring damage in sandwich structural materials containing a honeycomb core. This technique is based on a methodology of damage assessment using knowledge about the most probable deformation modes resulting in the damage. Optical fibers have been employed for in-situ monitoring: Fiber Bragg Grating (FBG) and Fiber Fabry-Pérot Interferometry (FFPI) are used.

The FBG technique is based on a photosensitive mechanism; the wavelength of a beam propagating in the optical fiber embedded or mounted on a sample may change with strain and temperature and hence this can be used for indications of the change. The strain and temperature sensitivities of this technique are coupled. The FFPI technique is based on the effect of changing position between the fibers' ends embedded in a material structure. This technique is sensitive to acoustic wave propagation through the FFPI gauge.

9.10 THERMOGRAPHY

Infrared thermography is a non-contact, non-intrusive technique of surface temperature mapping. The infrared thermography detects the energy radiated from objects in the infrared band of the electromagnetic spectrum and converts it into a visible image. Defects such as disbands, delaminations and fiber/matrix variations have distinctly different infrared response, and this is used as a basis for the inspection [139-141].

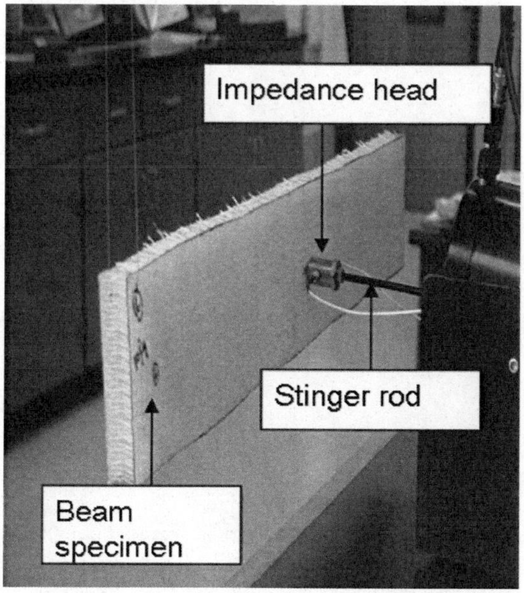

Figure 68. Vibration NDE of a 3D glass/epoxy sandwich panel. Arrangement shows the specimen, impedance head and stinger rod [129]

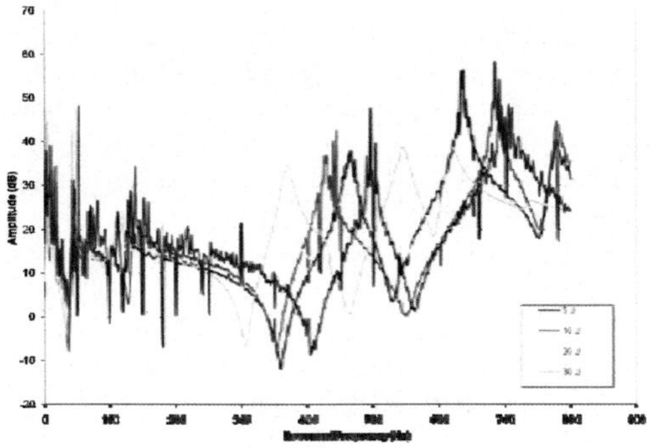

Figure 69. Vibration NDE of a 3D glass-hybrid/epoxy sandwich panel. Increasing impact damage For 5, 10, 20 and 30 J impact energy causing 6%, 12%, 24% and 30% reduction in resonant frequency[133]

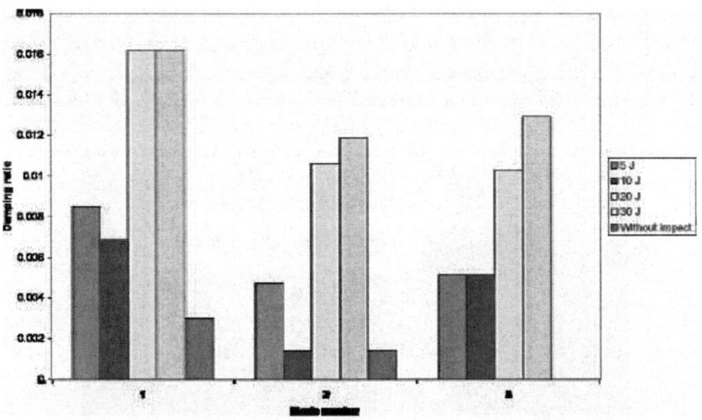

Figure 70. Vibration NDE of a 3D glass-hybrid/epoxy sandwich panel. Increasing impact damage. For 5, 10, 20 and 30 J impact energy causing 6%, 12%, 24% and 30% reduction in resonant frequency [133]

10.0 SUMMARY

This chapter presented current state of the art in test methodologies, recent and past work in the area of quasi-static, low, intermediate and high velocity (ballistic) impact and high strain rate response of composites; (i) Low velocity impact – The theory of LVI, Charpy, Izod and instrumented impact testing, impact damage mechanisms and structural response in relation to material type (fiber, matrix, laminate construction etc) was discussed. The role of fiber length scales, impactor type and geometry, and residual strength was discussed. The LVI section complements the Chapter 3 in this book, which covers the impact mechanics in greater detail; (ii) Intermediate velocity impact (IVI) – the conditions of testing that fall under the IVI impact regime, test equipment and recent work in IVI of composites using blunt and sharp impactor on laminates and sandwich composites was discussed. The IVI regime is the least standardized and more studies are needed in the intermediate velocity range, because the IVI impact regime is most common to structures in real life scenarios; (iii) High velocity impact - The test equipment, types of projectiles, past and recent work in ballistic impact, also referred to as high velocity impact (HVI) was discussed. Chapter 7 complements the ballistic impact in the experimental aspects. The interface along with the fiber/matrix constituents plays a prominent role in ballistic damage energy absorption and damage accumulates by progressive failure; (iv) Progressive failure - The damage accumulation in an impact damaged laminated or sandwich composite is modeled accurately with the aid of material models accounting for progressive damage. The correlation of the simulation to experimental data is largely dependent upon reliable material data over a range of loading and boundary conditions. These include standard quasi-static properties (tension, compression, shear) in different directions. In addition, the strain rate dependent behavior is required. The lack of all required material properties is a large gap in impact modeling, where new materials are being developed, but there limited data for modeling; (v) Quasi-static punch shear (QS-PST) - The quasi-static punch shear test is an intermediate means of generating sufficient input data required for ballistic modeling. The testing and correlation of quasi-static punch shear to ballistic behavior were also described in this paper; (iv) High strain rate impact (HSR) – The theory of HSR impact, typical test equipment, and HSR response of laminated and sandwich composites is described. The Hopkinson bar used for HSR is an effective tool to generate some of the input data required for modeling, particularly the strain softening parameters. Yet, the availability of HSR data is still largely inadequate; and (v) Nondestructive evaluation (NDE) methods including visual inspection, ultrasonics, X-ray radiography, acoustic impact, vibration NDE, thermography and smart sensor for evaluating impact damage induced matrix cracking, fiber damage, debonding and delaminations in composites were discussed.

A wealth of information is has been published over the last two decades with respect to impact damage in composite laminates. However standardized information on impact to composite materials over a range of strain rates, operating conditions and constituent materials is largely inadequate. The emergence of new materials with different relative constituent content, architecture, material forms and designs further complicates the standardization of data. The innovations in materials processing and design, has resulted in a host of material options to the designer - in terms of material composition, architecture, weaving styles, hybrids and so on. Each of these variations exhibits different impact properties and there is a large gap in terms of understanding of the impact response of newer composite material forms and structures.

11.0 ACKNOWLEDGEMENT

The author wishes to thank Dr. Serge Abrate for this support and helpful hints in writing of this chapter. The help provided by Dr. Mahesh Hosur in the high strain rate impact section is gratefully acknowledged. Several aspects of the work included were possible due to funding from the Office of Naval Research (Dr. Yapa Rajapakse), Army Research Laboratory and Department of Energy (Graduate Automotive Technology Education). The author is grateful to these organizations and individuals.

12.0 REFERENCES

1. Abrate, S. (1991). Impact on laminated composite materials. Applied Mechanics Review. 44 (4):155-190.
2. Abrate, S. (1998). Impact on Composite Structures. Cambridge University Press.
3. Richardson, M. O. W and Wisheart, M. J. (1996). Review of low-velocity impact properties of composite materials. Composites Part A 27A:1123-1131.
4. Sjoblom, P.O., Hartness, J.T. and Cordell, T.M. (1998). On low-velocity impact testing of composite materials. Journal of Composite Materials. 22:30-52.
5. Olsson, R. (2000). Mass criterion for wave controlled impact response of composite plates. Composites: Part A, 31:879-887.
6. Cantwell, W.J. and Morton, J. (1991). The impact resistance of composite materials-a review. Composites. 22(5):347-362.
7. Liu, D. and Malvem, L.E. (1987). Matrix cracking in impacted glass/ epoxy plates. Journal of Composite Materials. 21:594-609.
8. Robinson, P. and Davies, G.A.O. (1992). Impactor mass and specimen geometry effects in low velocity impact of laminated composites. International Journal of Impact Engineering. 12(2):189-207.
9. Instrumented Impact Testing of Plastics and Composite Materials. (1987). ASTM STP 936, American Society for Testing Materials.
10. Foreign Object Impact Damage to Composites. (1975). ASTM STP 568, American Society for Testing Materials.
11. Jang, B.Z. (1994). Advanced Polymer Composites: Principles and Applications. ASM International.
12. Liu, D. (1988). Impact-Induced Delamination-A View of Bending Stiffness Mismatching. Journal of Composite Materials. Vol. 22, No. 7:674-692.
13. Hufenbach, W., et al. (2008). Charpy impact tests on composite structures – An experimental and numerical investigation. Composites Science and Technology. Volume 68, Issue 12, September: 2391-2400.
14. Singleton, A.C.N., et al. (2003). On the mechanical properties, deformation and fracture of a natural fibre/recycled polymer composite. Composites: Part B 34:519–526.
15. Wardle, M.W. and Tokarsky, E.W. (1983). Drop Weight Impact Testing of Laminates Reinforced with Kevlar Aramid Fibers, E-Glass and Graphite. Composites Technology Review, Vol. 5, No. 1: 4-10.
16. Mitrevski, T., Marshall, I.H., Thomson, R. (2006). The influence of impactor shape on the damage to composite laminates. Composite Structures. 76:116–122.

17. Daihua, Zheng. (2007). Low Velocity Impact Analysis of Composite Laminated Plates, PhD Dissertation, University of Akron.
18. Rotem A. (1988). The Strength of Laminated Composite Materials under Repeated Impact Loading. Journal of Composites Technology Research, Summer:74-79.
19. Dorey, G. (1988). Impact damage in composites-development, consequences, and prevention. In Proc. 6th Int. Conf. on Composite Materials and 2nd European Conf. on Composite Materials', Imperial College, London, Vol. 3: 3.1-3.26.
20. Choi, H.Y., Wu, H.-Y.T. and Chang, F.K. (1991). A new approach toward understanding damage mechanisms and mechanics of laminated composites due to low-velocity impact: Part II, Analysis, Journal of Composite Materials. 25:1012-1038.
21. Jih, C.J. and Sun. (1993). C.T. Prediction of delamination in composite laminates subjected to low velocity impact, Journal of Composite Materials. 27(7): 684-701.
22. Cantwell, W.J., Curtis, P.T., and Morton, J. (1986). An Assessment of the Impact Performance of CFRP Reinforced with High-Strain Carbon Fibers. Composites Science & Technology. Vol. 25:133-148.
23. Thomason, J.L. and Vlug, M.A. (1996). Influence of fibre length and concentration on the properties of glass fibre-reinforced polypropylene: 1. Tensile and flexural modulus. Composites, 1996, **27A**:477-484.
24. Thomason, J.L. and Vlug, M.A. (1997). Influence of fibre length and concentration on the properties of glass fibre-reinforced polypropylene: 4. Impact properties. Composites, 1997, 28A:277-288.
25. Ticona Inc. (2008). www.ticona.com, last accessed, September 2008.
26. Belingardi, G., Vadori, R. (2002). Low velocity impact tests of laminate glass-fiber-epoxy matrix composite material plates. International Journal of Impact Engineering. 27:213-229.
27. Resnyansky, A.D. (2006). The Impact Response of Composite Materials Involved in Helicopter Vulnerability Assessment: Literature Review Parts I and II. Australian DSTO Defense Science and Technology Organization.
28. Jenq, S.T., Wang, S.B., and Sheu, L.T. (1992). A Model for Predicting the Residual Strength of GFRP Laminates Subject to Ballistic Impact. J. Reinforced Plast. Composites, 1992, 11: 1127 – 1141.
29. Williams J.G., and Rhodes M.D. (1982). Effects of Resin on Impact Damage Tolerance of Graphite/Epoxy Laminates. Composite Materials: Testing and Design, STP 787, Daniel I.M., Ed. ASTM: 450-472.
30. Cartie, D.D.R. and Irving, P.E. (2002). Effect of resin and fibre properties on impact and compression after impact performance of CFRP. Composites Part A, 33:483-493.
31. Tita, V., Carvalho, J., Vandepitte, D. (2008). Failure analysis of low velocity impact on thin composite laminates: Experimental and numerical approaches. Composite Structures 83:413-428.
32. Offringa, A. (2005). Thermoplastics, From Press-Forming to Co-Consolidation., 37th ISTC, Seattle, WA Oct 31-Nov. 3, CD-ROM Proceedings.
33. Cole, K.C, Nole, D. and Hechler, J.J. (1990). Crystallinity in PPS-Carbon Composites. A Study Using Diffuse Reflection FTIR Spectroscopy and Differential Scanning Calorimetry. Journal of Applied Polymer Science. 39: 1887-1902.

34. Pillay S., Serrano, J.C., Vaidya, U.K. (2005). Low Velocity Impact of Single Diaphragm Formed Carbon/PPS Laminates., Session on Transportation Applications. May 2005, Long Beach, CA, Session on Thermoplastic Composites, SAMPE Technical Proceedings, CD-ROM Proceedings.
35. Hosur, M.V., Adbullah, M., Jeelani, S. (2005).Studies on the low-velocity impact response of woven hybrid composites. Composite Structures 67:253–262.
36. Pillay, S. (2006). Thermoplastic Composite Resin Infusion., PhD Dissertation, University of Alabama at Birmingham.
37. Zenkert, D. (1997). The Handbook of Sandwich Construction. Published by Engineering Materials Advisory Services Ltd. (EMAS).
38. Anderson, T, Madenci, E. (2000). Experimental investigation of low-velocity impact characteristics of sandwich composites. Composite Structures, 50:239±247.
39. Johnson, H.E., et al. (2006). Health and Safety Executive. Damage modelling of large and small scale composite panels subjected to a low velocity impact. RR520 Research Report.
40. Ulven, C.A, Pillay, S, and Vaidya, U.K. (2003). Flexural and low velocity impact response of E-glass/balsa core sandwich composites. ASME Chicago, (CD-ROM Proceedings).
41. Mines, R.A.W, Worrall, C.M, Gibson, A.G. (1994). The static and impact behaviour of polymer composite sandwich beams Composites. Issue 2, Volume 25:95-110.
42. Vaidya, U. K et al. (2006). Impact and post-impact vibration response of protective metal foam composite sandwich plates. Materials Science and Engineering: A. Issues 1-2, Volume 428: 59-66.
43. Serrano-Perez, J.C, Vaidya, U.K, Uddin, N. (2007). Low velocity impact response of autoclaved aerated concrete/CFRP sandwich plates. Composite Structures, 80:621-630.
44. Vaidya, A.S. (2008). Design and Manufacturing of Composite Structural Insulated Panels (CSIPS), PhD Dissertation, University of Alabama at Birmingham.
45. Drzal, L.T, Rich, M.J, and Lloyd, P.F. (1982). Adhesion of graphite fibers to epoxy matrices: I. The Role of Fiber Surface Treatment, Journal of Adhesion, Vol. 16: 1-30.
46. Walter, R.W et al. (1977). Designing for integrity in long-life composite aircraft structures., STP 636, ASTM. 228-247.
47. Ko, F.K, Chu, H and Ying, E. (1986). Damage tolerance of 3-D braided intermingled carbon/PEEK composites. Advanced Composites: The Latest Development, ASM International. 75-88.
48. Krieger, Jr R.B. (1987). An adhesive interleaf to reduce stress concentration between plys of structural composites. Advanced Materials Technology. 87, SAMPE Symposium, Vol. 32:279-286.
49. Curtis, P.T. (1984). An initial evaluation of a high-strain carbon fiber reinforced epoxy., TR84004, Royal Aircraft Establishment.
50. Bartus, S. (2004). Impact Response of Long Fiber Thermoplastics., MS Thesis, University of Alabama at Birmingham.
51. Bartus, S.D. and Vaidya, U.K. (2005). Performance Of Long Fiber Reinforced Thermoplastics Subjected to Transverse Intermediate Velocity Blunt Object Impact, Composite Structures, Issue 3, Vol. 67: 263-277.

52. Grow, D. (2006) Protective Metal Foam Sandwich Composites, NSF SBIR Phase II Technical Report.
53. Bartus, S.D. (2006). Near Simultaneous and Sequential Impact of S2-glass/epoxy composites., PhD Dissertation, University of Alabama at Birmingham.
54. Coppa, A.P., Selden, G.F. and Mehan, R.L. (1979). Mach one impact test apparatus utilizing an 18.4 mm-bore gas gun with a novel decelerating device. Review of Scientific Instruments, 50(7), 824-828.
55. Bourne, N.K. (2003). A 50 mm bore gas gun for dynamic loading of materials and structures. Measurement Science and Technology, 14: 273-278.
56. Czarnecki, G.J. (1998). Estimation of the V_{50} using semi-empirical (1-point) procedures. Composites 29B: 321-329.
57. Jones A. Ed. (1998). Speer Reloading Manual Rifle and Pistol. Blount, Inc. Sporting Equipment Division, Lewiston, ID.
58. Jones R. (1994). Residual strength of composites with multiple impact damage. Composite Structures 28:347-356.
59. Chian-Fong, Y. (2002). Ballistic impact modeling of composite materials. In: proceedings of 7th International LS-DYNA Users Conference, Detroit, Michigan, 15-25.
60. Tsai, SW. (1965). Strength characteristics of composite materials. NASA CR–224.
61. Tsai, SW and Wu, EM. (1971). A general theory of strength of anisotropic material. Journal of Composite Materials. 5:58.
62. Hashin, Z. (1980). Failure criteria for unidirectional fiber composites. J of Appl Mech;47:329-334.
63. Cheng, FK and Chang, KY. (1987). A progressive damage model for laminated composites containing stress concentration. Journal of Composite Materials. 21:834-855.
64. Puck, A and SchÜrmann, H. (1998). Failure analysis of FRP laminates by means of physically based phenomenological models. Composites Science and Technology. 58:1045–68.
65. Puck, A, SchÜrmann, H. (2002). Failure analysis of FRP laminates by means of physically based phenomenological models; Part B. Composites Science and Technology 62:11633–68.
66. LS-DYNA theoretical manual, Livermore software technology. (1998).
67. Schweizerhof, K., et al. (1998). Crashworthiness analysis with enhanced composite material models in LS-DYNA-Merits and Limits. LS-DYNA world conference, Detroit, Michigan, USA.
68. LS-DYNA keyword user's manual, Livermore software technology. (2003).
69. Xiao, JR, Gama, BA, and Gillespie, Jr J.W. (2007). Progressive damage and delamination in plain weave S-2 glass/SC-15 composites under quasi-static punch-shear loading. Compos Struct. 78(2):82-196.
70. Matzenmillar, A, Lubliner, J, Taylor, R.L. (1995). A constitutive model for anisotropic damage in fiber composites. Mech of Mater 20:125–152.
71. Brown, K, Brooks, R, Warrior, N. (2005). Numerical simulation of damage in thermoplastic composite materials. In: proceedings of the 5th European LS-DYNA Users Conference, Birmingham, UK.

72. Deka, L. (2008). Multi-site Impact Response of Laminated and Sandwich Composites., PhD Dissertation, University of Alabama at Birmingham.
73. Goldsmith, W, Dharan, CKH, Hui, C. (1995). Quasi-static and ballistic perforation of carbon fiber laminates. Int J Impact Eng. 32(1):89–103.
74. Sun, CT. and Potti, SV. (1996). A simple model to predict residual velocities of thick composite laminates subjected to high velocity impact. Int J Impact Eng.18(3):339-353.
75. Potti, SV, Sun CT (1997). Prediction of impact induced penetration and delamination in thick composite laminate. International Journal of Impact Engineering. 19(1):31-48.
76. Lee, SWR, Sun, CT. (1993). A quasi-static penetration model for composite laminates. Journal of Composite Materials. 27:251-71.
77. Lee, SWR, Sun, CT. (1993). Dynamic penetration of graphite epoxy laminate impacted by a blunt- ended projectile. Compos Sci Technol. 49:369-380.
78. Vlot, A. (1996). Impact loading on fiber metal laminates. International Journal of Impact Engineering. 18(3):291-307.
79. Ursenbach, DO, Vaziri, R and Delfosse, D. (1996). An engineering model for deformation of CFRP plates during penetration. Composite Structures. 1995. 32:197-202.
80. Gama, B.A and Gillespie, J.W. Jr. (2008). Punch shear based penetration model of ballistic impact of thick-section composites. Composite Structures. Volume 86, Issue 4: 356-369.
81. Johnson, W. (1986). Historical and present-day references concerning impact on wood. International Journal of Impact Engineering. 4:161-174.
82. Johnson, W. (1986). Mostly on oak targets and 19th century naval gunnery. International Journal of Impact Engineering.. 4:175-183.
83. Reid, SR and Peng, C. (1997). Dynamic uniaxial crushing of wood. International Journal of Impact Engineering. 19(5-6): 531-570.
84. Buchar, J, Rolc, S, Lisy, J. and Schwengmeier, J. (2001). Model of the wood response to the high velocity of loading. In proceedings 19th Int. Symposium on Ballistics., Interlaken, Switzerland. 7-11.
85. Vural, M. and Ravichandran, G. (2003).Dynamic response and energy dissipation characteristics of balsa wood: experiment and analysis. International Journal of Solids and Structures. 40: 2147-2170.
86. Tagarielli, VL, et al. (2005).A constitutive model for transversely isotropic foams, and its application to the indentation of balsa wood. International Journal of Mechanical Sciences. 47: 666-686.
87. Murray, YD, et al. (2005). Evaluation of LS-DYNA Wood Material Model 143, U.S. Department of Transportation, FHWA-HRT-04-096.
88. Chan, S, et al. (2007). Ballistic limit prediction using a numerical model with progressive damage capability. Composite Structures. 77: 466–474.
89. Tabiei, A and Wu, J. (2000).Three-dimensional non-linear orthotropic finite element material model for wood. Composite Structures. 50:143-149.
90. Meyers, MA, (1994). Dynamic Behavior of Materials, John Wiley & Sons, Inc.
91. Nemat-Nasser, S. (2000). Introduction to high strain rate testing, ASM Handbook. Mechanical Testing and Evaluation, ASM Int, Materials Park OH. Vol. 8:427–428.

92. Hopkinson, B. (1914). A method of measuring the pressure produced in the detonation of high explosives or by the impact of bullets, Philos. Philos. Trans. R. Soc. London, Ser. A 213, 437–456.
93. Davies, RM. (1948). A critical study of the Hopkinson pressure bar, Philos. Trans. R. Soc. London, Ser. A 240(821), 375–457.
94. Kolsky, H. (1949). An investigation of the mechanical properties of materials at very high rates of loading, Proc. Phys. Soc. London, Sect. B 62(II-B). 676–700.
95. Gray III, GT. (2000). Classic split-Hopkinson pressure bar testing. ASM Handbook, Mechanical Testing and Evaluation, ASM Int, Materials Park OH. Vol. 8:462–476.
96. Nicholas, T. (1980). Tensile testing of materials at high rates of strain, Experimental Mechanics 21:177–185.
97. Gray III, GT, (2000). Split-Hopkinson pressure bar testing of soft materials. ASM Handbook, Mechanical Testing and Evaluation, ASM Int, Materials Park OH. Vol. 8:488–496.
98. Nemat-Nasser, S., Issacs, JB., and Starrett, JE. (1991). Hopkinson techniques for dynamic recovery experiments, Proc. R. Soc. London Ser. A 453, 371–391.
99. Couque, C., Albertini, C., and Lankford, J. (1993). Failure mechanism in a uni-directional fiber-reinforced thermoplastic composite under uniaxial, in-plane biaxial and hydrostatically confined compression, J. Mater. Sci. Lett. 12: 1953–1957.
100. Hosur, M.V. et al. (2003). High Strain Rate Compression Characterization of Affordable Woven Carbon/Epoxy Composites under off-Axis Loading, Polymers and Polymer Composites. n 7, v 11: 527-539.
101. Tsai, J. and Sun, C. T., Dynamic compressive strengths of polymeric composites, International Journal of Solids and Structures, Volume 41, Issues 11-12, June 2004, Pages 3211-3224.
102. Hosur, M.V., et al. (2003). High strain compression response of affordable woven carbon/epoxy composites, Journal of Reinforced Plastics and Composites, v 22, n 3: 271-296.
103. Hosur, M.V., et al. (2004). Dynamic punch shear characterization of twill weave graphite/epoxy composites, International SAMPE Technical Conference, SAMPE 2004. 739-748.
104. Hosur, M.V.; et al. (2004). Experimental studies on the high strain rate compression response of woven graphite/epoxy composites at room and elevated temperatures, Journal of Reinforced Plastics and Composites, n 5, v 23:491-514.
105. Gomez-del Rio, T et al. (2005). Dynamic tensile behaviour at low temperature of CFRP using a split Hopkinson pressure bar. Composites Science and Technology, n 1, v 65:61-67.
106. Bing, Q. and C.T. Sun., (2005). Modeling and testing strain rate-dependent compressive strength of carbon/epoxy composites., Composites Science and Technology, Issues 15-16, Volume 65: 2481-2491.
107. Tsai, J., and Sun, C.T. (2005). Strain rate effect on in-plane shear strength of unidirectional polymeric composites., Composites Science and Technology, Issue 13, Volume 65:1941-1947.
108. Ninan, L., Tsai, J. and Sun, C. T. (2001). Use of split Hopkinson pressure bar for testing off-axis composites, International Journal of Impact Engineering, Issue 3, Volume 25:291-313.

109. Zhao, H. (1997). Testing of polymeric foams at high and medium strain rates, Polym. Test. 16, 507–516.
110. Hembs, S., et al. (2007). Strain rate effects in phenolic composites and phenolic-impregnated honeycomb structures, Composites Science and Technology. 67: 2827-2837.
111. Tagarielli ,V.L., Deshpande, V.S. and Fleck, N.A., (2008). The high strain rate response of PVC foams and end-grain balsa wood., Composites, Part B. 39: 83-91.
112. ASNT Handbook Volumes 4, 6 and 7, American Society for Nondestructive Testing, Columbus,Ohio.
113. Chocron S., et al. (2001). Impact of the 7.62-mm APM2 Projectile Agains the Edge of a Metallic Target, Int. J. Impact Eng., v. 25:423-437.
114. Bhardwaj, M. (2010). Ultrasonic Evaluation of Materials, Ultran Co, www.ultran.com, last accessed, June 2010.
115. Chang, J., Zheng, C., and Qing-Qing, Ni. (2006). The ultrasonic wave propagation in composite material and its characteristic evaluation. Composite Structures. Issues 1-4, Volume 75:451-456.
116. Ultrasonic evaluation of matrix damage in impacted composite laminates Composites Part B: Engineering, Issue 1, Volume 31:1-6. (2000).
117. Saito, H., and Kimpara, I. (2006). Evaluation of impact damage mechanism of multi-axial stitched CFRP laminate. Composites: Part A. 37:2226–2235.
118. Růžek, R., Lohonka, R., and Jironč, J. (2006). Ultrasonic C-Scan and shearography NDI techniques evaluation of impact defects identification. NDT & E International. Issue 2, Volume 39:132-142.
119. Rosalie, S.C., et al.(2004). Variation in the group velocity of Lamb waves as a tool for the detection of delamination in GLARE aluminium plate-like structures. Composite Structures. no 1-4, v. 66: 77-86.
120. Toyama N. and Takatsubo J. (2004). Lamb wave method for quick inspection of impact-induced delamination in composite laminates. Composites Science and Technology, n. 9, v. 64:1293-1300.
121. Bourne, N.K., Forde, L., and Field, J.E. (1997). High-Speed Photography and Stress-Gauge Studies of the Impact and Penetration of Plates by Rods. In: Proc 22nd International Congress on High-Speed Photography and Photonics. Ed. Paisley, D.L. v. 2869, SPIE Press, Bellingham, Washington, USA:626-635.
122. Jain A. K., and Dubuissonnal, M.P. (1992). Segmentation of X-ray and C-scan images of fiber reinforced composite materials. Pattern Recognition. Issue 3, Volume 25: 257-270.
123. Roye, W. (1995) The X-ray backscatter technique for the nondestructive inspection of composite materials. NDT & E International, Volume 28, Issue 3 :191-195.
124. Bernardi, R.T. (1989). Computed tomographic and laminographic x-ray inspection of composites. Proceedings of the 12th World Conference on Non-Destructive Testing, Amsterdam, Netherlands. Vol. 2:1646–1648. Edited by Boogaard, J. and van Dijk, G.M. , Elsevier.
125. Wang, W., Su, C., and Liu, Po-Wen. (2008). Full-field non-destructive analysis of composite plates., Composites Part A: Applied Science and Manufacturing. Issue 8, Volume 39: 1302-1310.

126. Grédiac M. (2004). The use of full-field measurement methods in composite material characterization: interest and limitations, Composites Part A: Applied Science and Manufacturing., ns. 7-8, v. 35:751-761.
127. Dickinson L.P. and Fletcher N.H. (2008). Acoustic detection of invisible damage in aircraft composite panels. Applied Acoustics, In Press, Corrected Proof, Available online 21 February.
128. Raju, P.K, Patel, J.R. and Vaidya, U.K. (1993). Characterization of defects in graphite fiber based composite structures using the acoustic impact technique (AIT), Journal of Testing & Evaluation, n 5, v 21:377-395.
129. Vaidya, U.K. (1993). Nondestructive Evaluation of Carbon Fiber Based Composites using Acoustic and Vibration Techniques., PhD Dissertation, Auburn University.
130. Li, Zhuang and Crocker, M.J. (2005). A review on vibration damping in sandwich composite structures. International Journal of Acoustics and Vibrations. n 4, v 10: 159-169.
131. Karjatkar, Ravi, et al. (2003). Experimental determination of modal densities of foam filled honeycomb core sandwich panels., Proceedings of the Tenth International Congress on Sound and Vibration. 2325-2332.
132. Whittingham, B., Li, H.C.H., Herszberg I. and Chiu, W.K. (2006). Disbond detection in adhesively bonded composite structures using vibration signatures., Composite Structures. 75:351–363.
133. Vaidya, A.S. (2006). Multifunctional Sandwich Composites for Transportation Applications. MS Thesis, University of Alabama at Birmingham.
134. Guild, F. J. et al. (1976). Application of acoustic emission to fibre-reinforced composite materials, Composites, n 3, v 7 : 173-179.
135. Yu Y et al. (2006). A study on the failure detection of composite materials using an acoustic emission., Composite Structures 75:163–169.
136. Kwon, O-Y. and Hong, D.-H. (1997). Non-destructive Evaluation of Impact Damage CFRP Laminates by Acoustic Emission During Flexural Loading, In: Proc. 11th Int. Conf. on Composite Materials, Gold Coast, Australia, ACSS, Woodhead Publ Ltd. V.6:223-229.
137. Bar, H.N., Bhat, M.R., and Murthy, C.R.L. (2004). Identification of failure modes in GFRP using PVDF sensors: ANN approach, Composite Structures, n.2, v. 65: 231-237.
138. De Oliveira, R et al. (2004). Optic fibre sensor for real-time damage detection in smart composite, Computers & Structures, ns. 17-19, v. 82: 1315-1321.
139. Brown, J, Anderson, R and Visser, D. (2007). Detecting low-velocity impact damage in composite plates using infrared thermography. Proceedings of SPIE - The International Society for Optical Engineering, Thermosense XXIX: 6541-6550.
140. Dutton, A.G. (1996). Flaw detection in composite materials using infra-red thermography by the method of external heating., Proceedings of the Institution of Mechanical Engineers, Part C: Journal of Mechanical Engineering Science, n 5, v 210: 399-407.
141. Connolly, M.P. (1992). Measurement of porosity in composite materials using infrared thermography. Journal of Reinforced Plastics and Composites, n 12, v 11:1367-1375.

Vehicle Crashworthiness Design – General Principles and Potentialities of Composite Material Structures

Giovanni Belingardi and Giorgio Chiandussi

Department of Mechanics, Politecnico di Torino, Torino, Italy

1 - Introduction

Increasing demands from customers, government concern and national and international regulations put occupant safety at the forefront of vehicle design considerations. The safety level of a passenger vehicle, put in evidence by the number of stars (up to five stars) obtained in the European New Car Assessment Program (EuroNCAP) or in the US New Car Assessment Program (USNCAP) rating tests, are at present a very important point in the market promotion strategies for automotive manufacturers. The USNCAP is under mandate from the National Highway Traffic Safety Administration (NHTSA).
From the design point of view there are at least three different perspectives to look at the safety problem, therefore commonly (see Belingardi (2002), Prasad and Belwafa (2004), Seiffert and Wech (2003)) we use to distinguish between

- **passive** safety that includes all the car body parts that are devoted to dissipate the vehicle kinetic energy and automatically protect passengers from possible injuries during the crash events. Passive safety includes also all the devices that equip the passenger compartment and are activated automatically, when impact conditions are detected, in order to increase the passenger protection. It is clear that these safety devices - for example the seat belts or the airbag - are passive because they work during the accident event without command inputs from the driver;
- **active** safety that includes all the devices that are devoted to avoid the impact event, mainly by helping the driver during the vehicle riding. An example of this type of devices is the ABS that is intended to assist the driver during the braking phase by avoidance of the wheel locking. It is clear that these safety devices are active because they work with command inputs from the driver before the accident event;
- **preventive** safety that includes all the devices that are devoted to assist the driver's riding surveillance. These types of devices include for example driver vision enhancement in case of fog or insufficient road lighting, obstacle recognition, classification and warning. Driver tiredness evaluation and warning are presently under development. These devices are preventive because they are intended to help the driver avoid dangerous situations.

The concept of safety therefore includes both the likelihood that a collision will occur and the probability, type and rank of both injury to passengers and other road users, and damage to vehicles resulting from the collision.

The cause of a traffic accident are usually found in three different, often concurring, factors: the road, the vehicle and the driver. In order to give a rough idea where priorities should be in addressing this problem, we can consider that recent statistical data (see Seiffert and Wech (2003)) attribute about 7 - 10% of the causes of road accidents to the road itself and its surroundings, 3 - 5% to the vehicle and the remaining 85 -90% to driver lack of prudence and errors.

The **road** or, more generally, the environment, strongly affect safety. At first, we have to consider the influence of different aspects of the infrastructure such as the type of road, the separation between traffic lanes, the visibility at road crossings and corners, the existence of lateral guards and traffic signals, etc. Moreover, we have to consider the influence of the atmo spheric environment that includes visibility, the tire-road adherence, the presence and direction of wind and other factors.

The **vehicle** affects safety from at least two different points of view. On one hand, as will be discussed later, the vehicle structure is designed to give a good level of passive protection and is equipped with passive, active and preventive safety devices. On the other hand the vehicle itself can be the origin failure situations, that can be potentially dangerous. The main systems of the vehicle mechanics are involved as they determine the directional behaviour of the vehicle (suspension, wheels, tires, steering) and the braking performance. Even situations of failure of the vehicle systems related to the external visibility (windscreen, windscreen wipers, headlights) or to drive signals (position lights, direction signal lights, etc.) may have consequences with respect to safety.

Finally, as it is often written is accident analysis reports, by his/her conduct, the **driver** affects the safety of the vehicle as a consequence of dangerous actions, imperfect health and physical conditions or simply for not having foreseen a danger or not having avoided it through proper manoeuvres.

The level of safety is expressed by a probability. Design for safety does not mean to design a safe vehicle in absolute terms (it does not exist) but a vehicle that is, at some extent, safe when submitted to a certain type of accidental events, that are seen as more frequent and relevant in the real world scenario. Moreover, even with reference to these accidental events, the engineering solution does not lead to the total annulment of injury and/or damage, but to their reasonable limitation. This means, first of all, passenger survival and passenger non invalidating injuries.

In this chapter the attention is concentrated on the passive safety perspective.

It is really impressive that in the year 2000 in the US (NHTSA 2001) there were still about 37500 fatalities and about 2 million people injured (at different level of injury from permanent lesion to short lasting disease) as a result of vehicle

accidents. About the same number of fatalities and one million people injured in the European Community.
But this, on a statistical basis, is a relevant result obtained enforcing the safety design perspective through the national and international regulations.
A strong impulse to vehicle safety was given by the book "Unsafe at any speed" published by Ralph Nader in the late 1960s. Since that days the fatality rate per 10^8 Vehicle and Miles Travelled (VMT), as it is shown in Figure 1.1, decreased from about 5.3 (value of the ratio for the year 1967) to 1.7 (value of the ratio for the year 1991) with a nearly constant (negative) slope and over the last ten years has remained nearly constant at the value 1.6.

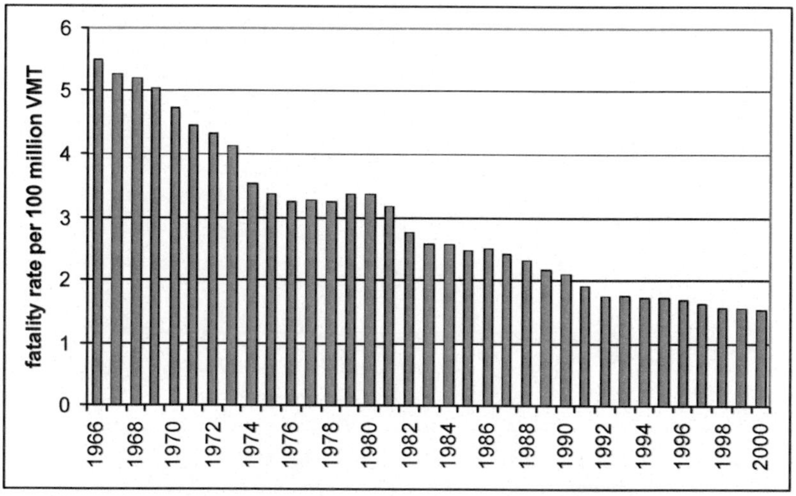

Figure 1.1 – Fatality rate in the US – Belingardi (2002).

This constant value does not mean that safety has stopped its improvement trend: both the number of registered motor vehicle and the number of travelled miles had a large increment in recent years so that still a constant ratio means a decrement of the fatality risk. Figure 1.2 shows the trend of the number of fatalities with the number of resident population, that is one of the risk exposure factors. While the resident population increases almost linearly in the years (NHTSA 2001), there is a progressive decrease of the number of fatalities. The figure also puts in evidence that for some periods the decrement is more evident. These can be related to the enforcement of safety regulation that became more and more severe.
Furthermore, of a total number of 57403 vehicles involved in fatal crashes in the year 2000, 27496 were passenger cars (47.9% with a fatality rate equal to 1.7) and 20295 were light trucks (35.4% with a fatality rate equal to 2.2), as shown in

Figure 1.3 together with other data. If we take in consideration the corresponding numbers of the year 1980: of a total number of 62312 vehicles involved in fatal crashes 39059 were passenger cars (62.7% with a fatality rate equal to 3.5) and 12680 were light trucks (20.3% with a fatality rate equal to 4.3) we can appreciate in much more detail the safety increment obtained through the regulations enforcement. We can notice that, although in absolute numbers the passenger car safety is increased while the light trucks safety is decreased, the fatality rate had the same trend: in thirty years both rates have been halved.

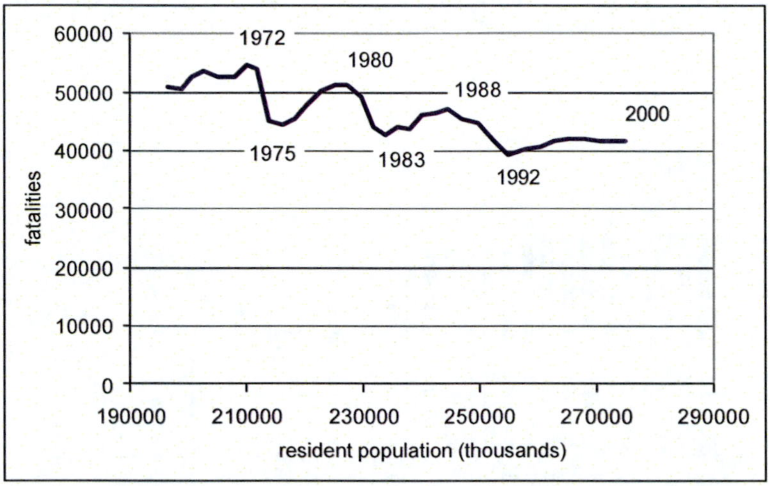

Figure 1.2 – Trend of the number of fatalities in the US – Belingardi (2002).

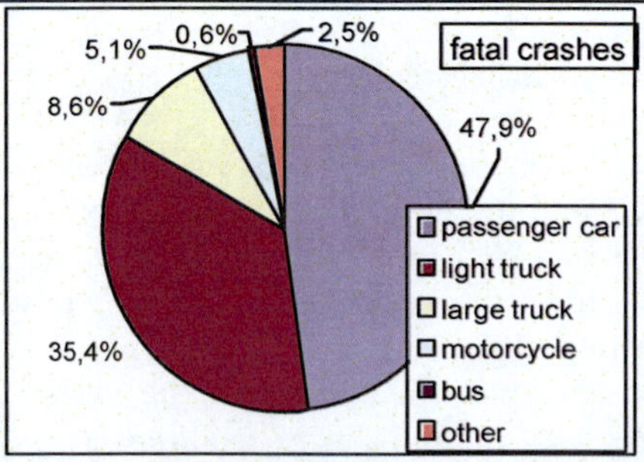

Figure 1.3 – Type of vehicles involved in fatal crashes in the year 2000 in the US – Belingardi (2002).

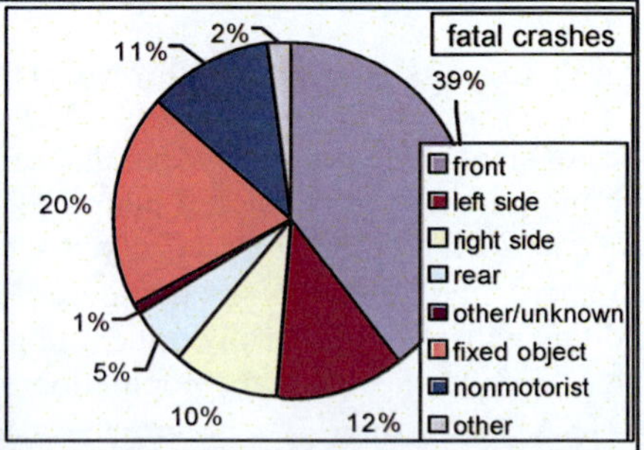

Figure 1.4 – Type of impact for passenger cars in fatal crashes in the year 2000 in the US – Belingardi (2002).

For what concerns passenger cars in the year 2000, the 27496 fatal crashes can be subdivided according to the initial point of impact in the following way:

60.6% were frontal impacts (that include car to car impacts and the majority of impacts against fixed objects and non motorists), 25.7% were side impacts (both left and right, and some impacts against fixed objects such as pole), 6.2% were rear impacts (that include car to car impacts and some impacts against non motorists). Thus frontal impacts are still the majority of the passenger car impact (see Figure 1.4), although a certain diversification should be done between straight impact with full overlap, straight impact with partial overlap and oblique impact. An impressive number of fatal crashes involves a vehicle and one, or more, non motorist, mainly a pedestrian. As we will se later (see sub-subsection 2.1.4) attention in regulations toward the Vulnerable Road Users (VRU) is increasing in the recent years, in particular toward pedestrians more than toward pedalcyclists.

The passenger survival space concept was first introduced in the 1970s. It required that, during standard crash tests, the passenger compartment remains nearly non deformed and non intruded by other structural parts of the vehicle itself, while the front structure deforms and thus absorbs the needed amount of energy (see Seiffert and Wech (2003)). Particular attention was paid to the steering wheel backward motion (that was a major cause of fatal injuries to the driver by his thorax deep compression), to the front hood (that must not penetrate the passenger compartment but must bend and remain anchored to the front structure) and to the windscreen (that must not be detached from its site and must not fragment).

Then the passenger movements immediately before and after the impact event have been considered: the safety belts have been introduced to maintain the driver and the other passengers fastened to their seats, avoiding their expulsion from the passenger compartment through the windows and, in particular, avoiding the impact of the driver chest against the steering wheel.

But the preservation of the survival space and the simple use of the safety belt proved to be insufficient: biomechanic studies pioneered by Dr. Stapp in the US and developed by an increasing number of engineers and researchers all over the world (from Wayne University in the US, TNO in the Netherlands and INRETS in France to many others) made clear that human beings can survive only if the head deceleration do not exceed well defined limits and human beings do not experience invalidating injuries only if the forces and bending moments applied to the neck, the chest, the abdomen, the pelvis hip joints and the femurs do not exceed well defined limits. The use of anthropomorphic instrumented dummies during the cash tests became mandatory.

Injury criteria have been introduced to correlate measurements made with these instrumented dummies and some biomechanical parameters that express the capability of the human body to survive the impact event and the level of the possible injuries of different types that can be caused by the impact event.

The national and international regulations prescribe the measurements that should be conducted during cash tests with these instrumented dummies and state the limits of the above mentioned biomechanical parameters in order to reduce gradually but definitely the fatal and invalidating injury risks.

In order to meet these quite strict requirements safety belts are generally not enough in case of frontal impact against a fixed barrier at a speed of only 50 km/h. Air bags have been introduced and nowadays all new cars are equipped with airbag devices for the driver and for the front passengers.

The main attention of the vehicle safety regulation and of the car manufacturers, at first, was devoted to the frontal impact accidents. This is justified by the statistical evidence that fatal crashes have in about 60% of the cases their initial impact point at the vehicle front. The frontal impact test is therefore the principal test, but progressively attention has been paid also to side impact and then rear impact and rollover events. For the vehicle rear structure designer can take advantage from the experience gained dealing with the vehicle front structure, that can be rather similar to the rear structure in terms of layout and structural elements. However the rear impact requires some specific design developments and tests related to the fuel tank (that is generally positioned in the rear part of the vehicle) and to the whiplash event for the passenger head-neck articulation.

Much more problematic are the side impact and the rollover events. The lateral body structure as well as the upper structure (roof) of the passenger compartment are generally constituted by thin panels with contour beams and therefore have a limited capability of absorbing energy by plastic deformation (compared to the front and rear structures). The door and the roof have been reinforced with specifically designed beams in order to obtain a sufficient energy dissipation, more over the contribution of the seat structure has been optimised.

In the analysis of frontal impacts attention is increasingly paid to partial over-lap frontal impacts and to oblique impacts. In these cases, only one half of the front structure dissipates energy and this can be insufficient. The designer cannot simply increase the strength of each single part because an overstiff structure may result dangerous in case of full over-lap impact; he has to put at work all the vehicle frontal structure even when there is a partial overlap impact.

Pedestrian deaths in car accidents in Europe (see the European Road Accident Database - CARE 2005) have decreased from 7000 victims in 1993 to 5000 in 2002 as shown in Figure 1.5. Even if a decreasing number of deaths can be perceived in the last years, better results can be achieved by improving the vehicle active and passive safety. The statistical analysis (Viano et al. 1997) of the human body regions that suffered major injuries in pedestrian/car accidents is shown in Figure 1.6 (data from Europe, United States, Japan and Australia). The chart shows that the legs and the head are the most critical body parts. A Japanese study (Mizuno Y. 2003) on car accidents involving pedestrians shows also that the main cause of death is the impact between the head and the vehicle.

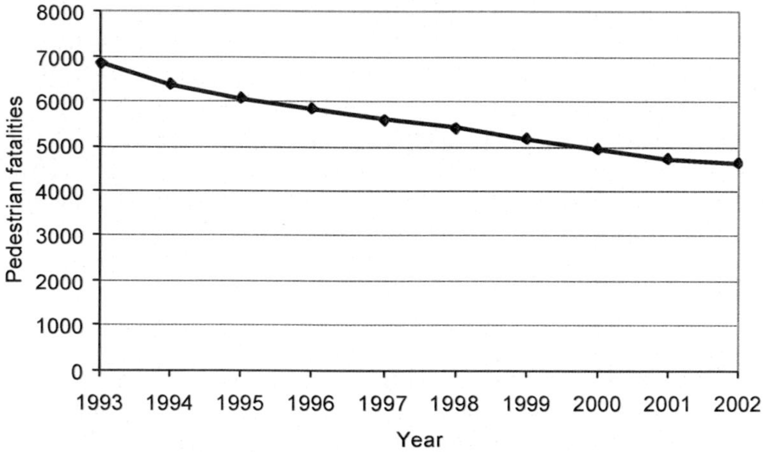

Figure 1.5 - Pedestrian fatalities in Europe since 1993.

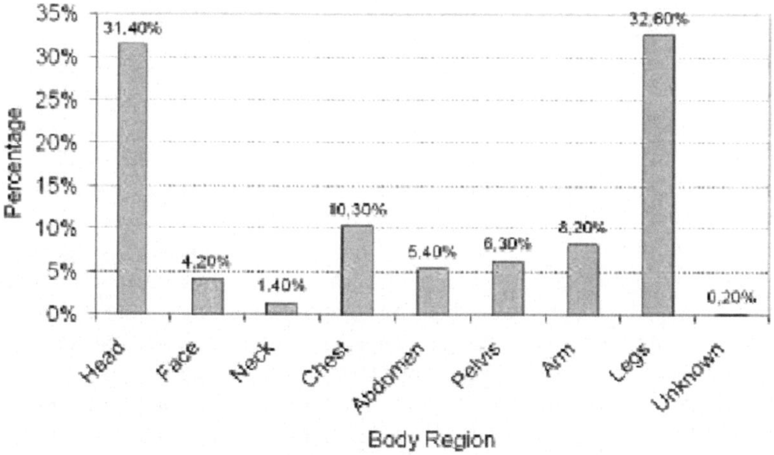

Figure 1.6 - Injury distribution over the human body.

As a consequence, particular attention has recently been paid to the study of pedestrian head impact against several car components like the hood, the windscreen and the pillars (Yang 2005). By taking into consideration the impact of a pedestrian against the frontal part of a vehicle, the impact against the hood is the most critical if a child is involved whereas the impact against the windscreen is the most critical if an adult is involved (Viano et al. 1997 and Nishimoto 2003). Particular attention has been paid to the hood design.

Concluding this introduction it is important to underline the growing importance of virtual (computer aided) simulation in the development of passive safety research. The possibility of performing realistic virtual tests using numerical simulation makes it possible to:

- analyse different architectural solutions in the very early stages of the vehicle structure design, while in the past it was only possible to try to correct the eventual passive safety inadequacies of the already drafted structure by local patches;
- analyse the vehicle behaviour submitting its mathematical model to a number of different type of impact events in the design development phase (i.e. before the availability of prototypes for tests that, taking into account their very high cost, were always less than desired for a complete evaluation);
- get an extraordinary detailed insight of each different part of the structure and therefore to put in evidence the role played and the contribution given by each different part of the structure in obtaining the global result;
- perform a number of different prescribed tests to demonstrate the conformity of a new vehicle to the national or international regulations with a reasonable level of costs.

This chapter consists of six sections. In the section 2 the mechanics of the vehicle collision will be introduced, after some equations for the general description of the impact phenomenon, the principal experimental tests prescribed by the regulations and those used in the safety rating of vehicles will be addressed. In the section 3 the typical layout of the vehicle structure will be described and some considerations about the advantage in the use of composite materials instead of metallic materials will be developed. The section 4 will deal with the passenger restraint systems and the dynamic behaviour of passengers inside the compartment. In the section 5 the main types of dummy used in standard tests will be described. Then the biomechanical parameters, identified from the measurements made during standard impact tests, will be introduced. For each of these parameters, limit values established by the approval regulations will be reported. These limit values try to evaluate the threshold between severe injuries (or even death) and light injury for human beings submitted to impact events. The section 6 will present and discuss some applications of composite materials for vehicle structures that are designed and tested against crash because they may be involved in impact events.

2 – Mechanics of the Vehicle Collision.

To understand how the vehicle collision phenomenon takes place, we can use a very simplified model. This model is usually called the impulsive model of impact. The impact phenomenon has an extremely short duration, of the order of magnitude of 100 ms. In our model the impact duration can be assumed equal to nearly zero. The forces that originate in consequence of the impact, i.e. those exchanged between impacting vehicle (bullet vehicle, BV) and impacted vehicle (target vehicle, TV) are so large to make negligible those normally exchanged between the vehicle and the surrounding environment (rolling, aerodynamics, acceleration or braking); so they can be neglected. The described phenomenon is of the impulsive type and the principle of conservation of momentum of the two vehicles has to be applied.

Let us assume that the collision is direct central impact, i.e. that the speed of the two vehicles, applied at their center of gravity, are perfectly aligned. With reference to Figure 2-1, let t be the generic time, t_1 the instant in which the two vehicles begin their contact, t_2 the time at which the impact finishes. Let d be the distance between the two vehicle centers of gravity, V_{TV} and V_{BV} the speeds of the two vehicles and V_{REL} the relative speed of the two vehicles, i.e. V_{TV} - V_{BV} .

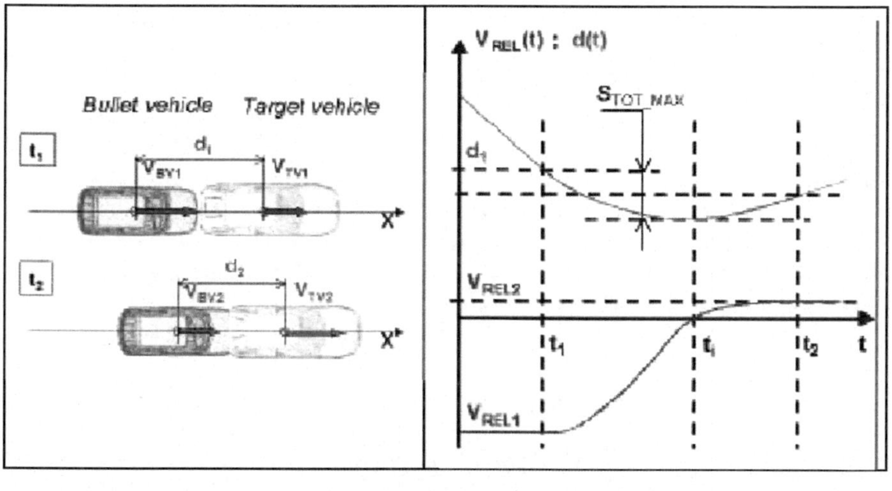

Figure 2-1 – The rear impact case; simplified history of distance d and relative velocity.

The subscript 1 refers to the beginning time, i.e. when the two vehicles begin their contact, while subscript 2 refers to the final impact time, i.e. when the impact finishes. If before the collision V_{BV} and V_{TV} have the same sign, we have the rear impact case, otherwise, if the velocities have opposite signs, we have the frontal collision case. The sign of V_{REL} is negative when the distance d decreases over time, and, on the contrary, is positive, when d increases.

During the collision distance d becomes evidently smaller than its initial value and we indicate with the subscript i, every quantity that relates at the time at which distance d assumes its minimum value. At the end of impact, at time t_2, distance d will be slightly increased with respect to d_i because of the elastic spring back of the structure of the two vehicles; time t_2 is assumed to coincide with the end of the spring back. The coefficient r of restitution is defined as the ratio V_{REL2} / V_{REL1} , that is generally called the Newton's law of impact. For a completely inelastic collision the coefficient r equals 0, while in case of a completely elastic collision it equals 1; in real cases, the value of r is between 0.05 and 0.20.

Taking into account that the velocity vectors are perfectly aligned in the x direction, the principle of conservation of momentum, that is a vector equation, can be written in its scalar form:

$$m_{BV} \; V_{BV1} + m_{TV} \; V_{TV1} = m_{BV} \; V_{BV2} + m_{TV} \; V_{TV2} \qquad (2.1)$$

i.e. the linear momentum of the two vehicle system does not change during the collision. The value of the coordinate x_G of the center of mass of the system is defined in terms of the coordinates of the center of mass of the two vehicles as

$$x_G = \frac{m_{BV} \; x_{BV} + m_{TV} \; x_{TV}}{m_{BV} + m_{TV}} \qquad (2.2)$$

while the other two coordinates y and z of the center of mass of the two vehicles have the same values y_G and z_G of the coordinates of the center of mass of the system. The following equation, obtained by taking the derivative of x_G with respect of time, shows that the velocity of the center of mass of the system does not change before and after the collision

$$v_G = \frac{m_{BV} \; v_{BV} + m_{TV} \; v_{TV}}{m_{BV} + m_{TV}} \qquad (2.3)$$

This is a direct consequence of the principle of conservation of momentum. We are therefore fee to eliminate the subscript 1 and 2 of V_G and to write the velocities of the two vehicles only as a function of the masses, the velocities of center of mass and the relative velocity:

$$v_{G1} = v_{G2} = v_G = \text{constant} \qquad (2.4)$$

$$v_{BV} = v_G - \frac{m_{TV}}{m_{BV} + m_{TV}} v_{REL} \qquad v_{TV} = v_G + \frac{m_{BV}}{m_{BV} + m_{TV}} v_{REL} \qquad (2.5)$$

With reference to the same time interval the comparison between the kinetic energy of the system before and after the collision has to be written in the form of inequality, the system energy after the impact is less than that before the impact as a result of energy dissipation due to plastic deformation of metals or fracture of composite and brittle structure during the collision

$$[m_{BV}\ (V_{BV1})^2 + m_{TV}\ (V_{TV1})^2\ ]/2 > [m_{BV}\ (V_{BV2})^2 + m_{TV}\ (V_{TV2})^2\ ]/2. \quad (2.6)$$

The amount of the energy dissipated during a collision can be calculated as the difference of the kinetic energy of the system before and after the collision and the following expression is obtained

$$E_D = E_{C1} - E_{C2} = \frac{m_{BV}\ m_{TV}}{2(m_{BV} + m_{TV})}(1 - r^2)\ V_{REL1}^2 \quad (2.7)$$

The dissipated kinetic energy is equal to the kinetic energy of a virtual body, whose mass M_{virt} is the ratio between the product of the two vehicle masses and the total mass of the system and whose velocity V_{virt} is the product of the relative velocity and the square root of $1\text{-}r^2$. Therefore, the dissipated energy is maximum when $r = 0$, i.e. for perfectly inelastic collision, and is zero when $r = 1$, i.e. for perfectly elastic collision

$$M_{virt} = \frac{m_{BV}\ m_{TV}}{m_{BV} + m_{TV}} \qquad V_{virt} = V_{REL1}\sqrt{1 - r^2} \quad (2.8)$$

Figure 2-1 describes qualitatively the evolution of the phenomenon; S(t) is equal to d_1-d(t), i.e. the total amount of the vehicles crush.

2.1 Crash tests and main international regulations

The main crash tests considered by the international regulations and by the rating procedures are briefly described in this subsection.

Safety regulations prescribe those crash tests on which depend the safety approval of the vehicle and therefore the possibility to put it on the market. The vehicle must have at least a safety level considered the minimum needed to ensure adequate protection both to automobile users and to other road users.

Crash test ratings have a different aim, they help customers evaluate the safety level of different vehicles, that have already passed the approval tests. Rankings are usually based on a combination of the results from various types of tests. The result of each test contributes to the calculation of a final score that summarizes the degree of safety offered by the vehicle. The evaluation result, in the EuroNCAP and USNCAP procedures, is expressed through a number of levels that ranges from sufficient to excellent. These results are visualized through a number of stars and / or different colours, the interested reader can see in the web site of the two organisations.

In the rating procedures there are both tests similar to those required by regulations and tests that analyse different aspects not yet considered into regulations (e.g. child safety). Usually the crash tests that are similar to those required by regulations are more stringent in their implementation (e.g. collision against the same kind of barrier but at an increased impact velocity) and in the procedures for the result evaluation.

One of the main features of these rating tests is that they are done by independent institutes (independent from vehicle makers) and that the results are publicly disclosed, in order to inform consumers so that, having reviewed the performance obtained in the field of safety by the various tested vehicles (usually sold in some different versions), can better orient their choice.

As safety rating is an important marketing factor, it has become a standard to be followed for manufacturers. Manufactures are pushed to improve their designs so that their vehicles are rated at the excellent safety levels. This contributes to the increase in the minimum safety level of vehicle fleets, and to the improvement of the existing legislation. Therefore, ratings can be considered a sort of driver for the legislation. The following description of crash tests takes into account the distinction between regulation crash tests and rating crash tests, and describes briefly the main impact configurations.

It should be noted that the study of human behaviour during impact is much more difficult than, for example, the mechanical behaviour of materials. This is due to two main reasons: first, it is not possible to carry out "destructive" tests with live persons on board and, secondly, there is a very large dispersion of responses between the persons due to their very large differences in age, weight, size, readiness and other physical and emotional factors.

Impact tests can be done only with anthropomorphic dummies (that have the feature of repeatability) and, in few cases, with corpses. However tests with cadavers (that are permitted only in few countries) provide incomplete and rather dispersed information and therefore do not represent a suitable solution. On the other hand the use of anthropomorphic dummies in these tests poses some questions about their bio-fidelity, the completeness of their instrumentation and the interpretation of the measurements that are still open questions. Both the safety legislation and rating procedures suffer from lack in establishing a precise relationship between the mechanic and physic characteristics of impact, the measurements made on dummies and the corresponding injury to human beings resulting from the vehicle crash. All these points ask for deeper insight and, consequently, are still under development.

Finally, standard crash tests can be considered as symbolic cases, quite particular between the nearly infinite possible modes of accident. Therefore all the rankings based on these test results, represent only an incomplete and approximate assessment of the safety performance of the cars examined. In this field, know-how is constantly evolving and future revisions are very likely to be introduced.

2.1.1 - EC 96/79 front impact test

The European Union directive 96/79/EC states the test procedures and requirements on the results to ensure protection of occupants of motor vehicles in the event of a frontal collision. According to this directive, the vehicle that is submitted to the impact test against a fixed deformable barrier (see Figure 2.2), must be representative of the series production, must include the equipment usually found in that vehicle and must be in the normal riding configuration. The restraint systems should be the standard ones provided by the manufacturer. A Hybrid III dummy (see subsection 5.1) is seated on each front seats. The position of dummies inside the vehicle and the application of the restraint systems on them are imposed by specific rules. The propulsion system of the vehicle can be its engine, or another device. At impact no steering or external propulsion device must be active on the vehicle. The speed of the vehicle at impact must be 56 km/h with a maximum allowed deviation of ± 1 km/h. However, if the test has been performed at a higher speed and the vehicle is found to comply with the requirements, the test is considered valid.

The trajectory of the vehicle must be such that 40% ± 20 mm of the width of the vehicle is superimposed on the barrier front (40% "overlap"). The barrier is positioned so that the first contact with the vehicle happens on the driver side.

The standard deformable barrier consists of several elements. In the back there is a thick metallic plate that supports the deformable part of the barrier and gives its positioning and fixture at a rigid wall. The main structure of the barrier consists of a large block of aluminium honeycomb, in front of which, in its lower part, a bumper-like element, once again made of aluminium honeycomb, is applied, with the aim of simulating the bumper structure of the antagonist impacting vehicle. The barrier front has to remain perpendicular (with a tolerance of ± 1°) to the trajectory of the vehicle. The force-displacement characteristic of the deformable barrier is prescribed by the directive and is intended to reproduce the impact deformation and energy absorption capability of the other vehicle when a "car-against-car" impact is considered (Sugimoto et al. (1998), Ultra-Light Steel Auto Body - ULSAB (1999), EuroNCAP (2008)).

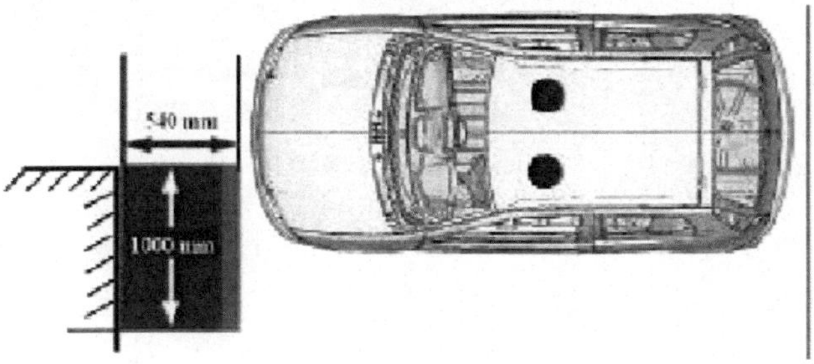

Figure 2.2 – Frontal impact test according to the EU regulations.

The barrier is constrained to a mass (rigid wall) of not less than $7*10^4$ kg, that is anchored to the ground or placed on the ground making use of other additional fixture devices to constrain its movement (a block of concrete is generally used). The size of deformable barrier is reported in Figure 2.2. During the test, measurements are executed to get the quantities needed to verify the necessary performance with respect to the approval criteria; needed parameters are recorded by independent data channels.

Results are evaluated by comparison of some biomechanical parameters with their limit values as specified in the regulation. In particular the following parameters are considered: the head injury criterion (HIC - see sub-subsection 5.3.1), the resulting acceleration of the head (see sub-subsection 5.3.2), the neck injury criteria (NIC - see sub-subsection 5.3.3), the thorax compression criterion (TCC - see sub-subsection 5.3.5) and the thorax viscosity criterion (VC - see sub-subsection 5.3.6), the compression force of the femur (FFC - see sub-subsection 5.3.7), the tibia index TI and the compression force of the tibia (TCFC - see sub-subsection 5.3.8), and, finally, the patella slide motion.

2.1.2 - EC 96/27 side impact test

The European Union directive 96/27/EC states the test procedures and requirements on the results to ensure protection of occupants of motor vehicles in the event of a side. According to this directive, the tested vehicle must be representative of the series production, must include the equipment usually found in that vehicle and must be in the normal riding configuration.

In this case the vehicle to be tested is motionless and the test is carried out on the driver's side. The dummy used in the test European Side Impact Dummy

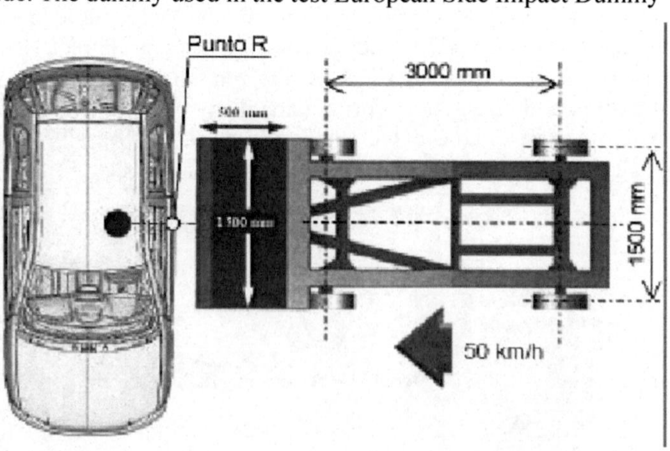

Figure 2.3 – Lateral impact test according to the EU regulations,

(EuroSID) has bio-fidelity performance only when loaded in the lateral direction, has the size and the mass of 50 percentile male. This dummy has not the forearms. See subsection 5.2 for a brief description of the EuroSID dummy.
The seat belt and/or other restraint system are dressed on the dummy as prescribed by the manufacturer. The speed of moving deformable barrier at the moment of impact is 50 ± 1 km/h; however, if the test is conducted at a higher impact speed and the vehicle meets the requirements, the test is considered valid. The speed is stabilized at least 0.5 m before the impact. The trajectory of the mean longitudinal vertical plane of mobile deformable barrier must be perpendicular to the mean longitudinal vertical plane of the impacted vehicle (see Figure 2.3).
The mobile deformable barrier consists of a cart and a device to prevent a second impact on the already crashed vehicle. The mobile deformable barrier is made with aluminium honeycomb and consists of six blocks, linked together (see Figure 2.4) and prepared in two rows of three items each. All the elements have the same width (500 ± 5 mm) and the same height (250 ± 3 mm); the elements of the upper row are made of aluminium 3003 and have a depth of 440 ± 5 mm while those in the lower row are made of aluminium 5052 and have a depth of 500 ± 5 mm. Also for this deformable barrier the directive prescribes the force-displacement characteristic as it is intended to reproduce the impact deformation and energy absorption capability of the other vehicle when a "car-against-car" impact is considered. During the test, measurements are performed to get the quantities needed to verify the necessary performance with respect to the approval criteria; needed parameters are recorded by independent data channels.

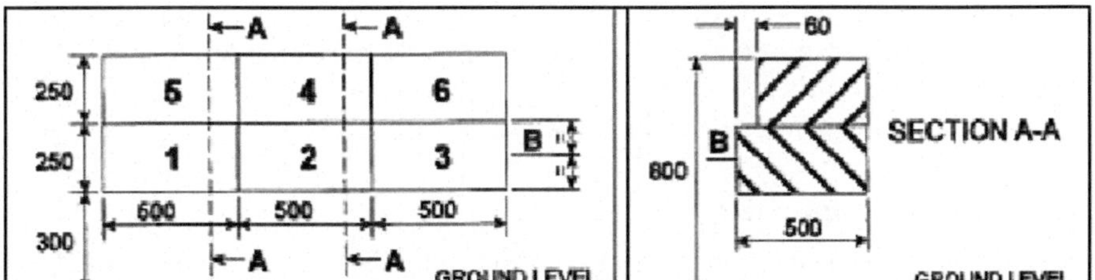

Figure 2.4 – Dimensions of the Mobile Deformable Barrier (MBD) used in EU side impact tests; see also EuroNCAP side impact test protocol (2004).

Results are evaluated by comparison of some biomechanical parameters with their limit values as specified in the regulation. In particular the following parameters are considered: the head injury criterion (HIC - see sub-subsection 5.3.1) in case of head impact against the vehicle interior, the thorax safety criterion according the following two parameters: the ribs deformation (CDC) and the soft tissue criterion (CT) or viscous criterion (VC), the maximum force on the pubic symphysis (FMSP) and the maximum internal force on the abdomen (FMA).

2.1.3 - ECE 32/34 rear impact test

In this type of test (see Figure 2.5) the vehicle, initially at rest, is impacted at its back by a 1100 kg cart equipped with rigid barrier at a speed of between 35 and 38 km/h. The rigid mobile barrier, 2500 mm wide and 800 mm high, has a clearance with respect to the

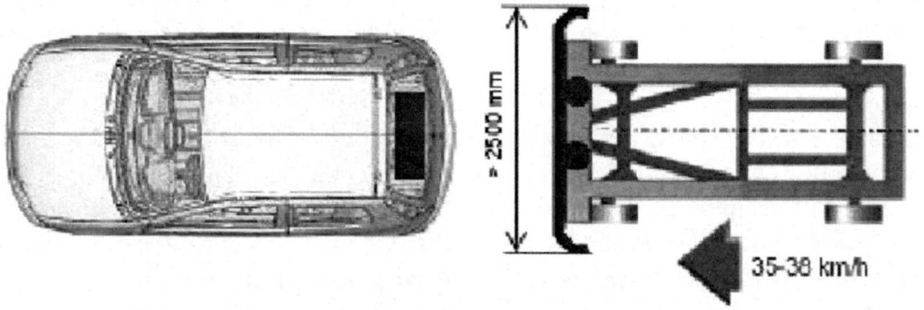

Figure 2.5 - Rear impact test according to the EU regulations

ground of 175 mm and should impact completely the rear zone of the vehicle, in order to have the condition of "full-overlap" impact.

At present, no dummies are on board of the vehicle as in this impact case no check of biomechanical type are required. Requirements are on the maximum intrusion in the passenger compartment (forward stroke of the back seat H point[1] less than 75 mm), on the maximum allowable loss of fuel (fuel spillage less or equal to 30 grams per minute), on the strength of the battery case (battery should remain in its site), on the possibility to open lateral doors after the collision (at least one door must open to allow the escape of occupants).

2.1.4 – Pedestrian impact test

Pedestrian impact tests are performed according to the legislative proposal made by the European Enhanced Vehicle Committee (EEVC) and has recently become the European Union directive 2003/102/EC.

The prescribed tests are intended to evaluate the protection offered by the vehicle against road users other than its occupants and the occupants of other vehicles, the so called Vulnerable Road Users (VRU) such as pedestrians and bicyclists. With particular reference to the pedestrian situation, it takes into account three different areas of the front part of the car (see Figures 2.6 and 2.7):

---

[1] the **H-point** (or **hip-point**) is the location of an occupant's hip, specifically the pivot point between the torso and upper leg portions of the body, relative to the floor of the vehicle

1 – the Bumper, 2 – the Bonnet leading edge (including the headlights and grille area), 3 – the Bonnet top (bonnet surface).

The pedestrian is represented by three segments of the human body:

1 – the adult complete leg or Legform, 2 – the adult pelvis or Upper Legform,

3 – the head in two versions: the adult head or Adult Headform and the child head or Child Headform

Tests to be carried out include the impact of the Bump er against the Legform, of the Bonnet leading edge against the Upper Legform, of the Child and Adult headform against the Bonnet Top. While in the bumper test the vehicle is moving against the legform, in the bonnet test it is the headform that is thrown against the bonnet of the car at rest.

The vehicle to be used during the impact may be the whole structure or only the front of the car. During the test, the height of the vehicle with respect to the ground should be that of normal running conditions (tires inflated at the recommended pressure, not steered front wheels, fluids necessary for the vehicle operation at their maximum level; two occupants in the car for a total weight of 150 kg, vehicle suspensions positioned as during the vehicle riding at 40 km/h).

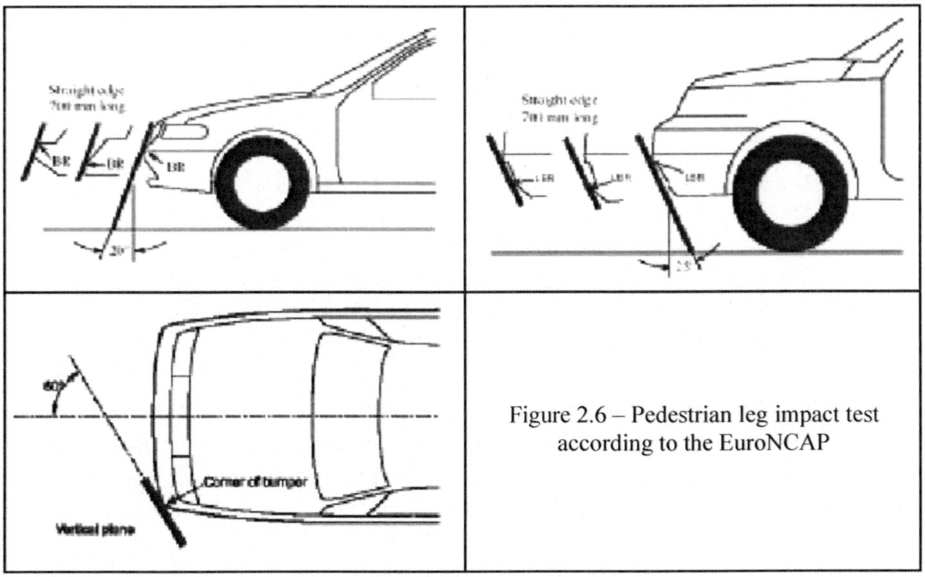

Figure 2.6 – Pedestrian leg impact test according to the EuroNCAP

The bumper area, according to this European directive, is the lower front structure of the vehicle, including all those structures that protect the vehicle in case of frontal collision at low speed, including the connections of the bumper to the frame of the car.

For what concerns the impact head - bonnet, the situation is as follows (see Figure 2.7):

- In the impact performed with the Child Headform, an angle of 50° ± 2° from the horizontal line and a headform speed of 40 km/h have to be used;
- In the impact performed with the Adult Headform, instead, an angle of 65° ± 2° from the horizontal line and a headform speed of 40 km/h have to be used.

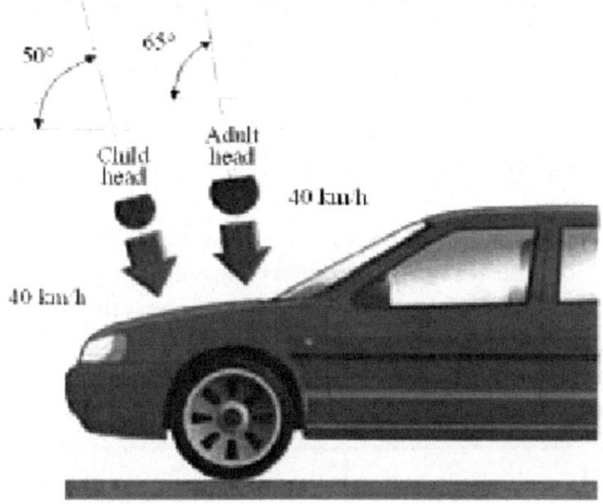

Figure 2.7 – The head – bonnet impact according to the EU pedestrian protection regulation.

Nine are the tests to be carried out: three for each third of the bonnet. The chosen impact points should be those that are expected to be the most dangerous for the head and must be in structurally different areas of the vehicle.
The criterion used to assess the severity of the possible injury is once again the Head Injury Criterion (HIC - see sub-subsection 5.3.1), both for the "child" impactor and for "adult" impactor.

2.1.5 - The EuroNCAP program

EuroNCAP (European New Car Assessment Program) tests are performed by an independent organisation and are devoted to supply costumers with information about the safety performance of new model of cars sold in the European market. This organisation performs tests that are similar to those performed for international regulation approval, but under heavier testing conditions. Results are published as rating in a web site, by means of a synthetic evaluation (a number of stars up to five) and some sketches that give information on the resulting protection level for every body parts. EuroNCAP performs mainly front and side impact tests, moreover information on pedestrian protection are supplied.

2.1.5.1 - EURO-NCAP rating - front impact test

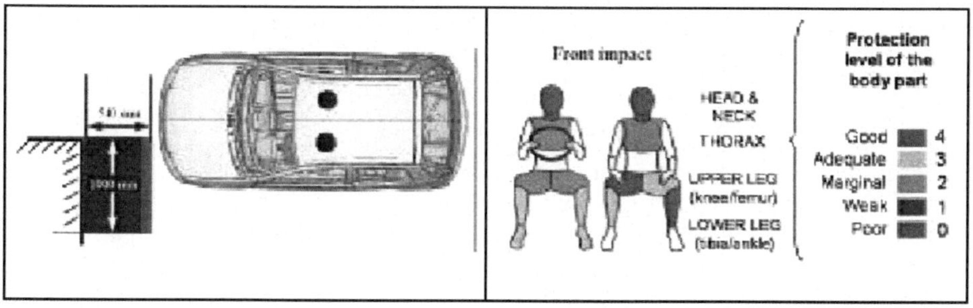

Figure 2.8 – Frontal crash test according to the EuroNCAP protocol (2008).

During the frontal crash test, the car is thrown against an offset deformable barrier (ODB) so that 40% of the width of the vehicle is superimposed on the front width of the barrier itself (see Figure 2.8). The barrier consists of a deformable aluminium honeycomb mounted on a rigid block, as that used for frontal tests in European approval procedure (see Figure 2.2).
The speed of the vehicle at the moment of impact must be 64 km / h, therefore 8 km / h higher than that prescribed in the approval test (see sub-subsection 2.1.1).
On the front seats of the car two anthropomorphic dummies of the Hybrid III type are seated, while two children dummies restrained by appropriate Child Restraint Systems (CRS) are seated on the rear seats. On the seat behind the driver a dummy with the size of a three years old child (TNO child dummy P3) is placed, while in the seat behind the passenger a dummy with the size of one and half year old child (TNO P1½ child dummy) is placed.
With the exceptions of the impact speed and the presence of children dummies, the test is identical to that prescribed by the European directive 96/79/EC for the front impact.
The good quality of vehicle crash behaviour is expressed mainly through the biomechanical response of front occupants: their body is subdivided into 4 areas or subgroups (head & neck, chest, upper legs and lower legs). To each of these areas a score is assigned out of five levels (from 0 to 4, i.e. from poor to good, as shown in Figure 2.8), depending on the level of the related indicators that has been achieved during the test by each areas of the body.
Resulting scores can be corrected (reduced) with appropriate coefficients or modifiers, if special situations occur such as chest-steering wheel contact, lack of integrity of the passenger compartment and so on. Each corrective coefficient can reduce the score on a region of the body by a maximum of two points. The partial scores of each body segment are added together and constitute a first subtotal the front impact contribution to the final score. The final score includes

the partial scores from other types of impact. The behaviour of systems intended for children protection, as well as the overall behaviour of the vehicle from various viewpoints, is summarized in the text that comments the publication of the above scores.

2.1.5.2 - EURO-NCAP rating - side impact test

The Euro-NCAP side test is similar to that specified in the European directive 96/27/EC for approval tests (see sub-subsection 2.1.2).
During the test, the vehicle is at rest and is impacted at the door on the driver side by a moving deformable barrier (MDB) that is travelling at a speed of 50 km/h (see Figure 2.9).

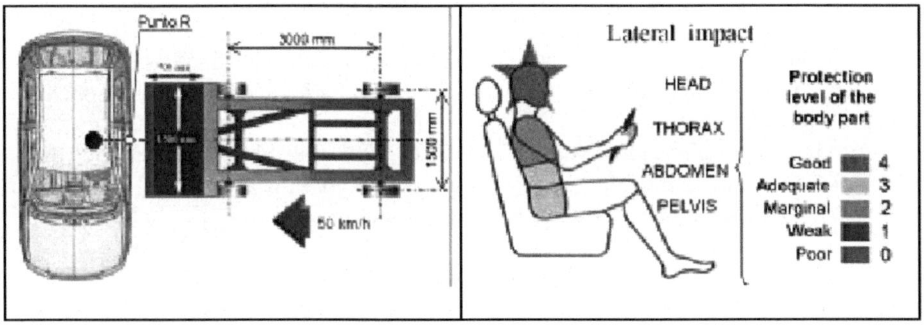

Figure 2.9 - Lateral crash test according to the EuroNCAP protocol (2004).

The dummy used in this test is the EuroSID (European Side Impact Dummy), it is quite different from the Hybrid III dummy used in the frontal impact tests as will be discussed in subsection 5.2. The dummy is placed on the driver seat, while at the rear seats two child dummies (one P 1½ and one P3) are placed and restrained by suitable child restraint systems (CRS). As in the front impact case, the body of the dummy is subdivided into 4 segments (head, thorax, abdomen and pelvis). To each of these areas a score, that ranges from 0 (poor) to 4 (good) is assigned, according to the level achieved by the appropriate biomechanical indicators; the scores obtained by these 4 body segments are summed together and the total is the second contribution to be considered for the final evaluation of the car.

2.1.6 - The FMVSS201 standard.

The FMVSS201 (Federal Motor Vehicle Safety Standard) and the companion European ECE 21 standards prescribe specific laboratory tests in order to assess the level of protection of the driver and the passengers of a car. During an impact situation they can hit one of the parts of the passenger compartment (for example the dash board or the pillars) causing severe injuries to the head. The mentioned standards prescribe that an instrumented headform, starting from a stated position relative to the driver or passenger front seats, is thrown and hits against any structure (see Figure 2.10) or object in the passenger compartment. The head acceleration history is measured and elaborated to obtain the value of the HIC index (see sub-subsection 5.3.1). It must be noted that the HIC index was developed with respect to a human being and the companion dummy with an appropriate level of bio-fidelity, in particular for what concerns the head support and guidance by the neck.

In the mentioned test procedure the dummy does not include a neck and the associated muscles, this results in a greater severity of the head stress. In order to obtain a meaningful value, comparable with those assessed from the whole human body, the FMVSS201 standard introduced a correction - HIC(d) (Dummy Equivalent HIC) - according to the following formula:

$$HIC(d)=0.75446\,(HIC)+166.4 \qquad (2.9)$$

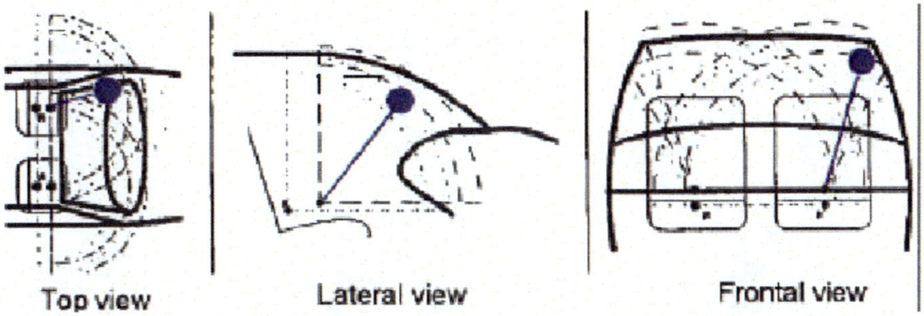

Figure 2.10 - Structure parts considered by the FMVSS201 standard (Belingardi et al 2002).

3 – Primer on the Structure Crash.

In case of collision with another vehicle or with a fixed obstacle, the kinetic energy of the vehicle must be completely dissipated. While the vehicle structure deforms to absorb the kinetic energy, the on-board protection devices, such as the safety belt pre-tension mechanisms and the air-bags, must be activated without commands by the driver for their activation (see Belingardi (2002), Prasad and Belwafa (2004), Seiffert and Wech (2003)).

The vehicle kinetic energy is absorbed by the plastic deformation of the material of the body structure if it is a metallic material, or by fracture and fragmentation if it is a brittle plastic material as generally are composite materials with polymeric matrix.

For passenger safety the progressivity of the energy absorption phenomena is essential. It is also very important that the accelerations and the forces to which the different parts of the human body are submitted do not exceed limits above which permanent injuries or even the death is caused.

3.1 – Vehicle front structure.

With this scope, the vehicle body structure (see for example Figure 3.1 where the characteristic components of the front structure of a recent vehicle are shown) is designed in order to get a vehicle deceleration history that has the desired characteristic of progressivity. The architecture of the structure shown in Figure 3.1 is nowadays generally adopted in passenger vehicles with small changes or adaptations. For formula racing cars, due to their smaller transversal dimensions, the front structure can be quite different. This is mainly due to the fact that the engine is placed in the rear part of the vehicle and the impact velocity that should be considered is higher that in the tests discussed in the previous section. Longitudinal and transverse beams are present also in the front structure of formula car, but generally the task of the energy absorption during impact is made by a specific front structure, the so called sacrificial front structure, see subsection 6.3

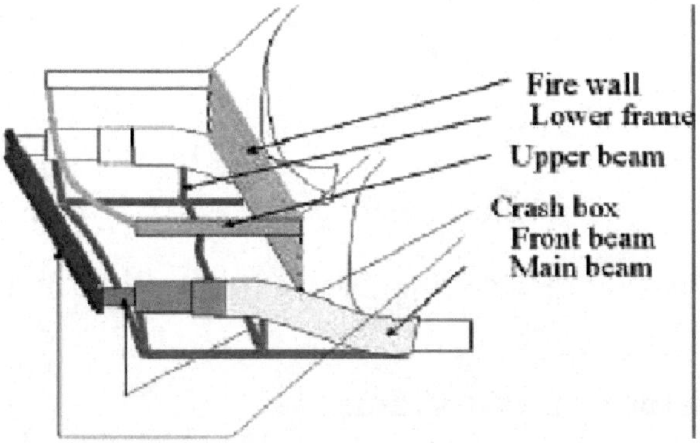

Figure 3.1 - Typical components of the front structure of a recent vehicle.

Figure 3.1 shows the front structure of the vehicle that is placed between the front bumper and the passenger compartment. The passenger compartment should remain essentially undeformed and not intruded into as possible, thus the structure that is behind the so called fire wall (that is the wall that separates the engine compartment from the passenger compartment) is assumed to give no contribution in the energy absorption. The front structure generally consists of four longitudinal thin-walled beams, two for each side of the vehicle, two are

positioned in the upper part of the engine compartment (upper beam), just below the hood, and two, generally of bigger dimensions with respect to the upper beams (main beams in Figure 3.1), at an intermediate height, behind the bumper. In recent vehicle, at the front extremity of these two main beams, two so called crash boxes are placed, one at each side. These crash boxes are intended to absorb energy in case of impact at low velocity thus avoiding large structural damage to the other parts of the front structure. At their rear extremities, the main longitudinal beams are connected to the strong longitudinal beams (the rocker rails) that are placed along the lower side of the passenger compartment and constitute its main structure. While the longitudinal upper beams, at their rear extremity, are generally connected to a quite complex structural joint that connects three other beams: the transverse beam below the windscreen, the so called A pillar aside the windscreen and the vertical column that is part of the frame around the fire wall. There is also a frame in the lower part of the engine compartment that has essentially the task of positioning and supporting the engine, also this frame gives a contribution, although quite small, to the front structure crash behaviour. Just behind the front bumper there is a transverse front beam, once again a thin walled beam, that has the task of positioning and supporting the bumper. The transverse front beam is connected at its extremities to the longitudinal main beams.
Figure 3.2 shows the typical history of the deceleration measured on the vehicle floor at the front seats anchoring points during a frontal crash done according to EuroNCAP testing procedure, as described in sub-subsection 2.1.5.1. In the graph, the different levels of the deceleration that are due to the different parts of the vehicle structure can easily be detected.

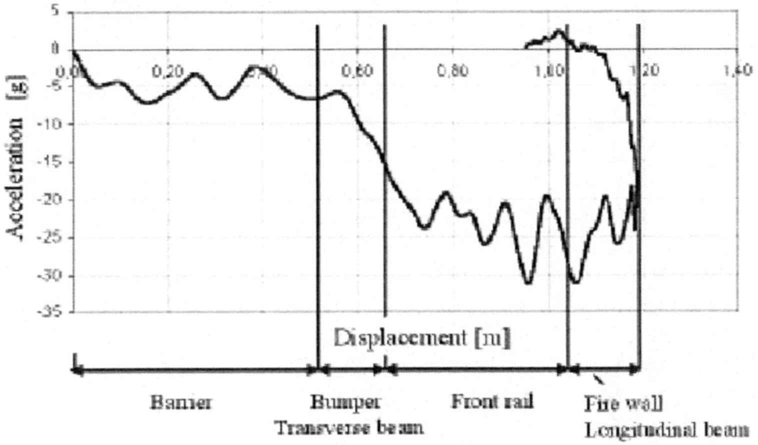

Figure 3.2 - Typical course of the deceleration during an EuroNCAP crash test.

The initial part of the curve (see Figure 3.2) shows the deceleration due to the compression failure of the deformable barrier (that simulates the other vehicle front structure, see sub subsection 2.1.1), thenthe deformation of, in the order, the transverse front beam, the so called crash boxes and finally the front structure longitudinal beams. Although all the longitudinal beams are involved the main contribution comes from the main beams. Usually, these structural beams are thin walled boxed beams made by drawn steel sheet and joined by spot welding. However, as it will be presented in subsection 6.1, solutions with composite material tubes and structures can be devised with interesting advantages.
The beam deformation mechanism of plastic axial collapse resulting in the formation of a sequence of folds is the most favourable mechanism in the case of metallic part, in order to guarantee the absorption of the desired amount of energy and in the meantime satisfies the progressivity requirements (Belingardi (2002), Prasad and Belwafa (2004), Seiffert and Wech (2003)). The initialisation of every new fold determines an increment of the beam axial antagonist force and, therefore, an increment in the vehicle deceleration. This increment is detectable in the deceleration history diagram as a peak. The plastic deformation evolves then through the development of localised plastic hinges in the beam walls. These localised plastic deformations of the metallic material give the desired energy absorption.

3.2 – Considerations on the use of UHS steels and composite materials,

In recent years the use of Ultra-High-Strength (UHS) steels has been proposed as this type of steel, due to its strength characteristics, makes it possible to reduce the weight of the vehicle, at least for those parts for which the design is based on tensile strength (ULSAB 1999). When the crash behaviour of the structure has to be evaluated, the advantage in using the UHS steels is not completely confirmed as the energy absorption mechanism is based on the bending plastic deformation of the thin walled components. The comparison between low carbon steels and UHS steels must therefore be based on the comparison between the resulting values of the plastic bending moment $M_y = \sigma_y h^2 / 4$, where σ_y is the material yield strength and h is the wall thickness that, being at the second power, plays a fundamental role. Moreover the UHS steels are generally less sensitive that mild steel to the strain rate effect and therefore the advantage in impact loading condition can be dramatically reduced.
Composite materials are considered as a possible alternative to UHS steels in order to match both lightweight and energy absorption in case of impact requirements (Avalle et al. 2004. Mangino et al. 2007).
Excellent review of composite material behaviour under impact loading are given by Abrate (1991, 1994 and 1998), Cantwell et al. (1991), Carruthers et al. (1998) and Thorton et al. (1998).

Table 3.1 collects typical values of the specific absorbed energy for some materials that can be of interest for car body structure.

Material	Absorbed energy per unit weight [kJ/kg]	Absorbed energy per unit volume [MJ/m^3]
Carbon – epoxy , lay- up $[0/\pm15]_3$	99	20
Carbon – epoxy , lay- up $[\pm45]_3$	50	6
Aramid – epoxy , lay- up $[\pm45]_8$	9	1
Aramid – epoxy , lay- up $[0/\pm15]_3$	60	22
Glass – epoxy , lay- up $[0/\pm15]_2$	30	13
Glass – epoxy , lay- up $[0/\pm75]_2$	53	26
Mild steel – AISI 1015	45	92
Aluminium alloy 6061	60	42

Table 3.1 - Typical values of specific absorbed energy of some materials for car body structure from Carruthers et al. (1998).

When composite materials are considered, the structure designer has to take into account that, even along the fiber direction, the laminate elastic modulus is usually lower than the corresponding values of steels while the tensile strength is comparable. Therefore, in order to match the structure targets in terms of stiffness, the use of composite materials leads to increase the wall thickness of the structural members with respect to the corresponding steel members.

The different collapse mechanisms of thin walled members made of composite material have to be taken into account. In particular the composite structure may fail due to compression strength or to bending strength, in both cases the local wall situation has to be considered. The limit compression load can be evaluated as $F_{comp} = \sigma_{lam,comp}\, h$, where $\sigma_{lam,comp}$ is the limit compression strength of the laminate, according to the considered loading direction, and h is the wall thickness. The limit bending moment can be evaluated as $M_{lim} = \sigma_{lam,bend}\, h^2 / 6$, where $\sigma_{lam,bend}$ is the limit bending strength of the laminate, according to the considered loading direction. The proposed formula for the limit bending moment takes into account that, generally, composite materials have not substantial plastic deformation, as steels have, but behave elastically up to fracture. For both collapse mechanisms, the increment in the wall thickness, due to the adoption of composite materials, is favourable for the crash behaviour, leading to an increment of the structure crush strength.

The energy absorption mechanism in the composite front rails is material progressive fragmentation with formation of fronds, as a consequence of the axial compression. Therefore the compression failure mechanism generally prevails on the global bending one.

Two other energy absorbing mechanisms are present and dissipate a significant portion of the energy: delamination and friction between different fronds and between fronds and the target. Figure 3.3 shows a synthesis of the failure mechanisms that take place in the axial collapse of a composite column. In the figure there are: a micrograph of the crushed tube (Mamalis et al. 2004), a draft of the crush zone (Mamalis et al. 2006) and a mechanical model for the analysis of the energy absorption (Solaimurugan et al. 2007).
According to this model the concurring energy terms are the following six:
(i) external work done by the crushing platen,
(ii) energy dissipated during circumferential delamination,
(iii) energy dissipated during frond formation (axial cracks),
(iv) energy dissipated on bending of fronds,
(v) energy dissipated through interlaminar shear deformation in the fronds during bending,
(vi) energy dissipated due to friction between crushing platen and fronds.

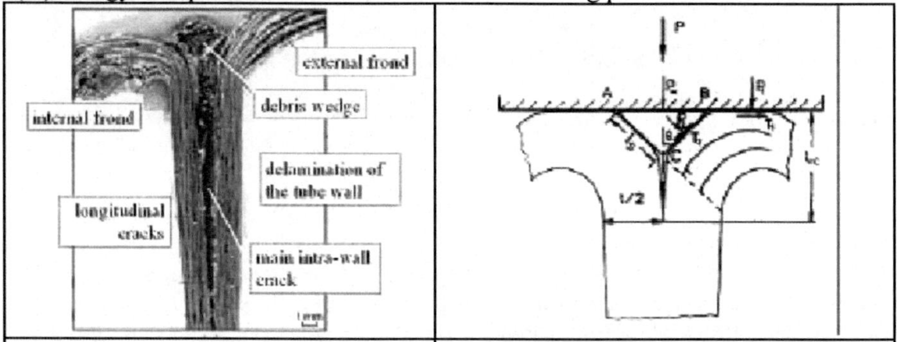

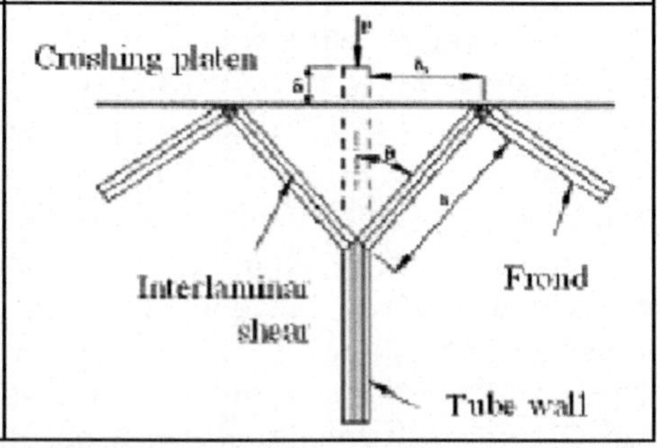

Figure 3.3 - Failure mechanisms with formation of fronds and debris wedge in the axial collapse of a composite column

A pioneering work to study the crush behaviour of composite tubes submitted to axial load has been done by Thornton (see Thornton et al. 1982), Farley (see Farley 1983) and Hull (see Hull 1983).
Thornton studied the behaviour of various composite tubes, taking into account difference in fiber type (glass, carbon and aramid fibers), different lay-ups and

different t/D thickness to diameter ratio. Energy absorption characteristic was found to be largely affected by the fiber type and by the lay-up, while results to be less sensitive to t/D ratio than in the case of metal tubes. Further rectangular and square section tubes were considered, experimental results showed that these section are less effective in energy absorption than circular section.
Farley and Jones (1991) studied the effect of crushing velocity on the energy absorbing characteristic of circular tubes made with aramid fibers and carbon fibers in epoxy matrix, with different lay-ups. They found that, in the case of carbon/epoxy tubes with [0/±θ] fibers, the specific energy does not vary with the crushing velocity, whereas with [±θ] fibers (they considered only two cases: θ equal to 45° and 75°) it increases with the increase of the crushing speed. With the aramid/epoxy tubes the energy absorption characteristic increases in all the considered cases.
Gupta et al. (1997) made an extensive experimental test program on the axial crush behaviour of composite tubes made of glass woven mats with epoxy resin. Different D/t diameter to thickness (from 8.5 to 54.4) have been considered. They found that the average crush stress decreases with the increase in D/t. They have done also some tests filling the tubes with polyurethane structural foam, results on the effect of the foam were not conclusive. Finally they proposed a model to predict the mean crushing load.

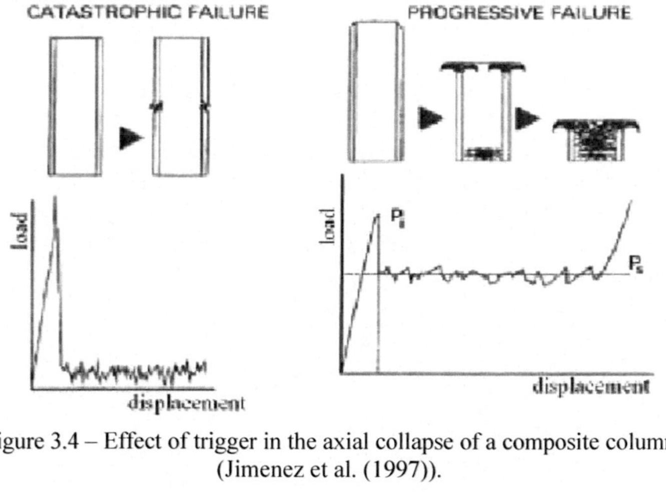

Figure 3.4 – Effect of trigger in the axial collapse of a composite column (Jimenez et al. (1997)).

Mamalis and co-workers have done an extensive research program on the axial crush behaviour of composite tubes, results have been presented in a number of papers and in one book (1998). They have initially done multiple experimental test campaigns, considering both cylindrical and conical (with different cone angles) tubes. They have analysed in depth the frond formation mechanism (see

figure 3.3a) both for circular and square cylindrical tubes. In a second phase they have developed a mechanical model to predict the axial crush behaviour of these tubes. Finally, in recent years, they have also develop finite element simulations, with a progressive refinement of the obtained results (Mamalis et al. 2002).
Composite front rail must have appropriate crush initiator both to initialise the right collapse mechanism and to reduce the value of first peak of the crush load (see Figure 3.4).
Siglas et al. (1991) made an extensive experimental campaign to study the effect of chamfer trigger on the energy-absorbing capabilities of glass fibre/epoxy composite tubes. Tubes with external chamfers ranging from 10° to 90° were submitted to axial crush load to different extents and the resulting microstructures were examined with a microscope. It was found that, for this material, the crushing process usually initiated by local bending of the chamfered parts of the tubes and by internal cracking. These events gave rise to wedges of crushed material which were pushed to the inside of the tube wall. The stresses generated at the root of the wedge gave rise to lateral cracks, which caused small rings of material to be sheared off. This sequence of events dominated the initial stages of the crushing process. At a later stage, the mode of crushing changed to one of frond-wedge-frond geometry.
Jimenez et al. (1997) made a systematic study of the effect of two different types of trigger geometry on the axial collapse of composite pultruded columns made of glass fiber in epoxy matrix. Figure 3.5 show the considered trigger types, namely a chamfer type (trigger B) and a so called "tulip" type (trigger T). The chamfer type has easier manufacturability than the tulip type.
Both the trigger geometry are effective in initialise the right collapse mechanism and in reducing the value of first peak of the crush load. Jimenez reports that all the considered chamfer angles, from 30° to 60°, reduced to a similar value the first load peak, however with the 60° solution has the highest value of the absorbed energy, close to 45 J/g, that is also higher than what he obtained with the tulip trigger.

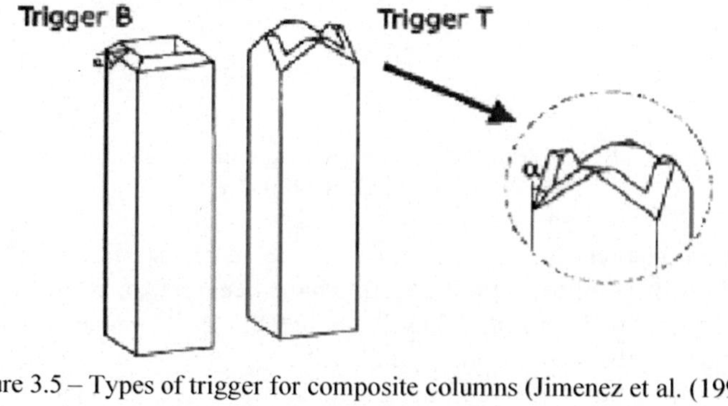

Figure 3.5 – Types of trigger for composite columns (Jimenez et al. (1997)).

3.3 – Vehicle body panels.

The vehicle structure is enveloped by some panels. The front and rear bumpers, the bonnet, the lateral and rear doors and the roof are of particular importance for the safety performance. All these parts are essentially made of a thin skin (a metal sheet or a composite laminate), that has a relevant aesthetic role and is stiffened and strengthened by means of a underlying structure. This structure consists of thin walled beams, mainly with a panel contouring layout but, especially in case of large panels such as the roof or the bonnet, some other beams are generally added as supporting ribs.

Lateral doors, together with the central pillar (the so called B pillar), are deeply involved in case of lateral impact. In order to give a contribution to the energy dissipation, all the lateral doors are strengthened with a longitudinal beam, placed inside the external panel, approximately at mid height of the external panel. During side impacts the door structure is loaded by the front longitudinal beams of the other vehicle. The result is affected by the structure of both vehicles, thus at present a large debate is on-going about vehicle compatibility.

In case of pedestrian impact the bonnet structure has a fundamental role as the pedestrian head hits against it. As it will be discussed in subsection 6.4, the main problem is to design a structure with a suitable compromise between the static stiffness target and the head impact target, that asks for a locally soft structure.

For all these external envelop panels a composite laminate solution can be devised. However some problems of colour matching between the composite and the painted metallic parts should be mentioned. This is, at present the main obstacle to a large adoption of composite.

Crash design criteria for these composite panel members are different from those discussed for the vehicle front structure. In all these cases, the main loading condition is bending. Once again the energy absorption mechanism is material progressive damage. This damage is mainly due to delamination. The other typical mechanisms of failure of a composite laminate, i.e. matrix cracking, matrix crushing and local buckling, fiber breakage, fiber debonding are also detectable but, in general, delamination appears to be the most important. (Abrate 1991 and 1998, Belingardi et al. 2002, 2003 and 2006).

3.4 – Passenger compartment.

The interior designer has to deal with the possible impact of the passenger head against the dashboard and the pillars.

For the dashboard, there are two main design solutions: to cover the metallic structure with an appropriate layer of foam or to make the structure with a material that is less aggressive than metals, for example a polymeric material, eventually reinforced with fibers. This type of material can offer interesting

solutions from the point of view both of the energy absorption capability and of structure light weight and appropriate stiffness.
For the pillars, the solution, usually adopted to pass the FMVSS201 standard tests, is to cover the metallic structures of the passenger compartment (the A, B and C pillars and the roof rails) with an appropriate layer of crushable foam (Barbat et al. 1995, Chou et al. 1995, Deb et al. 1997, Belingardi et al 2002).
The designer has to chose the most convenient type of foam and then he has to define the most appropriate values of the foam density (that means the stress-strain relationship and the energy absorption capability) and the foam layer thickness. These are the two design variables that can be considered in order to obtain a HIC value lower than the admissible limit value specified by the standard. Plastic foams are materials of great interest for the energy absorption problem in crashworthiness since they are light and at the same time able to absorb large amount of energy (Gibson and Ashby (1997), Sounik et al. (1997), Avalle et al. (2001) and (2007)).

4 - Primer on Restraint Systems.

In order to understand what happens during the collision, the first important point is on the structural behaviour of the involved vehicle. A second point is of at least equal or even higher importance: the control of motion of the occupants inside the vehicle passenger compartment, through appropriate restraint systems. The purpose of this section is to illustrate how this motion is connected with the global behaviour of the vehicle structure and to give a brief introduction to the design of restraint systems, regardless to their level of sophistication.
Let us consider the following two conditions, applied to the same vehicle:
- braking situation, an initial speed of 50 km/h is considered, the vehicle stop is gradually obtained after a certain distance X ;
- situation of frontal impact against fixed rigid wall, once again an initial speed of 50 km/h is considered, the vehicle stop is almost instantaneous.

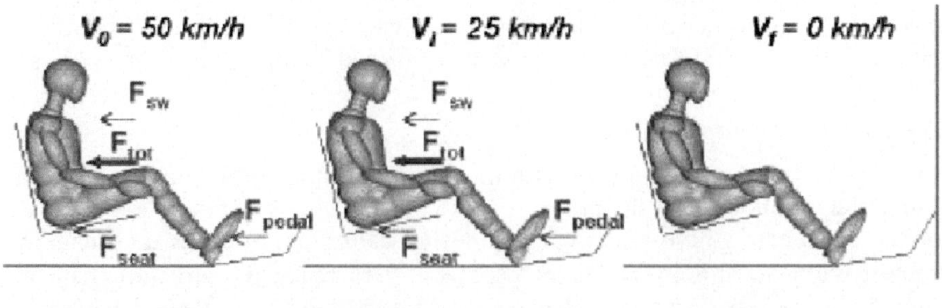

Figure 4.1 – Forces applied to the occupant during braking.

The two vehicles have therefore the same initial kinetic energy and both reduce their speed to zero, following two stopping processes that are very different. In the first case, the brake action generates a force at the tire-road contact points that, acting on the vehicle, reduces progressively its speed up to complete stop. This force between tire and road moves with the vehicle, makes a negative mechanical work that absorbs kinetic energy from the moving vehicle, up to produce its final stop.
In the second case, when the collision occurs against a non deformable (rigid) wall, the vehicle almost instantly loses all its speed relative to the ground and, consequently, all its kinetic energy. This energy is absorbed internally by deformation of the front structure of the vehicle.
In both cases, the impulse $\int_{t_1}^{t_2} F dt$ of force F(t) in the time interval $\Delta t = t_2 - t_1$ (the duration of the phenomenon) will be the same (as required by Newton's second law: $F = ma$ that can be rewritten as $\int_{t_1}^{t_2} F dt = m\,\Delta v$) but there are significant differences. First of all, the braking vehicle is submitted to relatively low value forces for a relatively long period of time while the impacting vehicle is submitted to forces much more intense for a much smaller time. Secondly, the point of application of the braking force (at tire-road contact) is moving on the ground, this generates a mechanical work with consequent removal of kinetic energy from the vehicle, while in the case of crash the force that stops the vehicle is exerted by the barrier, so it is not moving and does not generate work. Thus there is not an extraction of energy. The initial total energy of the impacting vehicle remains in the vehicle (the principle of conservation of energy) and the initial kinetic energy is converted into other forms of energy but is not extracted from the vehicle. Similar processes happened to the occupants of the vehicles.
In a braking situation, the occupant of the vehicle is submitted to a number of small forces that have slowed him down progressively together with the vehicle, by gradual extraction of the kinetic energy from his body, as a result of the movement of the points of application of these forces with respect to the ground. The friction force with the seat F_{seat} in Figure 4.1, the contact force between feet and footrest on the floor F_{pedal} in Figure 4.1 and the force applied by the hands and arms on the steering wheel F_{sw} in Figure 4.1 have to be considered.
If he is not restrained (see Figure 4.2), the occupant of the vehicle that impacts the barrier, faces a completely different situation much more similar to that of the vehicle in which he is located. Since the vehicle stops very quickly, the friction forces between the occupant and his seat are not sufficient to prevent his large motion. The occupant continues to move forward, into the passenger compartment, with his initial speed (50 km/h) until he hits the compartment surfaces (steering wheel, dashboard, etc.) that in the meantime have stopped their motion together with the vehicle against the rigid barrier. At this instant these

surfaces have already lost all their speed with respect to the ground. Since they are also rather stiff, the forces that they apply to the person to stop his free motion into the passenger compartment will be very high. During the very short time interval when forces are applied to the occupant by these surfaces, these latter move very little with respect to the ground and therefore they practically do not extract energy.

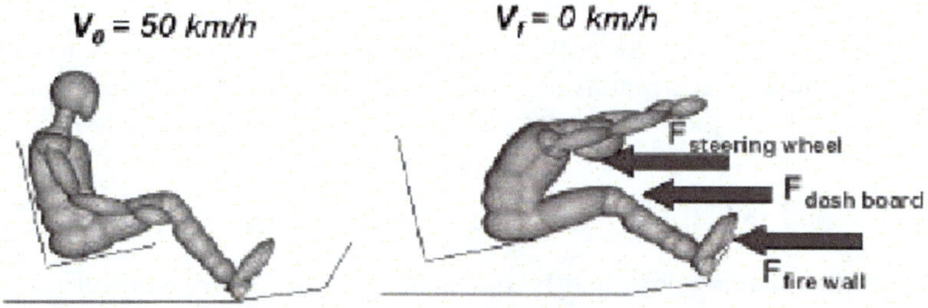

Figure 4.2 – Forces applied to the occupant without restraint systems during front impact.

Once again, since the kinetic energy of the occupant has to be dissipated and no part of it has been extracted by forces that make work, this initial kinetic energy must be converted into other forms of energy, remaining inside person body. It is this transformation of the kinetic energy that causes injuries suffered by the occupant during (essentially in the final part of) the collision. The introduction of a occupant restraint system significantly changes the sequence of events which he will be submitted to during the collision.

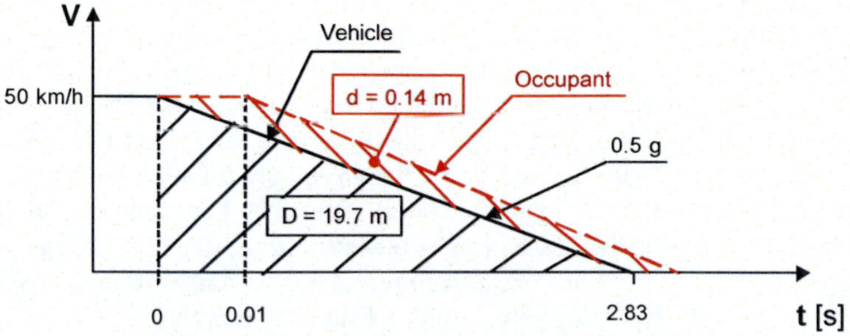

Figure 4.3 – Time – speed diagram in case of braking.

To better examine and explain the differences introduced by the presence of a restraint system, diagrams speed-time can be introduced. In these diagrams the speed with respect to the ground of each object is reported as a function of the time and the differences between the various situations are easily described at first, on these charts. In addition, a number of parameters are identified on these diagrams: the acceleration of each body is directly linked to the slope of the curve, the displacement of each object with respect to the ground is represented by area under its speed-time curve, and the relative displacement between two objects is equal to the area included between their two speed-time curves.

In Figure 4.3 an example of vehicle braking situation is shown: the initial speed is 50 km/h and the braking action is applied instantly at t=0.

The force at the interface tire-ground develops at time 0 (for sake of simplicity we consider instantaneous response to vehicle brake) and its value is such to produce an average deceleration of 0.5 g, which causes the vehicle to stop 2.83 s later. During this deceleration period, the vehicle travels 19.7 m on the ground: in the speed-time diagram, this absolute movement is equal to the area under the curve between 0 and 2.83 s.

If the various forces that act on the vehicle occupant through the seat, deck, etc.. , were applied exactly at the same time, the occupant would reduce his speed in the same way as the vehicle following the same curve in the speed-time diagram with perfect overlay. However, if those forces were applied with a small delay of 0.01 s (corresponding to a more realistic situation), the occupant curve is no more superimposed to the vehicle one and would include an area of 19.84 m (this is the overall displacement of the occupant on the ground, starting from the time 0). The area included between the two speed-time curves (now no more overlapping) amount to 0.14 m, that corresponds to the distance d travelled by the occupant with respect to the passenger compartment, inside the car, during braking.

In Figure 4.4 the situation of the vehicle crash is reported, in this case the occupant is not restrained inside the passenger compartment. The front structure of the vehicle collapses against the rigid barrier, let us suppose for sake of simplicity that deceleration has a constant value of 15.7 g. As a consequence the passenger compartment moves forward on the ground by a quantity equal to this overall deformation, that using the diagram of Figure 4.4, can be evaluated as the area under the deceleration line, equal to 0.625 m.

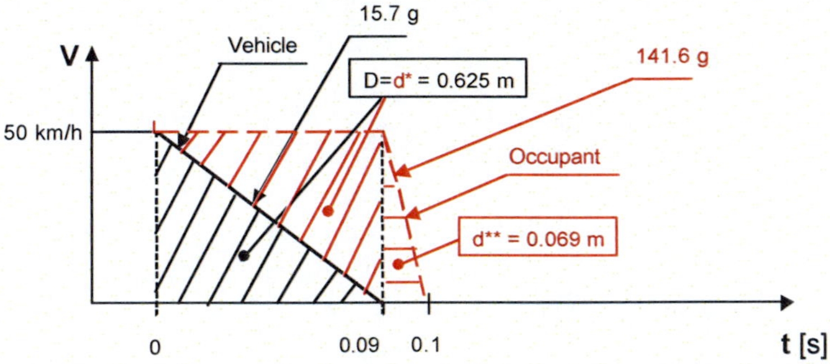

Figure 4.4 - Time – speed diagram in case of front impact.

The speed of the passenger compartment is shown in Figure 4.4 by the continuous line, with a slope of 15.7 g. The occupant that is not restrained and, as shown in Figure 4.4 by the dashed curve, will continue to travel at 50 km/h into the passenger compartment until he hits the steering wheel – dashboard group. Let suppose, for sake of example, that this group had a distance of exactly 0.625 m from the occupant initial position, the occupant hits the group at exactly the same time (0.09 s) at which the vehicle stops against the barrier, having translated a total distance of 1.25 m (0.625 + 0.625) on the ground. If the deformability of his body allows for only other 0.069 m absolute displacement on the ground, the occupant comes to a complete stop 0.01 s later, with an average deceleration of 141.6 g.
Figure 4.5 shows the speed-time curve of the same vehicle in the same crash conditions, in the case of occupant restrained by an appropriate restraint system. Let us assume that this device can not apply a restraint force to the occupant before he moves forward a distance d* = 0.156 m, with respect to passenger compartment (his seat), he will travel at a speed of 50 km/h for 0.045 s.
If the restraint system applies a force sufficient to zero the occupant speed, during the subsequent 0.045 s, he will decelerate at 4.31 g and will come to a complete stop exactly at the same time when the vehicle will be completely stopped against barrier (0.09 s). During this period of restraint, operated by the restraint system, the occupant has moved forward, with respect to the passenger compartment, of an additional distance equal to 0.156 m (the distance d** shown once again in Figure 4.5); therefore his total (or absolute) displacement on the ground will be of 0.937 m (0.625 m of the vehicle deformation plus 0.312 m of displacement inside the passenger compartment).

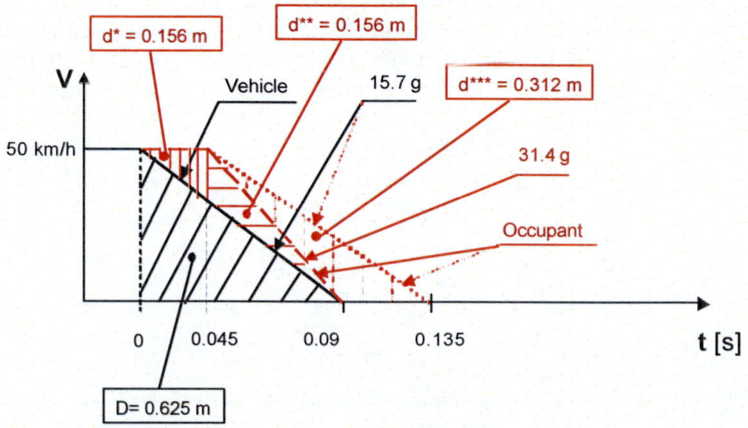

Figure 4.5 - Time – speed diagram in case of front impact with the occupant restrained by an appropriate restraint system

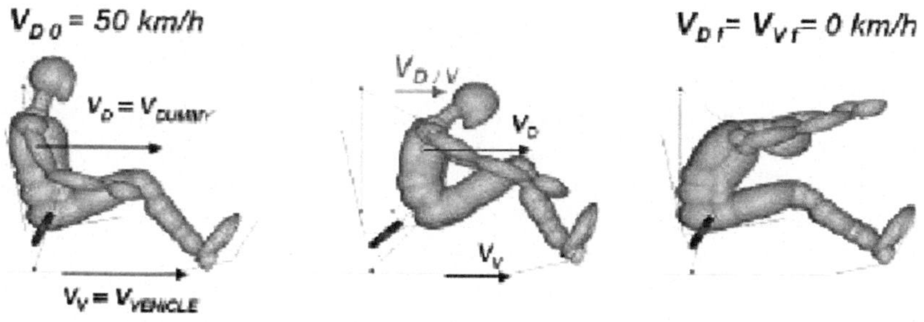

Figure 4.6 - Motion of a vehicle occupant with belt restraint system during front impact.

Assuming again that also in the described case, the steering wheel – dashboard group has initially a distance of 0.625 m in front of the occupant, he has not used to stop a space of 0.312 m that remains still available in passenger compartment. If the restraint system is modified to use the most of this remaining space, the occupant curve becomes the dotted one in Figure 4.5: the deceleration and hence the magnitude of forces applied by the restraint system to the occupant are halved.

In Figure 4.6 the restraining sequence of occupant with seat belt is shown.

The occupant has an initial kinetic energy (EC_{D0}), that must be extracted as much as possible from his body. As the occupant at the end collision is stopped, his final kinetic energy is zero ($EC_{Df} = 0$). The initial kinetic energy of the occupant, extracted from the occupant himself, is then transformed into internal energy of the vehicle system and the restraint devices ($E_{INTERNAL}$). A part is dissipated in the structure of the vehicle (through attachment points of the restraint system) and contributes to its deformation during the collision; the remaining part is dissipated by the deformation of restraint system, as a result of the relative motion of the occupant with respect to the passenger compartment, allowed by the restraint device during the collision.

In conclusion, the deceleration history of the vehicle and of the passengers inside the vehicle are deeply connected. This justifies the already mentioned design requirement that in case of impact the energy absorption has to be progressive as much as possible, in order to avoid very dangerous deceleration peaks. By a carefully considered design of the vehicle structure and of the restraint system the deceleration and the consequent load on the passengers has to be minimised to reduce the risk of injury. In the following section some information on the capability of human being to sustain loads, deformations and accelerations are briefly addressed.

5 – Dummies for Crash Test and Biomechanical Parameters.

During the development of the vehicle design, as well as in approval and rating tests, the evaluation of the vehicle crash performance is performed on the basis of the biomechanic response measured with the standard tests described in section 2 and by means of appropriate anthropomorphic dummies placed inside of the passenger compartment.
These dummies are made of systems of articulated metal masses, springs, dampers and joints and are covered by polymeric materials. They simulate the response of the human body in case of impact and allow properly positioned sensors to measure a number of physical quantities that are correlated with the type of biological damage (injury) that would occur to the passenger in the real conditions of the same impact.
The dummies must have a "bio-fidelity" behaviour for the types of impact in which they are used. They should serve to measure only those parameters that are considered related to biological damage in the considered specific load conditions. As a consequence, for different kind of impact (front, side, rear) dummies of various types should be used.
The various models of dummy differ also for the type of person of which they simulate the biomechanical response. Therefore there are child dummies of different age and adult dummies of different size. The definition of correlations between physical quantities measurable on the dummies and the corresponding levels of biological damage on the human body and the development of the dummies with bio-fidelity features, were based and are still based on studies conducted on corpses (or Post Mortem Human Subjects, PMHS) in a few research centers specialized in these types of tests (e.g. the Wayne State University in USA, INRETS in France, Heidelberg in Germany). In general, any type of dummy must fulfil, beside the bio-fidelity of its biomechanics response, the following requirements: durability, reparability by easy interchangeability of its parts, high repeatability (similar results using the same dummy in numerous repetitions of same test), high reproducibility (small variability of the results obtained with different dummies used in several repetitions of the same test).
In the following, three dummies will be briefly described. Two of them are for front crash test and one is use in Europe for side crash test. Further the main biomechanical injury parameters associated with these dummies will be discussed.

5.1- Dummy for frontal crash tests.

The Hybrid III 50th percentile dummy is the most widespread dummy for frontal crash test. It reproduces the size and the inertial and dynamic stiffness characteristics of an average American adult male (height 1.75 m, weight 78 kg). The development of this particular dummy is due to the research work conducted

by General Motors in the 1970s. Between 1971 and 1976 four generations of dummies for different crash tests were designed, culminating precisely with the Hybrid III dummy (that was preceded by Hybrid I, II and Hybrid ATD 502).
A proper anthropometry of the dummy is an important feature to be considered. Everything else being equal, results of crash tests are more realistic if the dummy has shapes, sizes and weights typical of a man. In particular, the seating position, the shapes, the location of the centers of the joints, the centers of gravity and

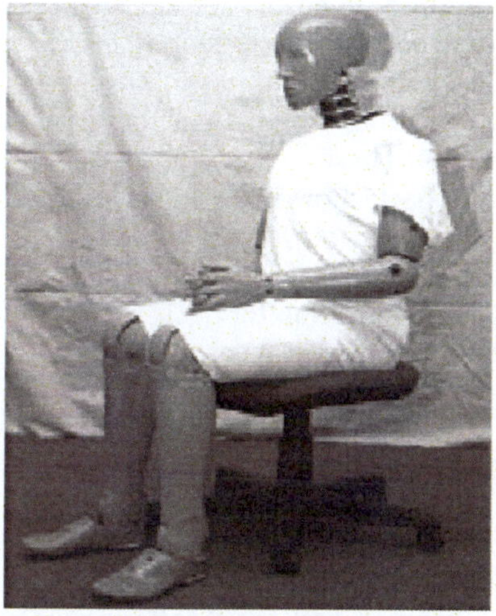

PART	WEIGHT [kg]
Head	4.54
Neck	1.54
Upper Torso	17.19
Lower Torso	23.04
Upper Arms	3.99
Lower Arms and Hands	4.54
Upper Legs	11.97
Lower Legs and Feet	11.34
Total Weight	78.15

DIMENSION DESCRIPTION	SPECIFICATION [cm]
Head Circumference	59.89
Head Breadth	15.49
Head Depth	20.32
Erect Sitting Height	88.39
Shoulder to Elbow Length	33.78
Back of Elbow to Wrist Pivot Length	29.72
Buttock to Knee Length	59.18
Knee Pivot Height	49.53

Figure 5.1 – Hybrid III dummy for front crash test; part weight and dimensions.

weights of the various segments constituting the dummy represent the main factors that determine the kinematics of the dummy during a crash test. The Hybrid III dummy has been designed to take naturally the typical posture of a man sitting inside the car (the so-called "automotive seated position").
Figure 5.1 gives the weight, the sitting height and the stature of the Hybrid III 50th percentile dummy, together with the weights and dimensions of most of the parts. A brief description of the characteristics of the most relevant part of the dummy is provided in the following.
The **head** (see Figure 5.2a) consists of an aluminium skull made using a net shape cast process. A thick skin is applied on this metallic skull, to ensure the biomechanic fidelity and repeatability of the response in the impact of the head against hard surfaces. Three orthogonal accelerometers are mounted at the center of gravity of the head to measure accelerations.
The **neck** is a flexible component that should have representative stiffness and damping characteristics both in flexion and in extension. It is made of three rigid

aluminium discs (that are the vertebrae) spaced with elastomeric elements (Butyl). The butyl was chosen for its large dumping properties, that allow to simulate with good approximation the biomechanics hysteresis of the neck. A single steel cable runs along the center line of the neck to ensure a great value of axial strength (see Figure 5.2b). The neck is manufactured by injecting the elastomer in a special mould, in which the three vertebrae and the two aluminium disks of extremity (bearing the links) are appropriately located: so, due to the adhesion developed by the elastomer, a single part is obtained. The cross-section of the neck is not symmetrical, in order to ensure greater stiffness for the flexion (forward head movement) rather than for the extension (backward head movement) as required by biomechanical data. Appropriate transducers measure shear loads, axial loads and bending moments at the upper link to the head (occipital condylus) and at the lower link to the chest.

The **chest** of the Hybrid III dummy consists of a back spine bone and a rib cage covered with a removable protective jacket. Each part has added-on weight to obtain the correct overall weight and the correct position of center of gravity. The chest spine is of welded steel and serves as a connection for neck, clavicles, ribs and lumbar spine. A triaxial accelerometer is mounted at its center of gravity (see Figure 5.2c). The thoracic cage consists of six ribs made with 1074 steel. The ribs are linked at the back to the spine by means of leaf springs; they are made with a shape that imitates the shape of the human ribs and have a damping material applied at their inner surface, in order to ensure the dynamic response of the cage characterized by energy absorption during the front cash. A Urethane bib is attacked on the rib external surface to distribute more uniformly the load (see Figure 5.2c). The deformation transducer (that measures the chest crush) is a rotation potentiometer placed at the end of a fork that extends its arms beyond the lumbar spine, while a rod is placed between the transducer and the sternum in order to provide the input signal to the potentiometer. This chest model allows for a maximum deflection of 90 mm.

The **lumbar spine** is a curved element of elastomer to which extremity plates are applied in order to anchor the chest spine and the pelvis (Figure 5.2d). Two cables pass through the lumbar spine, they are linked to the final plates and allow for lateral stability in seating while permit a suitable longitudinal flexibility. The lumbar curved spine allows the dummy to take a position not erected when placed on the seat of the vehicle or to get a posture more similar to a human being, thus obtaining a greater repeatability of the initial position (as the dummy is lying inside the vehicle and not forced into a position).

The **lower trunk and legs** of the Hybrid III dummy are shown in Figure 5-3e. They consist of two legs connected to the pelvis by a couple of hip joints. Each leg has one metallic beam of circular section, instrumented with two load cells, that represents the femur, one hinge that represents the knee, another metallic beam, again instrumented with two load cells, that represents the tibia and finally the ankle-foot system that is connected to the tibia by a hinge. The metallic parts are covered by polymeric parts that give the typical dimensions and elastic consistency and stiffness of the human parts where the bones are covered by muscles and skin.

The **knee** of the Hybrid III dummy (see Figure 5.2e) is made of a moulded aluminium part, an insert that simulates the knee padding and a vinyl coating. The rubber insert is intended to improve the dynamic response of the knee

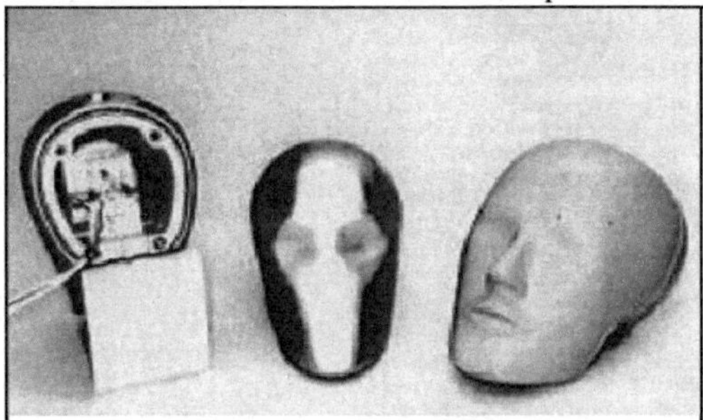

Figure 5.2a – Hybrid III head

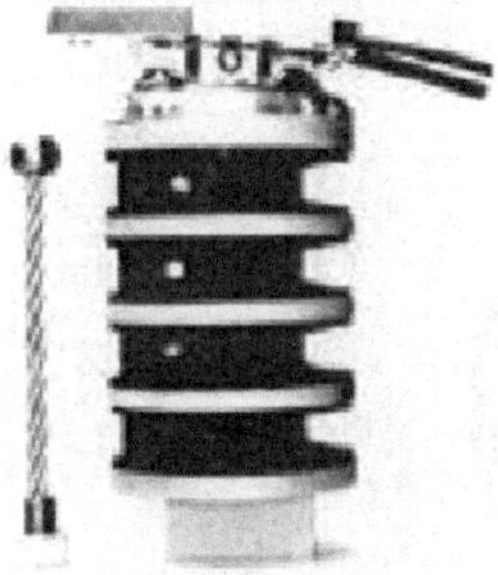

Figure 5.2b – Hybrid III neck

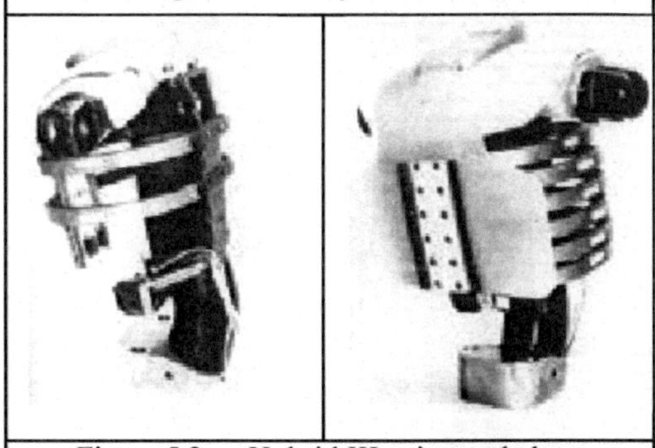

Figure 5.2c – Hybrid III spine and chest

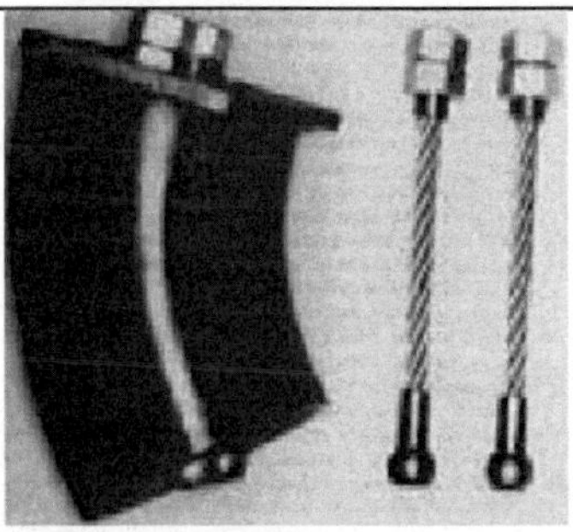

Figure 5.2d – Hybrid III lumbar spine

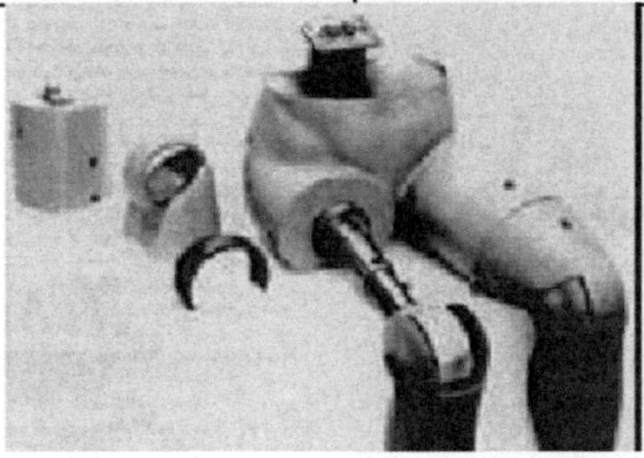

Figure 5.2e – Hybrid III lower trunk and legs

impacts against rigid surfaces.
The **foot-ankle** system of the Hybrid III dummy is not instrumented. The ankle is made of a spherical joints with angle limitator that allows for foot maximum excursions, in the various directions, similar to those of the human body. Therefore the Hybrid III dummy does not allow to investigate the damage types to the lower part of the legs in full.

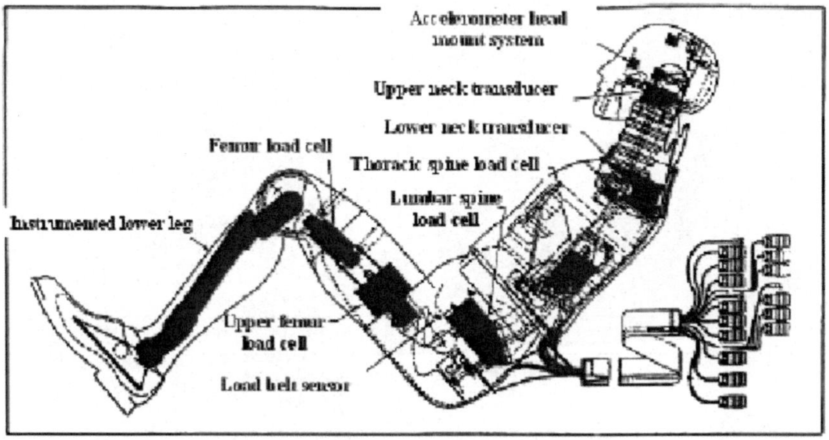

Figure 5.3 – Hybrid III transducer map.

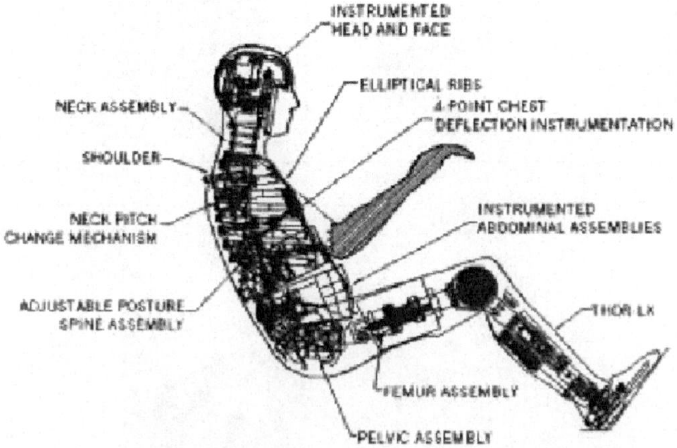

Figure 5.4 – THOR features.

Figure 5.3 summarizes the quantities measurable with the Hybrid III dummy and shows the location and nature of the instrumentation.

Figures 5.4 and 5.5 summarize the features of the new **THOR dummy** (Test device for Human Occupant Restraint), that has been recently developed as part of the NHTSA research program (begun in 1995). The THOR dummy is a extensive development of the Hybrid III dummy (that only keeps unchanged the arms) in terms of bio-fidelity, dynamic response and instrumentation.

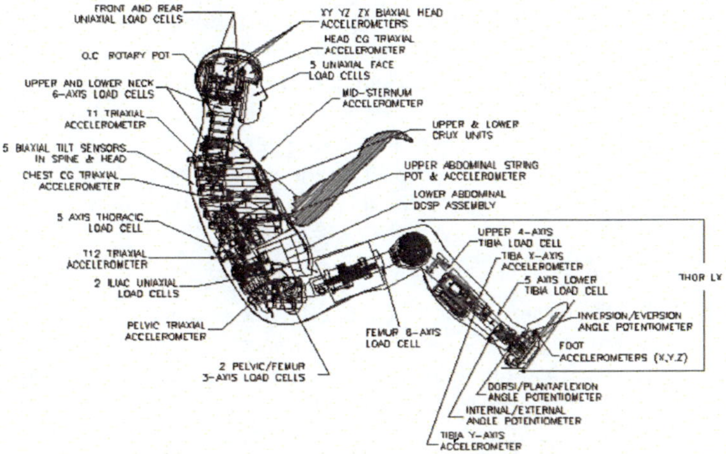

Figure 5.5 – THOR transducer map.

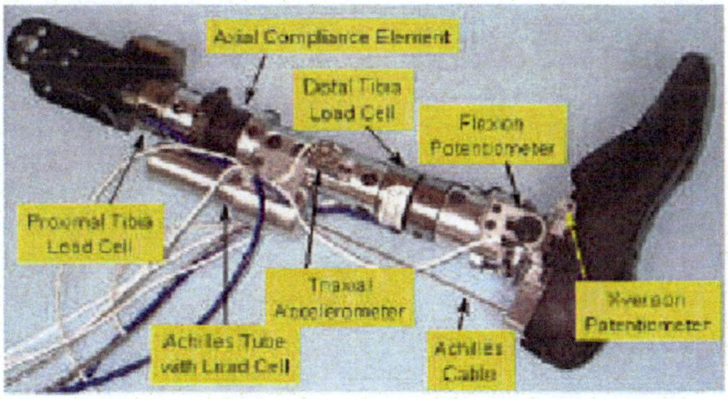

Figure 5.6 – THOR lower leg

The new dummy incorporates a completely new lower leg; a facial region instrumented with unidirectional load cells to verify the risk of fracture; a new neck with multidirectional bio- fidelity; a new adjustable spine; a new chest with elliptic ribs and a new crushing sensor, which detects the three-dimensional compression of the chest in four separate points; a new abdomen that in its lower part measures directly the three-dimensional intrusion of seat belts in two separate places, while in its high part measures the compression produced by the inflated airbag; an instrumented pelvis with a triaxial load cells at each of the two femur heads and load sensors for the belt pressure on the iliac crests. This dummy is quite complicated and is still the subject of studies. Therefore, at the moment, it is not required by law and / or rating tests, that continue to rely on the Hybrid III dummy.

One of the main change with respect to the Hybrid III dummy is a new lower leg (THOR-LX, new Lower eXtremity). This relevant change is motivated by the fact that this type of injury is quite frequent and, although not fatal, produces high costs to society due the disability associated with it. This leg is considerably more sophisticated (with a system that simulates the Achilles tendon) and enriched in instrumentation with respect to the previous one (see Figure 5.6). This leg can also be mounted as a "retrofit" on the Hybrid III dummy.

5.2 - EuroSID-1 Dummy

The EuroSID-1 is the dummy used in Europe for the evaluation of biomechanic response of the passenger in case of side impact. The dummy is designed according to the specifications laid down in Annex 6 (Technical description of side impact dummy) of the ECE Regulation 95. The EuroSid-1 represents, once again, a 50th percentile adult male, without the lower part of his arms.

The **head** is that of the Hybrid III dummy, while the legs are those of the Hybrid II dummy.

The **chest** consists of three identical ribs that can be mounted on the spine through three damper elements, indifferently on the left or on the right, in order to be able to perform crash tests in both the impact configurations (i.e. left-hand drive and right-hand drive versions) that may be of interest for the vehicle.

The **abdomen** is made of a metal moulded part covered with a polyurethane foam that includes some rubber parts, with well defined weight and curvature. The shape of the **pelvis** is representative of the corresponding human bone, with particular attention to the points that can be impacted during the side impact accident and to the points through which the interaction with the seat and the safety belt of the vehicle occurs. The two iliac crests are removable and are made with a special polyurethane plastic. The lumbar spine is straight. A jacket made of elastomeric material covers the chest, the two upper portions of the arms and the lower pelvis region.

The EuroSID-1 gives as output the quantities that are listed in Figure 5.7a; its main characteristics in terms of size and weights are given in Figure 5.7b.

In 1998 the American NHTSA took into consideration the possibility of incorporating the EuroSID-1 in FMVSS Std. 214, in order to replace the US-SID. For this purpose, many experimental tests were conducted with EuroSID-1, which pointed out significant mechanical problems. One of these problems, most likely linked to hysteretic phenomena due to friction effect in the rib-dumper system, leads to a dynamic force-compression curve that is characterised by a constant value of the chest compression for a non negligible period of time while the impact force decreases from its peak (this phenomenon is called "flat-top").

PART	WEIGHT [kg]
Head	3.6 - 4.4
Neck	0.9 – 1.1
Thorax	20.9 – 23.9
Arm	1.2 – 1.4
Abdomen	4.5 – 5.5
Pelvis	11.0 – 13.0
Leg	11.5 – 13.5
Total Weight	71.5 – 72.5

DIMENSION DESCRIPTION	SPECIFICATION [cm]
Sitting height	89.7 – 91.1
Seat to shoulder joint	55.2 – 56.2
Seat to H-point	9.6 – 10.0
Seat to arm	23.7 – 24.7
Buttock to knee lenght	60.5 – 61.5
Shoulder / Arm width	47.7 – 48.7
Pelvis width	35.0 – 36.0
Sole to seat, sitting	45.1 – 46.1

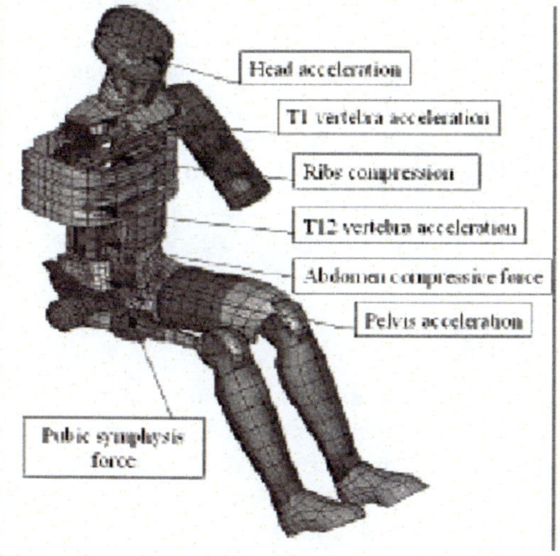

Figure 5.7 – EuroSID dummy for side crash test – (a) part weights and dimensions, (b) quantities measured by equipment sensors.

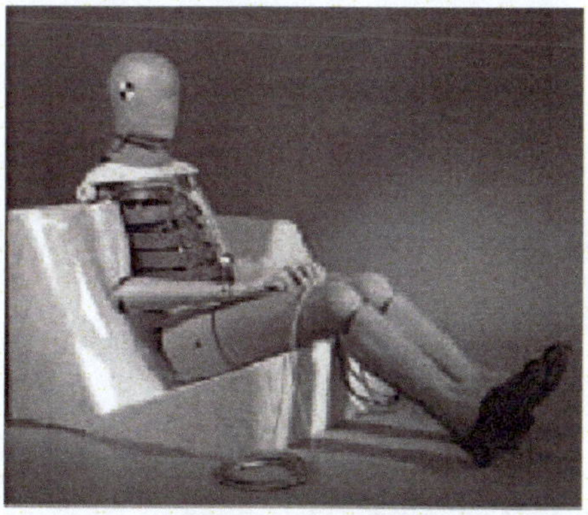

Figure 5.8 – WorldSID dummy prototype

After the development of a new test procedure that, thanks to a suitable pendulum, allow for impacts on the dummy with different angles in the horizontal plane and for hit contacts (in a manner similar to what happens inside of the car) on the dummy from the abdomen to the shoulder, it was found that the mentioned "flat-top" phenomenon increases passing from lateral angled impacts from the front (with respect to the chest) to similar angled impact from behind (range +/-20 degrees). Therefore design modifications have been developed to eliminate the "flat-top" phenomenon. This led to the design of the EuroSID-2 dummy: changes affect the shoulders, the "back-plate" (a fixture plate located in the back, that has been reduced in size), the pelvis and the rib-dumper system (where mechanisms have been implemented with friction reduction during the rib compression movement). The EuroSID-2 dummy is the descendant of the EuroSID-1 dummy for tests in legislation and / or rating, waiting for the WorldSID dummy. The WorldSID dummy is under development through a program for research and development of a single dummy that could replace, at the level of world standard, the various versions of "side impact dummy" used today. Figure 5.8 shows a prototype of this WorldSID dummy.

5.3 - Injury criteria

The physical parameters measured with the instrumented dummies are properly interpreted on the basis of the injury criteria that give the correlation between the output (acceleration, force, deformation, etc.) and the the corresponding damage on the human body, for the body segment of interest. In general, an injury criterion specifies both an appropriate signal processing that take into account the time evolution of the physical quantity under consideration, and the threshold limits or the range within which the result of this process must fall, in order to assess the injury amount. Injury criteria are usually the result of the experimental tests conducted on corpses. In the following sub-subsections a brief description of the most important injury criteria is given.

5.3.1 - Head Injury Criterion (HIC)

This index (also known as Head Performance Criterion, or HPC) is a criterion of the head injury based on the proper "reading" of the resultant acceleration history measured in the head center of gravity (Versace 1971, Clifford et al. 1974). In the formula of the HIC

$$HIC = \max\left\{\left[\frac{1}{t_2 - t_1}\cdot\int_{t_1}^{t_2} a(t)dt\right]^{2.5}\cdot\left(t_2 - t_1\right)\right\} \qquad (5.1)$$

a(t) is the time history of the resultant acceleration, expressed in g, while t_1 and t_2 ($t_2 > t_1$) are two instants that define the amplitude of a time window moving along the time interval covered impulse acceleration detected on the head. Currently mobile time windows with amplitudes t_2-t_1 $\leq$ 36 milliseconds are considered, as it has been noted that the maximum value of this indicator is obtained in such range of amplitudes (associated with typical durations of impact phenomena of the head). A practical interpretation of the 36 ms limit can be obtained by noting that if the acceleration is assumed constant over the time interval t_2-t_1 (that is actually not very realistic, but is a useful simplification), a HIC value of 1000 in the interval of 36 ms corresponds to a mean acceleration of 60 g. On the a(t) curve a moving time window (along the axis of time) with variable amplitude (up to 36 ms) is applied, obtaining for each window amplitude a corresponding value of HIC: the maximum of the obtained values is the actual HIC, that has to be compared with the limit of 1000. This value is a

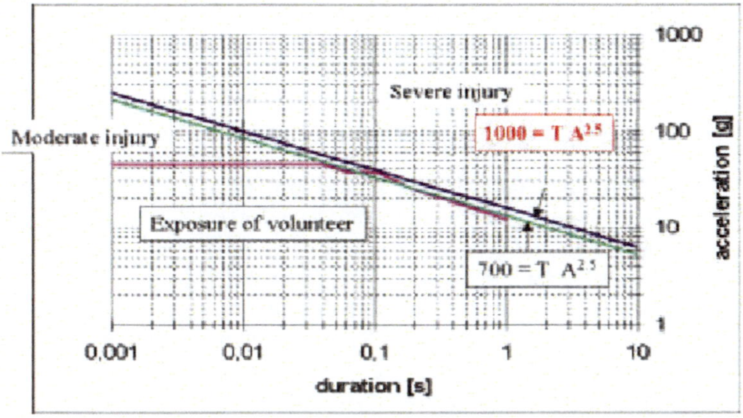

Figure 5.9 HIC – Limit values of the head linear acceleration, constant acceleration case.

boundary limit for a particular level of head injury (linear fracture of the skull) and originates from studies begun in the 1930's at Wayne State University in Detroit on the response of the skull (presence or absence of fractures as linear indicator damage up to risk of death) when impacted against a rigid wall (high acceleration levels for short durations). It should be noted that in recent years NHTSA (Eppinger et al 1999) has proposed a different limit value of 700 calculated in a interval reduced to 15 ms. It makes difference in case of constant acceleration history, as shown in Figure 5.9, but in typical acceleration history there is not a large difference.

The initial results were gradually up-graded with data from tests conducted in the'50s on animals and volunteers (lower levels acceleration for higher durations). The limit value of 1000 for the HIC comes from the equation of the line that, in a logarithmic diagram whose axis are the duration in seconds and the acceleration in "g", gives the best-fit (proposed by Gadd in 1966) of the edge between the area of absence and that of presence of linear skull fracture, obtained from the experimental test database (see Figure 5.9). One must always bear in mind that the HIC index, by its genesis, has a meaning only when there is an impact of the head against something, for example against some solid part inside the passenger compartment or an airbag. If the head does not hit a solid part, there is neither risk of skull fracture nor risk of other brain injuries, in these cases not only the HIC index has very little value but the application of equation 5.1 is questionable.

5.3.2 - Maximum Head Acceleration

The acceleration resultant value of the head, during its forward motion during the collision, should not exceed 80 g for a total (cumulative) duration of 3 milliseconds (criterion of the three milliseconds). The cumulative calculation of the time for which the acceleration curve exceeds the prescribed limit does not apply to the section corresponding to the bounce movement of the head.

5.3.3 - Injury criteria for the neck (NIC & Nij)

In case of collision, the Neck Injury Criterion (NIC) is defined in the EC Directive 96/79. The axial force (F_z) and the shear force (F_y) must comply with the limit curve of Figure 5.10 (a) and (b), while the bending moment (M_y) must not exceed the value of 67 Nm, in extension (Mertz et al. 1971).
Moreover, the limit curves are cumulative, this means that, for each level of force, they give the maximum time for which they may be applied, in order of decreasing size. Therefore the curve provided by the neck load cell is elaborated, in order to be in the conditions laid down by limit curve, as follows: moving along the ordinate axis, toward the decreasing loads, for each load it can be seen for how long this is cumulatively exceeded, so the coordinates from the cumulative curve to be reported on the diagram of the limit curve are obtained.

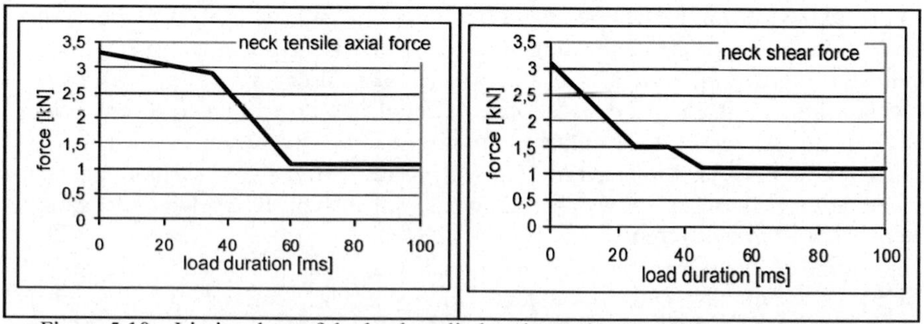

Figure 5.10 – Limit values of the load applied to the neck: a) axial force, b) shear force

The American Std. 208 specify the peak values that the axial force should never exceed (F_z <4170 N in tension and <4000 N compression). NHTSA (Eppinger et al. 1999) proposed the use of the N_j criterion. The motivation is that the capability of the neck to sustain an axial force and a bending moment at the same time cannot assessed as if the two loads are independent one from the other but a weighted mixture of the two loads is more appropriate. The definition of N_j index is

$$N_{ij} = (F_z / F_{zc}) + (M_{ocy} / M_{yc}) \quad (5.2)$$

For both the axial force and the flexure moment NHTSA recommended values that change according to the loading direction, the size of the person and his seating position (in-position or out of position) (Eppinger et al. 1999). In Figure 5.11 are reported the limit value for a mid sized male, seating correctly in position. Formula 5.2 states the relationship between the measured values and the corresponding values that are critical for the human body. During the collision, axial force F_z can be both in tension and compression, while condylus occipital moment (M_{ocy}) can be both in flexion (when the head rotates forward) and extension (when the head rotates backward).

This brings to four possible load conditions described with the N_j: tension-extension (N_{te}), tension-flexion (N_{tf}), compression-extension (N_{ce}) or compression-flexion (N_{cf}). At each time only one of the four loading conditions occurs and the corresponding value of N_{ij} is calculated, while the remaining three loading modes give zero. At any time during the collision, none of the four "N_{ij}" must exceed the value of 1.0, see Figure 5.11.

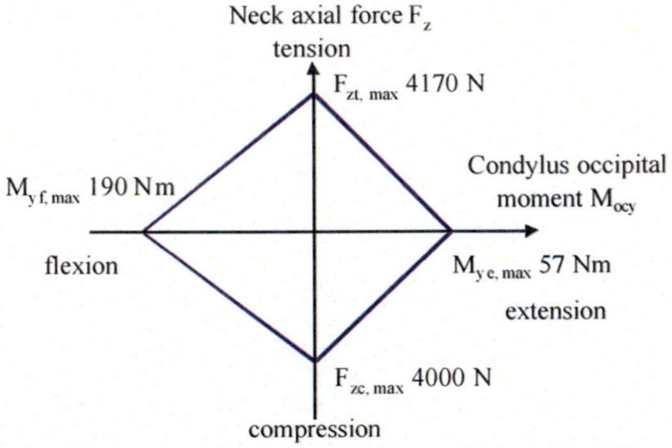

Figure 5.11 – Limit values of the load applied to the neck according to the proposed N_{ij} criteria

5.3.4 - Maximum Chest Acceleration

The Std. 208 prescribes a maximum chest acceleration of 60 g with the criterion of the three milliseconds. This limit comes not only from tests conducted on corpses but also from tests on volunteers (stunt-drivers, soldiers) and the data on accidents from free falls. The recommended limit of 60 g, the acceleration of gravity of the chest, is primarily the result of the research developed by Mertz and Patrick that has been proposed in 1971.

5.3.5 - Chest Compression Criterion

Another important parameter for possible thoracic injuries is the value of its compression in the sagittal plane (vertical plane passing through the spine and sternum, dividing the human body), that is the reduction of the distance between sternum and spine during the impact and that is primarily due to the restraint operated by the safety belts. The European standard for frontal impact (CE/96/79) prescribes a maximum compression of 50 mm, while the American standard (Std. 208) prescribes a maximum compression of 76 mm. For the lateral compression (or rib compression) the European legislation (CE/96/27) prescribes a maximum value of 42 mm.

5.3.6 - Viscous Criterion (VC)

The viscous criterion refers to injuries of the soft tissue chest (heart, lungs and blood vessels), that, as shown by studies conducted on corpses, beside to be produced by high chest compression, are also caused by high speed of impact with very small deformations of the chest (as it happens, for example, in the impact of a bullet against a safety jacket). An effective risk index of thoracic injury is therefore given by the maximum value, in function of time, of the product of instantaneous deformation speed by the instantaneous chest compression. Since the compression of the chest may occur in the two main modes of impact, front and side, actually there are two VC indexes, whose definitions are

$$VC = \max\left(\frac{D(t)}{0.140} \cdot \frac{dD(t)}{dt}\right) \quad \text{for the EuroSID dummy} \tag{5.3}$$

$$VC = \max\left(1.3 \cdot \frac{D(t)}{0.229} \cdot \frac{dD(t)}{dt}\right) \quad \text{for the Hybrid III dummy} \tag{5.4}$$

where D(t) is the rib relative displacement for the side impact and the sternum relative displacement for the front impact. In these formulas displacement D is normalised with the chest depth: 140 mm for the EuroSID dummy and 229 mm for the Hybrid III dummy (Lau and Viano 1981, 1986). In the case of frontal impact, for example, experimental studies on animals and corpses showed that an impact resulting in a VC value equal to 1.3 m/s has 50% probability to cause a serious chest injury. For a VC value equal to 1 m/s the injury probability drops to 25% and this is the VC value chosen as the limit by the European law 96/79 EC.

5.3.7 - Femur injury criterion.

The injury criterion for the femur consists essentially in a maximum value of the compression force not to be exceeded during the impact. It is based on the test results obtained from corpses aimed to determine the level of force corresponding to the fracture of the femur. Over time, there have been variations on this limit that initially was 5.43 kN (1200 lb) and has suffered a series of increments until reaching the current value 10.19 kN (2250 lb) of the Std. 208.

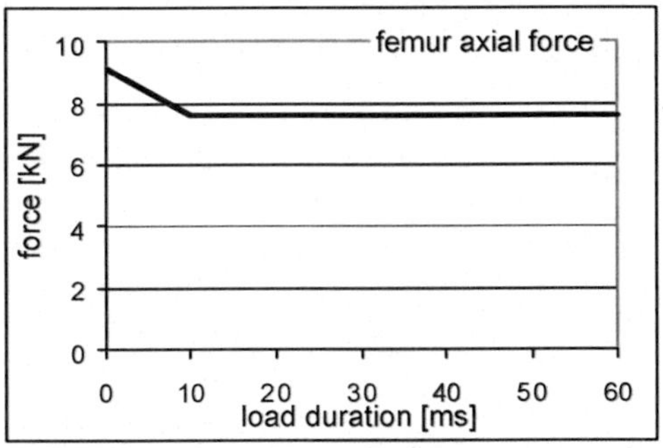

Figure 5.12 – Limit values of the axial load applied to the femur

This maximum force aims to establish a failure locus for the femur fracture, although generally for this value of the force fractures of the knee, femur and femur-pelvis hip joint occur jointly. While no one can say with certainty which is the load on the femur that is causing the fracture of each one of the mentioned parts, we assume the same value as the overall index. Indeed, if loads were measured on each part, each fracture load results to be different from the other (in the evolution of crash dummies it is now taken into consideration to measure loads in other points besides the femur, as in the case of THOR already mentioned, where load cells are placed also at the femur-pelvis hip joint). In European legislation, the injury criterion on the femur is provided in terms of cumulative force-time limit curve (see Figure 5.12).

5.3.8 - Tibia Index (TI)

The TI tibia injury criterion (defined by Mertz in 1993) is given by

$$TI = \left|\frac{M_r}{M_{Cr}}\right| + \left|\frac{F_z}{F_{Cz}}\right| \tag{5.5}$$

where M_r and M_{Cr} are the resulting bending moment and its limit value (equal to 225 Nm), F_z and F_{Cz} are the axial compression force and its limit value (equal to 35.9 kN).

This index is calculated at the top and the base of each tibia; however, the axial compression force can be measured (Hybrid III dummy) only in one of the two points (at the bottom where the load cell is placed) and this value is then used to calculate the tibia index both at the top and at the base of the tibia. The two bending moments M_x and M_y are measured separately in the two points. According to the European legislation, the EC 96/79, the Tibia Index should not exceed the value of 1.3 in either extremity.

6 – Composite Structures for Vehicle Applications.

To design the vehicle structure having in mind the potentials but also the peculiar mechanical behaviour of composite materials can result in innovative solutions. In order to take full advantage of all these opportunities, it is necessary to abandon the "substitution policy" with respect to the equivalent parts made of steel and to "think composite" from the very beginning of the design phase.
In the following subsections four applications of composite materials for vehicle structures are presented. In particular the front longitudinal beams of an all composite body for a passenger car (subsection 6.1), the rear impact absorbing structure for a Formula One racing car (subsection 6.2), the frontal sacrificial structure for a Formula One racing car (subsection 6.3) and the bonnet of a passenger car, optimised for pedestrian safety (subsection 6.4).

6.1 - Vehicle front beams.

This subsection summarises results from an experimental study performed on parts that are not simple specimens but are true scale components of the front structure (as shown in Figure 6.1) of a composite material vehicle designed with colleagues of the FIAT Research Center (CRF) within the framework of a national research program devoted to evaluate the suitability of such a type of component for the vehicle crash performance (Belingardi et al. 1993). Figure 6.1 shows the base geometry of the tested part that has been designed to have a driven destroying mechanism. It is a thin-walled beam composed of two prismatic square section tubes connected by a flat vertical plate. The two tubes are disposed with their axis converging toward the front of the vehicle so that there is a progressively increasing resisting cross-section starting from the first impact point and moving backward to the passenger compartment.
This tapered shape has been chosen to obtain also a proper value of the bending stiffness and strength: it should be taken into account that, in this application, this beam is loaded in bending and torsion both by the engine mounting reactive loads and by the vehicle suspension loads.
The parts were manufactured starting from a three-layer glass-fiber fabric, injecting the epoxy resin into the closed mould and, finally, heating at the curing temperature, according to the well known RTM (Resin Transfer Moulding)

technology. The fiber weight ratio was about 70%. The stacking sequence has the cross ply outer and inner layers with the fibers oriented at ± 45° with respect to the beam axis, while the internal layer has a unidirectional fiber reinforcement with the fibers oriented in the beam axis direction.

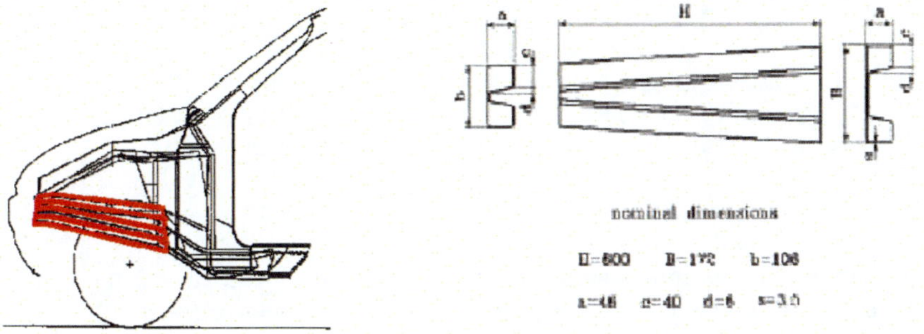

Fig. 6.1 – Composite front beam in the vehicle front structure; dimensions of the front beam.

The design of the beam has been completed with a trigger placed at the front extremity of the beam. This trigger consists in a very simple tapering (progressive reduction) of the wall thickness in order to reduce locally the resisting section. This trigger is intended both to reduce the value of the initial force peak and to initialise the beam collapse in a stable way.

Quasi-static axial loading tests were performed using a standard material testing machine. Quasi static tests offer the advantage of easy observability of the development of crushing collapse. The results are quite similar to those obtained with a dynamic impact test if

a – the material mechanical characteristics are not or, at least, are little influenced by the loading rate; in this perspective it is important to remind that the strain rate sensibility of this type of composite material is very little, and

b –inertia effects and stress wave propagation effects can be neglected, as they do not affect the global behaviour of the beam; in this perspective it is important to remind that the standard test are made at an impact speed of about 60 km/h and this testing condition causes neither relevant inertia effects nor noticeable stress wave propagation effects.

The beams to be tested have been placed between two rigid steel plates and the load is applied by means of the downward motion of the cross head. The test was conducted in the displacement control mode, and the cross head velocity set at a constant 15 mm/min. Signals from the load cell and from the cross head

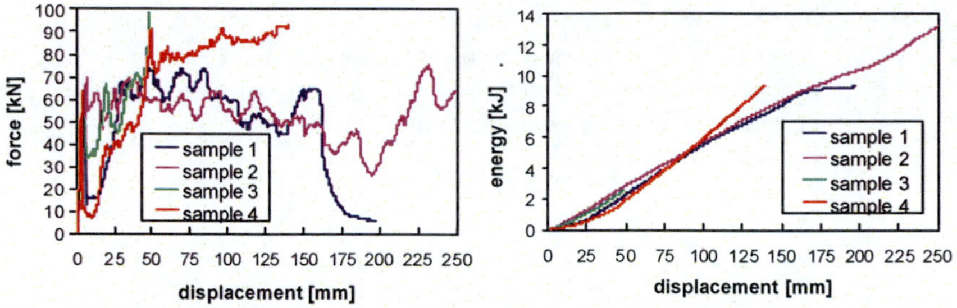

Figure 6.2 – Experimental test results: force – displacement curves on the left, absorbed energy – displacement curves on the right (Belingardi et al. 1993).

displacement sensor are acquired by means of a two-channels dynamic data acquisition board and stored in the computer memory.

Test on sample 3 was interrupted after 50 mm of cross head stroke due to the protection device of the testing machine: the crush load tried to exceed 100 kN.

Figure 6.2 shows (on the left) the curves of the force versus the displacement. These curves are quite irregular, due to the composite fracture mechanism together with friction of the fronds that slide over the plates (see also Hull (1991) and Farley (1992)), but at the same time the range of variation of the force values is quite narrow. The force at the collapse initialisation has been reduced approximately to the mean force value due to the proper design of the crush trigger. Without the application of the collapse initialisation trigger the initial peak value of the crushing force is larger than 100 kN. Such a large value of the collapse load could cause a completely different behaviour of the vehicle structure as the energy absorption could take place in a different part of the structure, even affecting the passenger compartment.

In the present situation, due to the proper effect of the trigger, the mean load is only a little lower than the maximum load and the load uniformity parameter (that is defined as the ratio of these two values $LU= L_{max}/L_{mean}$) is rather close to one (one is the value that characterize the ideal energy absorber). A narrow range of variation of the force opposed by the beam during the crush means both a quite stable, progressive collapse mechanism and a resulting nearly constant deceleration for the vehicle.

Figure 6.2 shows also (on the right) the history of the absorbed energy versus the displacement, obtained by simple integration of the previous force-displacement history. The linear trend of these curves confirms the progressive fracture of the composite material.

The mean stress value, obtained dividing the mean load by the beam section area, has also the meaning of the total energy absorbed per unit of material volume. It is of interest to note that the mean stress value is, in this case, about 43 MPa. This is a rather high value, higher than the values of the absorbed energy per unit volume reported in table 3.1. It compares well with values usually exhibited by metallic structures.

Pictures in Figure 6.3 shows the fracture mechanism of the beam at five successive stages of the collapse. The progressive formation of fronds is clearly visible.

Similar behaviours have been described by Mamalis and co-workers in their papers published in 1991, 1996 and 2003 and in the book published in 1998.

Figure 6.3 – Progressive fracture of the beam in a sequence of pictures.

6.2 – Rear impact absorbing structure for a Formula 1 racing car.

The 2005 FIA regulations for the Formula 1 racing cars prescribes that "an impact absorbing structure must be fitted behind the gearbox symmetrically about the car center line". "This structure must pass an impact test and be constructed from materials which will not be substantially affected by the temperatures it is likely to be subjected during the use". The impact test has to be performed according to the following prescriptions: "The structure and the gearbox must be solidly fixed to the ground and a solid object, having a mass of 560 kg and travelling at a velocity of 10 m/s (36 km/h), will be projected into it. The object used for this test must be flat." "The resistance of the tested structure must be such that during the impact:

1 – average deceleration of the object does not exceed 25 g;

2 – maximum deceleration does not exceed 60 g for more than a cumulative 3 ms in the direction of the impact;

3 – all the structural damage must be contained within the area behind the rear wheel centreline."

Results reported in the literature by Savage et al. (2006) and by Bisagni et al. (2005) indicate that hollow truncated cones can be an interesting solution to meet the above regulation requirements, provided that the wall thickness, the cross section dimensions and the semi-apical angle are chosen appropriately to obtain a stable progressive collapse and avoid catastrophic failure.

In the design reported by Feraboli et al. (2007) composite materials with either commercial grade or high modulus fibres and with commercial grade epoxy resins have been considered.

The selection of the epoxy resin has taken into account both the requirement of a tough resin, in order to optimise the energy absorption capability, and the requirement of quite high operating temperature, due to the proximity to the engine. Therefore a rubber modified epoxy has been discarded in favour of a more brittle, higher temperature epoxy matrix.

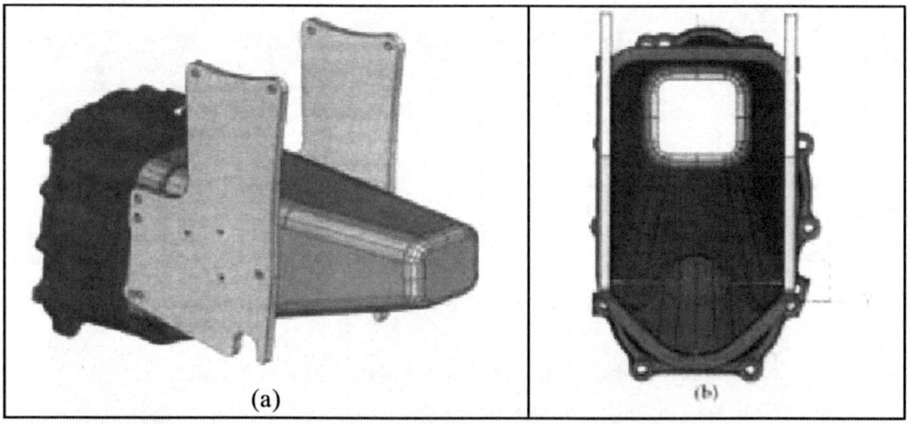

Figure 6.4 - Views of CAD model of the RIMP structure. Feraboli et al. (2007).

In the test condition, the kinetic energy at impact is $E_c = m\ V^2 / 2 = 28$ kJ, this energy has to be absorbed by the hollow truncated cone structure.

The average impact force can be calculated on the basis of the maximum allowable deceleration specified by the regulation (25 g), thus the mean crash force results $F = m\ a = 140$ kN.

Finally the crush displacement of the hollow truncated cone can be evaluated as $D = E_c / F = 0.200$ m. If more conservative value of the deceleration are considered, say 20 g or 15 g, the crush displacement increases to, respectively 0.250 and 0.333 m.

The final design of the rear impact absorbing structure (RIMP) has been defined by Feraboli and co-workers (2007) as a truncated pyramid having the minor base nearly parallel to the major base. The minor base has a square section of 81.3 mm side, while the major base is roughly rectangular in shape and has dimensions of 130.8 x 209.6 mm. The structure is visible in Figure 6.4.
Two different designs of the composite material laminate have been considered by Feraboli and co-workers (2007), both fabric and tape prepregs are used. The first design is based on a quasi-isotropic lay-up (it has been referred as FQI lay-up) with axial, angle and hoop plies, the second design has a greater content of angle plies but no fibres in the hoop direction (it has been referred as K0/45 lay-up). This latter laminate results to have a easier manufacturability thanks to its simpler staking sequence.
The FQI cone is subdivided in three segments with a progressively increasing thickness of the laminate: in the first segment, 76.2 mm long, near the minor base of the truncated pyramid, the laminate thickness is 1.5 mm, in the intermediate segment, 165,1 mm long, the laminate thickness is 2.3 mm, and, finally, in the third segment, 101.6 mm long, near the other extremity, the laminate thickness is 3.0 mm. In such a way that the laminate becomes progressively stronger along the pyramid axis. The lay-up of the first segment is $[(0/90)/90/(0/90)/0_2/(0/90)/(\pm45)/90]_T$, in the second and third segments other [0] tapes and [±45] fabrics are added up to the specified laminate thickness. The laminate is well balanced and alternate layers of unidirectional tape and fabric stacked at different angles through the thickness, avoiding to group together plies of the same orientation.

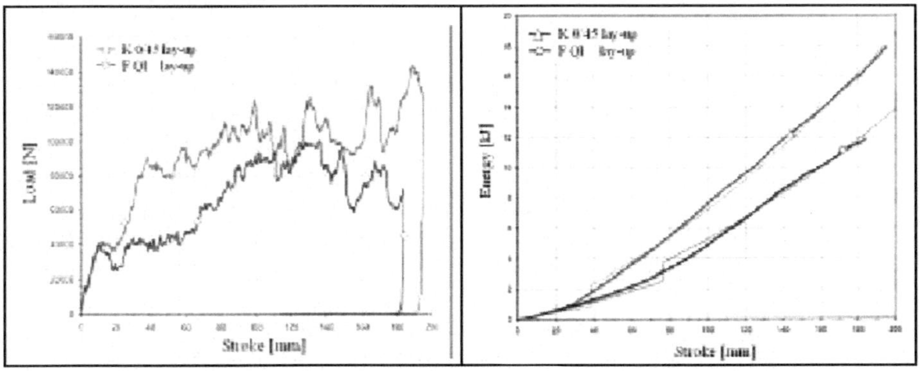

Figure 6.5 - Results of the quasi-static axial crush tests. Feraboli et al. (2007).

The K0/45 cone has a constant laminate thickness of 2.8 mm. The layers are stacked in a somewhat balanced way of the 0 and ± 45° fibres. About 54% of the fibres are unidirectional tape plies oriented in the cone axis direction, while the remaining plies are ± 45 plies that provide torsional stiffness. The initial trigger zone is 38.1 mm long and consists of wall thickness reduced to 2.0 mm and unidirectional ply content reduced to 35%.

Both designs have trigger in order to initialise the collapse of the structure by the conical shape of the component and also by the progressive reduction of the wall thickness along the cone axis from the major to the minor bases. However, the adopted strategies for thickness reduction are different in the two designs as already described.

The cones have been submitted to quasi-static and dynamic crush tests, the dynamic test was conducted according to the FIA regulation.

Figure 6.5 shows the results of the quasi-static axial crush tests in terms of force – displacement curves and of energy – displacement curves.

From the diagrams of Figure 6.5 it can be noted that

- the collapse evolution of both cones is progressive and stable;
- the initialisation trigger appears to behave properly but seems to induce a too large reduction of the force in the first part of the curve, particularly with the FQI cone. The force at first peak (that is the initialisation of the laminate fracture) is lower than the average force value. The energy absorption capability of both cones could be improved by decreasing the effect of the trigger and, as a consequence, by letting the force value at first peak increase up to the average force value, as it is expected with an ideal energy absorber.
- the maximum value of the crush force is essentially below the limit of 140 kN for both the cones, being only 100.5 kN for the FQI cone, while for the K0/45 cone it reaches 144.3 kN (i.e. slightly over the limit value) at the last peak.
- The average value of the crush force is 93.5 kN for the K0/45 cone and only 64.9 kN for the FQI cone.

The design of the K0/45 cone appears to be more effective than the FQI cone in terms of capability of energy absorption, however at the end of a 200 mm crush stroke the absorbed energy is only about 19 kJ, i.e. only two thirds of the required value. Assuming that the collapse evolution remains stable and with the same increment rate of the absorbed energy, a crush length of about 300 mm is needed to dissipate all the kinetic energy considered by the FIA regulation.

Figure 6.6 shows the results of the dynamic axial crush test of the K0/45 cone in comparison with the quasi-static one, in terms of force – displacement curves and of energy – displacement curves, it can be noted that

- the dynamic force curve results to be close to the quasi-static one; also in this case the macroscopic effect of the strain rate, at least for the considered material and for the particular loading condition, appears to be negligible.
- the force values in the dynamic test, up to a crush distance of about 200 mm, are lower than the quasi-static one, then, at a crush distance of 225 mm, it is visible a 200 kN peak, followed by a large load decrease, down to 40 kN;

- the cones after the dynamic test shows fracturing behaviour similar but not identical to the quasi-static specimen.

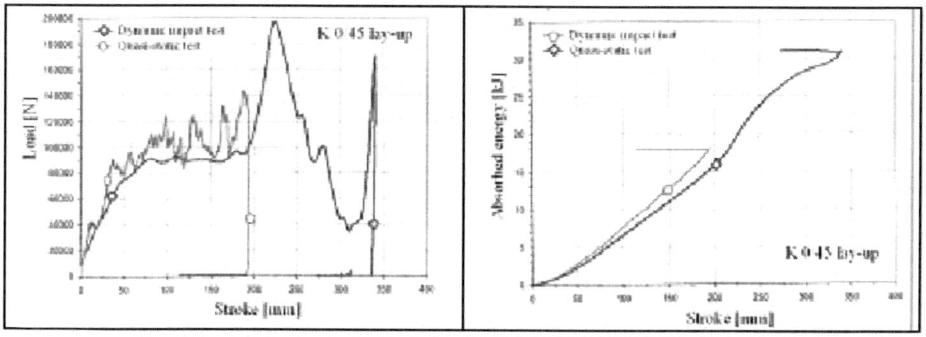

Figure 6.6 - Results of the dynamic axial crush test in comparison with the quasi-static one for the K0/45 cone. Feraboli et al. (2007).

To explain the somewhat unexpected difference in Figure 6.6 between the quasi-static and the dynamic curves Feraboli and co-workers (2007) mentioned three points that, a part from their particular study, are of general interest for this type of tests. The points are listed below in the following without reference to their relative importance.

The first point deals with the problems connected with the dynamic measurements, in particular the signal sampling rate, the signal filtering and elaboration. Inappropriate sampling rate and filtering can cause a hiding effect of the oscillations in the response curve.

The second point is that a relevant part of the absorbed energy is due to the work done by the friction force at the sliding surface between the fronds originated by the composite laminate fracture and the metallic counterpart, as already reported by Hull (1991) and Farley (1992). It should be pointed out that in the two tests done by Feraboli and co-workers the metallic counterparts were different and therefore the friction coefficient could be different.

The third point is that the friction coefficient, generally, is velocity sensitive and the velocity at the interface between the fronds and the metallic counterpart is clearly different in the quasi-static and dynamic tests.

In the Feraboli and co-workers paper there is not an attempt to explain the large peak-valley phenomena in the second part of the dynamic curve.

6.3 – Frontal sacrificial structure for a Formula One racing car.

Nowadays, in Formula one car, the main safety device in case of frontal impact is a sacrificial structure made of composite materials. Generally it is a sandwich structure with composite material skins and a structural polymeric foam core.

The numerical simulation of the crash behaviour of this structure by finite element modelling could be of great interest in order to optimise its performance and weight.

While the simulation of the impact response of a shell structure made of composite laminate when loaded in a direction perpendicular to the plate surface is nowadays feasible with commercial codes such as LSDYNA obtaining results that are in very good agreement with experimental test results (see for example Belingardi et al. (1998)), the simulation of a thin-walled beam loaded in the axial direction comes upon some non trivial modelling problems.

This is mainly due to the difficulty of modelling accurately the different collapse mechanisms that can take place and to decide which will be the correct one in the particular case of the considered structure and the considered loading condition. Some papers published by Farley (1983 and 1992) and by Mamalis et al. (1996) with reference to axial loading conditions, describe both the fragmentation with fronds formation, already mentioned in the two previous subsections, and a progressive folding mechanism that appears to be similar to what happens with metallic structures.

This latter behaviour can be modelled obtaining results in rather good accordance with the experimental test results, provided that one is able to supply the appropriate value to the parameters of the constitutive equation that describe the material behaviour in dynamic loading conditions.

Anghileri, Chirwa and co-workers, in their paper published in 2005, present an interesting procedure. The identification of the parameters of the constitutive equation with particular reference to the damage initiation and evolution for the considered composite material can be performed by means of an inverse approach base on an optimisation technique. The objective function of the optimisation is the minimisation of the distance of the load-displacement curve obtained through the numerical simulation with respect to the experimental one, the design variables are the parameters to be evaluated.

Anghileri and co-workers base the identification on the behaviour of a cylindrical composite tube with two different stacking sequences. Then a conical tube made of the identified material has been submitted to axial crushing load both in the laboratory and in the virtual test environment in order to get a confirmation of the validity of the identification approach and the suitability of the parameter values.

Figure 6.7 shows the obtained results in terms of the axial crushing force – displacement curves. Three curves are reported the reference curve that is the experimental test result, the curve obtained by numerical simu lation with the initial value of the material parameters and, finally the curve obtained by numerical simulation with the value of the material parameters determined by the application of the inverse approach. Although there are still some differences between the experimental and the numerical optimised curves, the quality of the

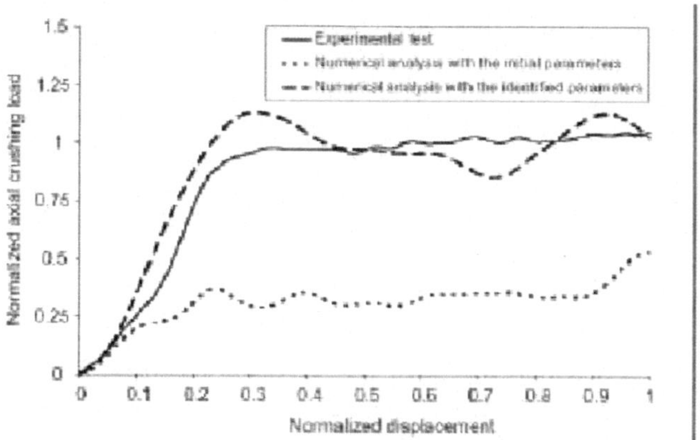

Figure 6.7 – Comparison of the experimental/numerical curves of the axial crush tests for the conical structure. Anghileri et al. (2005).

obtained results is remarkable.

Then Anghileri and co-workers come to the simulation of the axial impact crush of the frontal sacrificial structure for a Formula racing car.

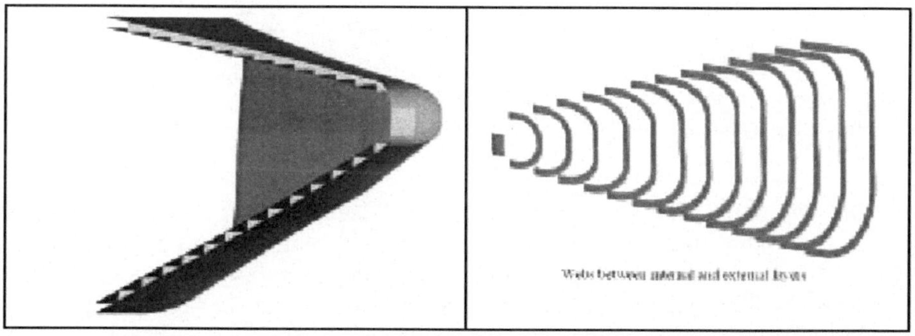

Figure 6.8 – Shape and modelling of the considered frontal sacrificial structure. Anghileri et al. (2005).

Figure 6.8 shows the shape and the modelling of the considered structure, it consists of two truncated pyramidal shells of carbon fibre / epoxy matrix composite with a foam core. The pyramid length is 660 mm and is oriented so that the cross-section dimensions increase along the pyramid axis when moving from the vehicle front toward the driver seat. The stacking sequences for each of the two shells include layers parallel to the pyramid axis, hoop layers and ±45 layers. In the finite element model the foam has been substituted by a number of webs (as shown in the right side of Figure 6.8) that act as spacers between the two shells, with an appreciable simplification of the model that thus avoid the well known problems connected with the numerical simulation of foams when very large compression deformations take place. The impact speed is 12 m/s.

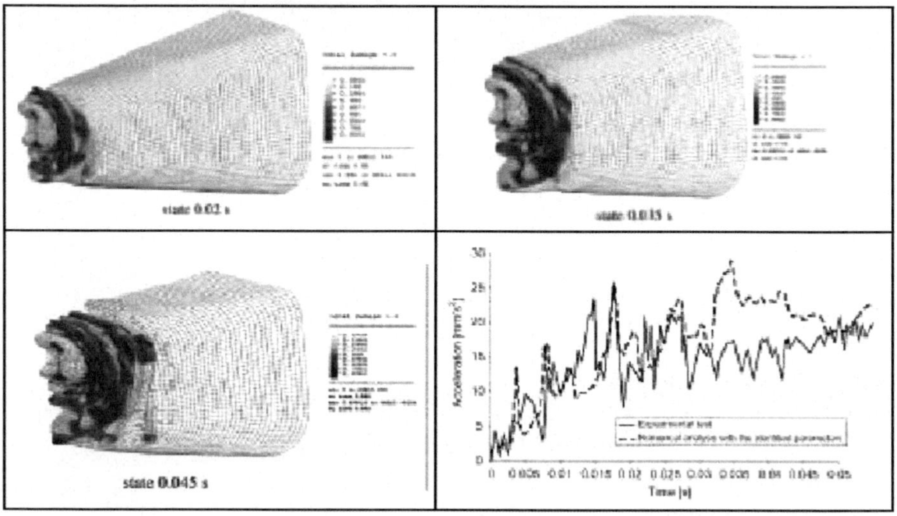

Figure 6.9 – Finite element simulation results: a,b,c) - sequence of the collapse shape of the structure at three times with evidence of the progressive damage of the composite material; d) - a comparison of the experimental and numerical deceleration time histories. Anghileri et al. (2005).

Finally Figure 6.9 shows the shape of the collapsed shells at three subsequent times. The figures include also information on the evolution of the damage level for the composite material. The figure include the diagram of the deceleration time history allowing for a comparison of the experimental test result with the numerical simulation one. The agreement between the two curves of Figure 6.9 appear to be quite good.

6.4 – A new hood composite structure optimised for pedestrian safety.[2]

6.4.1 – Introduction.

The hood design has to satisfy different requirements deriving from style, safety and structural needs. Kerkeling et al. (2005) described a new design of hood inner panel for pedestrian protection. The multi bulge structure is drawn into the inner panel instead of a traditional rib structure and offers a more uniform

[2] The present subsection is a summary of the paper authored by Giorda et al. (2006)

stiffness distribution. D.Y. Kwak et al. (1997) studied a one-piece fiberglass composite hood with reinforcing ribs obtaining a 30%÷40% weight reduction if compared to conventional steel hoods. Stammen et al. (2001) demonstrated the importance of the zones chosen for the impact test evaluation due to the presence of the underhood structure and its influence on the local stiffness of the hood. He showed that the HIC values are strongly influenced by the impact angle between the impact direction of the headform and the local tangent to the hood surface. Yang (2005) analysed the influence of the hood structure on the structural responses of the pedestrian head with varying car-front parameters. The results obtained from this parametric study indicate that the head injury severity is significantly affected by the local stiffness of the car front structure and that the local stiffness of the hood can be reduced by using energy-absorbing materials, by removing sharp edges, as well as by spreading impact forces over a body area as wide as possible.

6.4.2 - Concept design process

The hood design is usually characterised by the presence of an external metal sheet layer (that has to satisfy style and aerodynamic requirements) supported by an underlying frame-like (rib) reinforcement structure. The supporting structure is generally joined to the external metal sheet layer and has to guarantee the required global and local torsional/bending stiffness of the hood. The presence of the underlying reinforcement structure leads to a non uniform stiffness distribution of the hood and, unfortunately, the zones nearby the ribs are characterised by large stiffness values.

In order to keep the hood local stiffness uniform, the proposed concept design[3] is characterized by a thin sheet metal layer linked to a composite laminate structure. Bond is only along the edges to build a two shell structure. The reinforcement structure contributes to the global stiffness of the system without modifying the quite small local stiffness of the thin sheet metal layer. The thin sheet metal layer is not able to guarantee an adequate energy absorbing capability to the mechanical system even if very small HIC values can be obtained in a head impact simulation. The system consisting the external sheet metal layer and the underlying reinforcement structure is upgraded by the addition of a foam layer bonded to the sheet metal layer that will: (1) uniformly increase the local stiffness; (2) reduce the local elastic buckling load (denting) of the external sheet metal layer; (3) increase the total energy absorbing capability of the hood; (4) reduce the maximum displacement of the sheet metal layer during the pedestrian head impact. The proposed hood concept design introduces the presence of a sort of closed thin-walled structure and differs from similar

[3] patent n. TO2006A000574 authored by A. Giorda, I. Gaviglio, G. Chiandussi of the Mechanics Department of the Politecnico di Torino

sandwich concept designs proposed in several patents (see for example Devin et al. (2002), Cate et al. (2003), Kimoto et al. (2005), Engel et al. (2006) and Willinger (2006)) where both the upper steel (composite) layer and the reinforcement structure are tightly joined to the foam material layer.

6.4.3 - Hood geometry and requirements

The conceptual design problem concerns the hood for a small city car. The hypothetical external geometry of the hood (Figure 6.10) and the available space for an underlying reinforcement structure have been defined in collaboration with the vehicle designer. The design has to satisfy some requirements with respect to static and dynamic loading conditions.

For what concerns the static load conditions, requirements are essentially of minimum bending and warping stiffness. The hood is partially constrained in points A and B thus simulating the hinges that link it to the car body. C and D represent two points where displacement values are imposed and force values are calculated in order to evaluate the hood global stiffness. A maximum displacement value of 5.5 mm in C (D) by considering D (C) fully constrained has to be obtained due to the application of a 50 N force (corresponding to a K=9100 N/m minimum global stiffness).

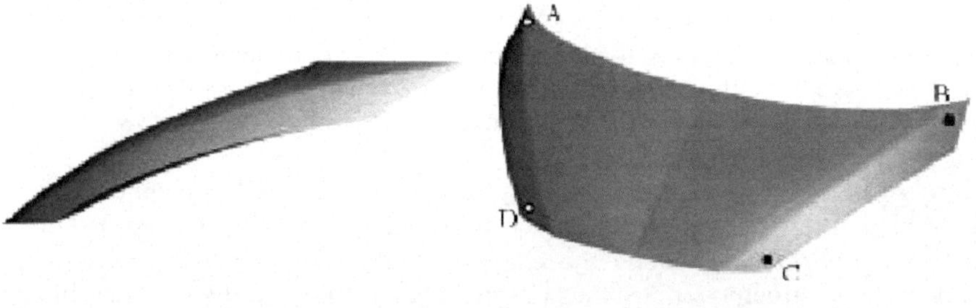

Figure 6.10 - Overall hood geometry and constraints.

For what concerns the dynamic load conditions, requirements are essentially in terms of pedestrian passive safety. The proposed hood represents a concept design. The hood structure has been considered in isolation and any under-hood structure representing, for example, an engine block has been taken into consideration. Due to regulation present and future requirements in terms of pedestrian passive safety, the distance between the hood and a possible structure lying beneath it has been considered as a hood behaviour dependent design variable.

6.4.4 - Reinforcement design problem

The optimal reinforcement structural layout has been identified by the topology optimization code OPTITOP developed by Chiandussi (2001) and (2006).

The solution of a structural topology optimisation problem leads to the identification of the material distribution that maximizes the elastic potential energy accumulated at the equilibrium. The structural members the optimal structure is made of, should be characterized by uniform normal stresses, to say, mono-axial stresses (tension or compression). Similar results should be obtained by using a composite material if the maximum material stiffness direction and the maximum normal stress direction are the same locally.

A carbon/epoxy composite structure with ribs has been hypothesized. Ribs have been laid out following the optimisation results in order to maximise their effectiveness (Figure 6-11). The composite structure is made of two carbon/epoxy composite layers coupled and oriented at 0° and 90°. Each layer is made of unidirectional carbon fibres in an epoxy matrix with the following mechanical properties $E_{11}=1.810 \cdot 10^5$ MPa, $E_{22}=1.028 \cdot 10^4$ MPa, $G_{12}=7.172 \cdot 10^3$ MPa, $\nu_{12}=0.280$, ?=1600 kg/m^3. Composite laminates with thickness ranging from 1 to 3 mm have been taken into consideration.

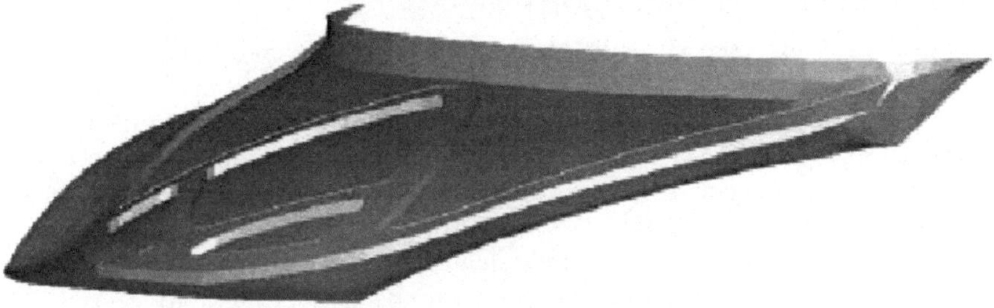

Figure 6.11 - Carbon/epoxy reinforcement structure with ribs.

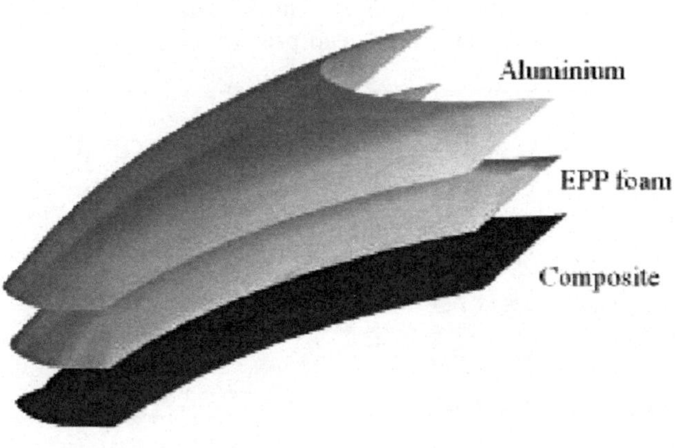

Figure 6.12 - Hood structure: a) external steel sheet layer, b) structural foam layer, c) carbon/epoxy reinforcement structure with ribs.

In order to increase the local stiffness of the hood, a layer of structural EPP (Expanded Polypropylene) foam has been introduced between the sheet metal layer and the carbon/epoxy reinforcement structure. The overall structural configuration of the hood is described in Figure 6.12 and is characterised by:

a) an external steel sheet metal layer with a 0.3 mm thickness,

b) a structural foam layer bonded to the sheet metal layer in order to increase the local stiffness and the energy absorption capability of the hood, see Avalle et al. (1999), (2001) and (2007),

c) a carbon/epoxy layer with ribs in order to guarantee the global stiffness of the structure and to limit the maximum displacement of the external steel sheet metal layer during the head impact.

The hood-pedestrian head impact simulation problem and procedure are described and the results obtained by taking into consideration different foam densities are discussed in the following.

6.4.4.1 - Hood-pedestrian head impact simulation.

The EuroNCAP test (see sub-subsection 2.1.4) has been considered to evaluate the impact response of a pedestrian head against a vehicle hood. This test requires the use of a specific headform hitting the hood at a speed of 40 km/h (Figure 6.13a). The adult headform is a 165 mm diameter aluminium sphere with a 4.5 kg mass covered by a 7.5 mm rubber layer reproducing the mechanical characteristic of the human skin.

The HIC (see sub-subsection 5.3.1) is calculated on the basis of the acceleration history measured by an accelerometer placed in the headform gravity centre. The EuroNCAP rating test prescribes the evaluation of the HIC value in three different areas on the hood as described in Figure 6.13b. The three areas can be identified on one half of the hood due to its longitudinal symmetry and are defined differently depending on which headform, the child (C) or the adult (A)

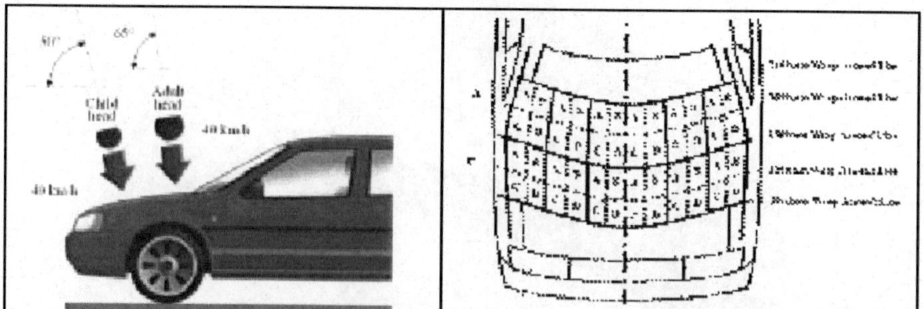

Figure 6.13 – a - Headform types and correlated impact zones; b - EuroNCAP rating tests impact areas for the adult (A) and the child (C) headforms.

headform, is used in the simulation. As prescribed by the EuroNCAP standards, the HIC has to be lower than 2000 in one of the three impact zones and less than 1000 in the remaining two zones.
The final design solution is characterised by foam layer that does not fill all the volume between the two (metallic and composite) shells. This solution has been analysed in detail by taking into account both the thickness of the carbon/epoxy layer and different foam densities searching for the optimal configuration that combines a suitable maximum hood deflection with allowable HIC values.
The foam layer (see Figure 6.14a) contributes to the hood global and local bending stiffness and helps to distribute the deformation field around the impact area reducing the local impact load.
Data have been obtained by the simulation of an impact of an adult headform against the hood in the position described in Figure 6.14b.

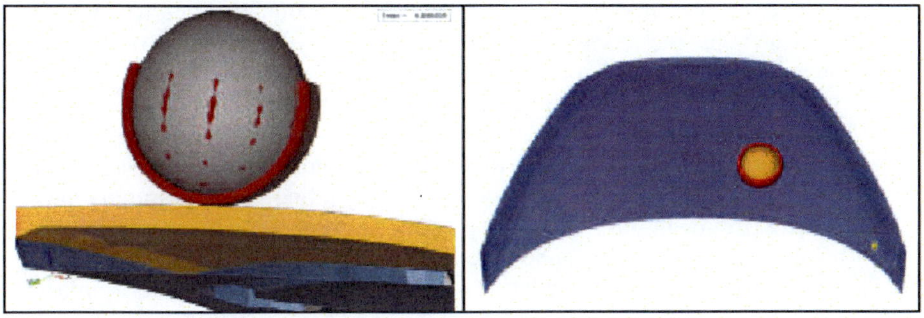

Figure 6.14 – a - Hood design with a little thickness foam layer; b - Impact point for deflection and HIC evaluation.

The role of the thickness of the carbon/epoxy laminate and of the density of the structural foam (that is given in g/l by the number in the material identification name) on the maximum hood deflection and on the HIC parameter are described in table 6.1 and 6.2, respectively.
The maximum deflection of the hood increases by decreasing the thickness of the carbon/epoxy material layer and by decreasing the structural foam density

Composite thickness	Foam type			
	EPP30	EPP45	EPP70	EPP145
1 mm	68	70	64	49
2 mm	42	43	39	32
3 mm	30	32	31	26

Table 6.1 - Maximum hood deflection (mm) for different foam densities and carbon/epoxy reinforcement thickness.

Composite thickness	Foam type			
	EPP30	EPP45	EPP70	EPP145
1 mm	958	926	1011	1296
2 mm	1360	1330	1494	1883
3 mm	1830	1829	1927	2257

Table 6.2 - HIC value for different foam densities and carbon/epoxy reinforcement thickness.

until the EPP45 foam (table 6.1). The results obtained by introducing a foam with a lower density (EPP30) show instead a little decrease of the maximum deformation of the hood. The HIC parameter increases by increasing the carbon/epoxy laminate thickness and decreases by decreasing the foam density (table 6.2). Once more, the introduction of the EPP30 foam leads to an HIC parameter larger than in the case of EPP45. The solution characterised by a 1 mm thickness of the composite laminate and an EPP45 structural foam guarantees an HIC value lower than 1000.
The conceptual solution characterised by the larger foam density value ?=70 g/l and a 1 mm composite laminate thickness has been selected due to the lower hood maximum deflection. Starting from this conceptual solution, the impact in other two different points on the hood has been studied. The test points have been selected following as strictly as possible the EuroNCAP guidelines. The half hood surface has been subdivided in small areas similar to the thirds indicated in the EuroNCAP regulation (Figure 6.13b) and the headform impact has been simulated in every third, i.e. near the hood boundary (position 3), in the middle in correspondence of a rib of the underlying reinforcement structure (position 1) and in an intermediate position not interested by the presence of underlying ribs (position 2) as shown in Figure 6.15.

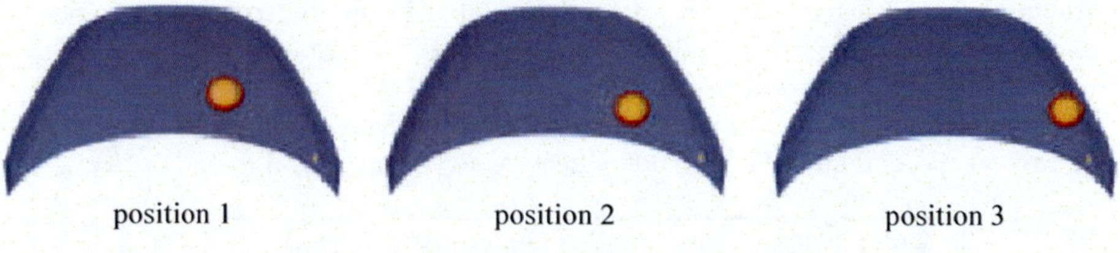

Figure 6.15 - The three headform-hood impact positions taken into account.

The results of the simulations are shown in table 6.3. The EuroNCAP requires an HIC value lower than 2000 in all positions and at least one impact point characterised by an HIC value lower than 1000. The HIC values for the selected conceptual design solution exceed the HIC value of 1000 only twice and only for a very little percentage showing a quite well distributed local stiffness on the hood.

position on the hood	design parameters	
	deflection (mm)	HIC
position 1	64	1011
position 2	47	1033
position 3	51	972

Table 6.3 - HIC and maximum hood deflection for the headform-hood impact in the three selected test points (1 mm carbon/epoxy composite structure and 70 g/l foam layer).

The total weight of the optimised design solution is 4.6 kg: a very good result compared with the total weight of about 8.6 kg that characterises the normal production metallic hood. The weight reduction is about 47%.
The global stiffness of the optimised structure has also been verified and found to satisfy the required values.

6.5 - Conclusions

Four applications of composite material to structural parts of a vehicle car-body have been presented and discussed in details.
The first application is of particular interest since the front longitudinal beams are intended for production vehicles. The results demonstrate that the adoption of composite materials not only leads to lighter structure but also makes it possible to design structures that have even better crash performance than steel structures. The particular type of energy absorbing mechanism, that was observed during experiments, allows to design component with a load uniformity parameter close to 1 and this means an excellent way of using the material.
The second and third applications come from the racing vehicle world. In these cases, the performance requirements are even more severe that for general production vehicles because the vehicle velocity is much higher and consequently the kinetic energy that must be dissipated to save the pilot life and to avoid invalidating injuries. Composite materials are commonly used for this type of vehicles because of the light weight structure that can be done, but at the same time they guarantee an excellent crashworthiness. With the third application the problems and the possibility of performing a structural analysis by simulating the crash event of a composite structure have been discussed. Some problems connected with the determination of the actual material parameters are still open and ask for refinement of material models. In particular it is of fundamental important in the impact simulations to describe in a proper way the progressive damage of the composite laminate and the consequent decrease in strength and stiffness.

The fourth application is an innovative multi-material bonnet, again intended for production vehicles. The use of a sandwich structure, made of aluminium, foam and composite material, allows to satisfy all the design targets. Aesthetic and paint problems are solved with the outer metallic skin, the stiffness problems are solved with the properly designed thickness of the foam core, the local denting problems are solved by the presence of the foam core that make stiffer the thin metallic sheet, finally the pedestrian head impact are solved with a structure characterised by a distributed stiffness.
The achieved result is significant: the designed bonnet is nearly half weight with respect to the steel normal production and has adequate crashworthiness performance.
Two more points have to be briefly addressed in these concluding remarks: the structural joints and the material recycling requirements at the end of the vehicle or component life.
The evolution of the adhesive capabilities, that has taken place in recent years, opens to hybrid structures made of different materials, without concerns about strength, fatigue and impact performance of the structural joints.
The material recycling problem is still of concern when dealing with composite material structures as it is not easy with an industrial process to separate back the fibers from the matrix and thus, at the moment the composite material and its constituent parts are not reusable. Research is on going and one promising possibility comes from the use of vegetal and biodegradable fibres.

7. - References

Abrate S. (1991) – "Impact on laminated composite materials", Applied Mechanics Review, vol. 44 n.4

Abrate S. (1994) – "Impact on laminated composites: recent advances", Applied Mechanics Review, vol. 47 n.11

Abrate S. (1998) – "Impact on composite structures", *Cambridge University Press*

Anghilleri M., Chirwa E.C., Lanzi L., Mentuccia F.(2005) - "An inverse approach to identify the constitutive model parameters for crashworthiness modelling of composite structures", Composite Structures, vol. 68, pp. 65-74

Avalle M., Belingardi G. (2004) - "Advanced materials for automotive applications", Mobility & Vehicle Mechanics, vol. 30, pp. 51-66

Avalle M., Belingardi G., Montanini R., Mangino E. (1999) - "Energy absorption characteristic of polymeric structural foams for passive safety applications", proc. of 6th Int. Conf. Florence ATA 1999 - Firenze, 17-19 November 1999

Avalle M., Belingardi G., Montanini M. (2001) - "Characterization of polymeric structural foams under compressive impact loading by means of energy -absorption diagram", Int. J. of Impact Engineering, vol. 25, pp. 455-472

Avalle M., Belingardi G., Ibba A. (2007) – "Mechanical models of cellular solids: Parameters identification from experimental tests", Int. Journal of Impact Engineering, vol. 34, pp. 3-27

Barbat S.D., Prasad P. (1995) - "Finite element modeling of structural foam and head impact interaction with vehicle upper interior", SAE paper 950885

Belingardi G., Cavatorta M.P., Duella R. (2003) - "Material characterisation of a composite-foam sandwich for the front structure of a high speed train", Composite Structures, n. 61, pp. 13-25

Belingardi G., Cavatorta M.P., Paolino D.S. (2006) – "Repeated impact behaviour and damage progression of glass reinforced plastics", proc. ECF 16 (European Conference on Fracture), Alexandroupolis (Greece), 3-7 July 2006

Belingardi G., Chiara A., Martinotti M., Vadori R. (1993) - "Experimental evaluation of the quasi-static collapse of automotive front beams made of glass fiber reinforced plastics", II Polish-Italian Seminar, Torino

Belingardi G., Duella R., Caminiti A. (2002) 'Optimal choice of the foam design parameters in order to meet the HIC index limit of the FMVSS201 standard', Proc. 2002 IBEC Conference.

Belingardi G., Grasso F., Vadori R. (1998) - "Energy absorption and damage degree in impact testing of composite materials", proc. of XI ICEM (Int. Conf. Experimental Mechanics), Oxford (UK), pp. 279-285

Belingardi G., Gugliotta A., Vadori R. (1998) - "Numerical simulation of fragmentation of composite material plates due to impact", Int. J. Impact Eng., vol. 21, pp. 335-347

Belingardi G., Vadori R. (2002) - "Low velocity impact tests of laminate glass-fiber-epoxy matrix composite material plates", Int. J. Impact Eng., vol. 27, pp. 213-229

Belingardi G., Vadori R. (2003) - "Influence of the laminate thickness in low velocity impact behavior of composite material plate", Composite Structures, n.61, pp. 27-38

Belingardi G. (2002) – "Some recent development and perspective in car passive safety", Mobility & Vehicle Mechanics, vol. 28, pp. 87-105

Bisagni C., Di Pietro G., Fraschini L., Terletti D. (2005) – "Progressive crushing of fiber reinforced composite structural components of a Formula 1 racing car", Composite Structures, vol. 68, no. 4, pp. 491-503

Cantwell W.J., Morton J. (1991) –The impact resistance of composite materials – a review, Composites, vol. 22(5), pp. 347-362

CARE Database (2005), General Directorate for Transport of the European Commission. http://ec.europa.eu/transport/road_safety/observatory/statistics/care_en.htm

Carruthers J.J., Kettle A.P., Robinson A.M. (1998) – "Energy absorption capability and crashworthiness of composite material structures: a review", Applied Mechanics Reviews, vol.51, pp. 635-649

Cate P.J., Naughton P.J. (2003) 'Energy absorption unit', WO 03047842.

Chiandussi G., Gaviglio I., Ibba A. (2001) – "Evolutionary method for topology optimisation with maximum stress control", Proc. European Conference on Computational Mechanics (ECCM 2001).

Chiandussi, G. (2006) – "On the solution of a minimum compliance topology optimisation problem by optimality criteria without a priori volume constraint specification", Computational Mechanics, Vol. 38, pp. 77-99.

Chou C.C., Zhao Y., Lim G.G., Patel R.N., Shahab S.A., Patel P.J. (1995) - "Comparative analysis of different energy absorbing materials for interior head impact", SAE paper 950332

Clifford C.C., Nyquist G.W. (1974) – "Analytical studies of the Head Injury Criterion (HIC)", SAE paper 740082,

Deb A., Calso S. Saha N. (1997) – "Effectiveness of countermeasures in upper interior head impact", SAE paper 970391
Devin J.M., Schmit F., Brun O., Deprez P. (2002) – "Reinforced and lightweight motor-vehicle bonnet", US 6398286.
DeWeese, R.L., and D.M. Moorcroft (2004) – "Evaluation of a Head Injury Criteria Component Test Device", US Department of Transportation, Federal Aviation Administration, Office of Aerospace Medicine, Document DOT/FAA/AM-04/18
Engel M., Itzen G., Thomas U. (2006) 'Bonnet', EP 1614590.
Eppinger R., Sun E., Bandak F., Haffner M., Kaewpong N., Maltese M., Kuppa S., Nguyen T., Takounts E., Tannous R., Zhang A., Saul R (1999) – "Development of improved injury criteria for the assessment of advanced automotive restraint systems" – NHTSA
EuroNCAP (2004), European New Car Assessment Program – Side impact testing protocol - May 2004
EuroNCAP (2008), European New Car Assessment Program - Frontal impact testing protocol v4.2, May 2008
EuroNCAP (2008a), European New Car Assessment Program - Pedestrian testing protocol, Version 4.2, June 2008.
EuroNCAP (2008b), European New Car Assessment Program – Assessment protocol and biomechanical limits, Version 4.2, June 2008
Farley G. L. (1983) – "Energy absorption of composite materials", J. of Composite Materials, vol. 17
Farley G.L., Jones R.M. (1991) – "The effect of the crushing speed on the energy absorption capability of composite tubes", J. of Composite Mat., vol. 25, pp. 1314-1329
Farley G. L. (1992) – "Relationship between mechanical properties and energy absorption trends for composite tubes", NASA TP 3284, ARL-TR-29
Feraboli P., Norris C., McLarty D. (2007) – "Design and certification of a composite thin-walled structure for energy absorption", Int. J. Vehicle Design, vol. 44, pp. 247-267
FIA (2005) – Art. 15.5 : safety structures" in Formula 3 technical regulations, Appendix J
Gadd CW. (1966) – "Use of a weighted-impulse criterion for estimating injury hazard." – Proc. of the 10th Stapp Car Crash Conference, SAE Paper 660793.
Gibson L. J., Ashby M. F. (1997) - "Cellular Solids: structure and properties", 2 ed., Cambridge (UK), Cambridge University Press
Giorda A., Gaviglio I., Chiandussi G. (2006) - "A numerical study on a new hood design", paper of the Mechanics Department of the Politecnico di Torino
Gupta N.K. et al. (1997) – An analysis of axial crushing of composites tubes", J. of Composite Materials, vol. 31, pp.1262-86
Hull D. (1983) – "Axial crushing of fibre-reinforced composite tubes", in *Structural Crashworthiness*, editors N. Jones and T. Wierzbicki, Butterworths, pp. 118-135
Hull D. (1991) – "A unified approach to progressive crushing of fibre-reinforced composite tubes", Composite Science and Technology, vol. 40, pp. 377-421
Jimenez M.A., Miravete A., Larrode E., Revuelta D. (2000) – "Effect of trigger geometry on energy absorption in composite profiles", Composite Structures, vol. 48, pp. 107 -11
Kerkeling C., Schäfer J., Thompson G.M. (2005) 'Structural hood and hinge concepts for pedestrian protection', Proceedings 19th International Technical Conference on the Enhanced Safety of Vehicles, paper n. 05-0304.

Kimoto Y., Kiyama H., Kawashima S., Enomoto Y. (2005) – "FRP panel for automobile", WO2005070747.

Kwak D.Y., Jeong J.H., Cheon J.S., Im I.T. (1997) "Optimal design of composite hood with reinforcing ribs through stiffness analysis", Composites structures, Vol. 38, No. 1-4, pp. 351-359.

Lau IV and Viano DC. (1981) - "Influence of impact velocity and chest compression on experimental pulmonary injury severity in an animal model", Journal of Trauma, Vol. 21, pp 1022-1028.

Lau IV and Viano DC. (1986) – "The viscous criterion – bases and applications of an injury severity index for soft tissue", Proc. of the Thirtieth Stapp Car Crash Conference, pp 123-142, SAE Paper No. 861882.

Locke D.J., Clark C.L. (1996) - "Energy-absorbing thermoplastics for head impact applications", SAE International Congress and Exposition, paper 960154

Mamalis A.G., Manolakos D.E., Viegelahn G.I. et al. (1991) – "On the axial crumpling of fibre-reinforced composite thin-walled conical shells", Int. J. of Vehicle Design, vol. 12

Mamalis AG, Manolakos MB, Demosthenous GA, Ioannidis MB (1996) – "The static and dynamic axial collapse of fibreglass composite automotive frame rails", Composite Structures, vol. 34, pp. 77-90.

Mamalis AG, Manolakos MB, Demosthenous GA, Ioannidis MB (1998) – "Crashworthiness of composite thin-walled structural components", *Technomic*

Mamalis AG, Manolakos MB, Ioannidis MB, Kostazos PK. (2003) – "Crushing of hybrid square and with composite vehicle hollow bodyshells with reinforced core subjected to axial loading: numerical simulation", Composite Structures, vol. 61, pp. 175–186

Mamalis AG, Manolakos MB, Ioannidis MB, Papapostolou DP (2004) – "Crashworthy characteristics of axially statically compressed thin-walled square CFRP composite tubes: experimental", Composite Structures, vol. 63, pp. 347–360

Mamalis A.G., Manolakos D.E., Ioannidis M.B., Kostazos P.K., Papapostolou D.P. (2002) – "Axial collapse of hybrid square sandwich composite tubular components with corrugated core: numerical modeling", Composite Structures, vol. 58, pp. 571–582

Mangino E., Carruthers J.J., Pitarresi G. (2007) – "The future use of structural composite materials in the automotive industry", Int. J. Vehicle Design, vol. 44, pp. 211-232

Mertz HJ., Patrick LM. (1971) – "Strength and response of the human neck", Proc. of the 15th Stapp Car Crash Conference, SAE Paper No. 710855.

Mertz H. (1993) - "Anthropomorphic Test Devices", in Accidental Injury, Biomechanics and Prevention, edited by Nahum, A., Melvin, J., Springer-Verlag

Mizuno Y. (2003) –"Summary of IHRA pedestrian safety WG activities (2003) – Proposed test methods to evaluate pedestrian protection afforded by passenger cars", Proceedings 18th International Technical Conference on the Enhanced Safety of Vehicles, paper n. 580.

Nader R. (1965) – "Unsafe at any speed: the designed-in dangers of the American automobile", Pocket Books

NHTSA (2001) - "Traffic Safety Facts 2000" published by the US Department of Transportation - National Highway Traffic Safety Administration

Nishimoto T. (2003) – "Introduction of the regulation of pedestrian head protection in Japan", Proc. 18th Int. Tech. Conf. on the Enhanced Safety of Vehicles , paper n. 503.

Prasad P., Belwafa J.E. (2004) – "Vehicle crashworthiness and occupant protection", AISI, Southfield (MC-USA)

Savage G., Bomphray I., Oxley M. (2006) – "Exploiting the fracture properties of carbon fibre composites to design lightweight energy absorbing structures", Engineering Failure Analysis, vol. 11, pp. 677-694

Seiffert, U., and L. Wech (2003) – "Automotive Safety Handbook", Warrendale, SAE International

Sigalas I., Kumosa M., Hull D. (1991) – "Trigger mechanisms in energy-absorbing glass cloth/epoxy tubes", Composites Science and Technology, **40**(3), pp.265-287

Solaimurugan S., Velmurugan R. (2007) – "Influence of fibre orientation and stacking sequence on petalling of glass/polyester composite cylindrical shells under axial compression", Int. J. of Solids and Structures, vol. 44, pp. 6999–7020

Sounik D.F., Gansen P., Clemons J.L., Liddle J.W. (1997) - "Head impact testing of polyurethane energy-absorbing (EA) foams", SAE International Congress and Exposition, paper 970160

Stammen J.A., Saul R.A., Ko B. (2001) – "Pedestrian head impact testing and PCDS reconstructions", Proc. 17th Int. Technical Conf. on the Enhanced Safety of Vehicles, paper n. 326.

Standard No. 201 (1998) – "Occupant protection in interior impact", 49 CFR, Ch. V, Parts 571.201, Federal Motor Vehicle Safety Standards, National Highway Traffic Safety Admin., Dept. of Transportation

Sugimoto, T., Y. Kadotani, S. Ohmura (1998) – "The offset crash test - a comparative analysis of test methods", Proc. 16th Int. Tech. Conf. Enhanced Safety of Vehicles (ESV), Windsor, Ontario, Paper Number 98-SI-O-08

Thornton P.H., Edwards P.J. (1982) – "Energy absorption in composite tubes", J. of Composite Materials, vol. 16, pp. 521-545

Thornton P.H., Jeryan R.A. (1998) – "Crash energy management in composite automotive structures" – Int. J. of Impact Engineering, 1998, vol. 7, pp. 167-180

ULSAB – Ultra-Light Steel Auto Body (1999) - AVC Technical Transfer Dispatch #6, CAE Analysis for Crashworthiness

UNECE (2006) World Forum for Harmonization of Vehicle Regulations (WP.29), Working Party on Passive Safety (GRSP): Proposal for a global technical regulation on uniform provisions concerning the approval of vehicles with regard to their construction in order to improve the protection and mitigate the severity of injuries to pedestrians and other vulnerable road users in the event of a collision, Document ECE/TRANS/WP.29/GRSP/2006/2

Versace J. (1971) – "A review of the severity index", Proc. of the 15th Stapp Car Crash Conference SAE Paper No. 710881.

Viano D., Von Holst H., Gordon E. (1997) 'Serious brain injury from traffic-related causes: priorities for primary prevention', Accident Analysis and Prevention, Vol. 29, pp. 811-816.

Willinger R. (2006) 'Multilayer hood with frangible outer skin and reduced impact in case of collision with a pedestrian', WO2006016053.

Yang J. (2003) 'Pedestrian head protection from car impacts', Int. J. Vehicle Design, Vol. 32, No. 1-2, pp. 16-27.

Yang J. (2005) 'Review of injury biomechanics in car-pedestrian collisions', Int. J. Vehicle Safety, Vol. 1, No. 1-2-3, pp. 100-117.

The Impact Resistance of Fiber Metal Laminates and Hybrid Materials

W.J. Cantwell[1] and R. Day[2]
[1] Department of Engineering, University of Liverpool, Liverpool, L69 3GH, U.K.
[2] School of Science and Technology Glyndŵr University, LL11 2AW U.K.

This chapter discusses the impact properties of fiber-metal laminates, stacked arrangements of thin metal layers and composites plies. Initial attention focuses on the strain rate sensitivity of the constituent materials, including the rate sensitivity of the composite-metal interface. This is followed by a review of the low velocity impact response of FMLs where such things as failure mechanisms and scaling effects are considered. Finally, the high velocity impact response of FMLs is considered.

1- Introduction

Fiber metal laminates (FMLs) are multi-layered materials based on stacked arrangements of aluminum alloy and fiber-reinforced composite materials. Currently, FMLs such as GLARE (glass fiber/aluminum) and ARALL (aramid fiber/aluminum) are attracting the interest of a number of aircraft manufacturers (see e.g. Vlot 2001). For example, ARALL, was used in the manufacture of the cargo door of the American C-17 transport aircraft, whilst GLARE is being used in the manufacture of the upper fuselage of the Airbus A380, an aircraft that is capable of carrying up to 525 passengers.

A number of different types of GLARE have been developed and these are summarized in Table 1, GLARE 3 is based on a 2024-T3 aluminum alloy and a 50/50 crossply (i.e. 50% of the fibers in the longitudinal direction and 50% in the transverse direction) glass reinforced epoxy. GLARE 4 has the same aluminum alloy, but has a 67/33 crossply composite structure and GLARE 5 which has additional $0^{\circ}/90^{\circ}$ plies compared to GLARE 3. GLARE 6 is based on a 2024-T3 FML with its composite plies oriented at $+/-45^{\circ}$. The mechanical properties of the FML can also be varied by changing the stacking sequence of the laminate. For example, FMLs are often denoted by their aluminum/composite arrangement, where a 3/2 laminate refers to a five layer system based on three layers of aluminum alloy and two layers of composite (i.e. A/C/A/C/A where A refers to an aluminum sheet and C refers to a composite ply).

Table 1. Summary of the different types of GLARE laminates.

Laminate	*Sub-group*	*Aluminum type*	*Orientation of fibers*
GLARE 1		7475-T761	0°/0°
GLARE 2	GLARE 2A	2024-T3	0°/0°
	GLARE 2B	2024-T3	90°/90°
GLARE 3		2024-T3	0°/90°
GLARE 4	GLARE 4A	2024-T3	0°/90°/0°
	GLARE 4B	2024-T3	90°/0°/90°
GLARE 5		2024-T3	0°/90°/0°/90°
GLARE 6	GLARE 6A	2024-T3	+45°/-45°
	GLARE 6B	2024-T3	-45°/+45°

It is well documented that FMLs combine the durability of metals with the impressive fatigue and fracture properties of fiber-reinforced composite materials (see e.g. Vlot 1991 and Vlot et al. 1998). Krishnakumar (1994) showed that the tensile strength of many fiber-metal laminates is superior to that of traditional aerospace-grade aluminum alloys as shown in Table 2. From the table, it is clear that GLARE-1 offers a tensile strength that is almost three times that of the plain aluminum alloy. Although the modulus of the FML is lower than that of the plain metal, this is more than offset by other key impressive properties under both static and long-term fatigue loading.

Vogelesang and Vlot (2000) investigated the tension-tension fatigue response of GLARE and showed that crack growth rates in these multi-layered systems were between one tenth and one hundredth of those measured in a monolithic aluminum alloy. Other works (see Bagnoli et al. 2009) have successfully demonstrated FMLs for use as crack retarder systems in damaged metallic structures.

In recent years, there has been growing interest in understanding and characterizing the dynamic response of FMLs. A number of workers have shown that FMLs offer attractive properties under impact and blast loading conditions and attempts have been made to evaluate the response of these hybrid materials over a wide range of strain rates. The following chapter presents a summary of published work investigating the dynamic response of FMLs.

Initially, the effect of strain rate on the properties of FMLs and their constituents will be considered. This is an important aspect and needs to be considered when designing FML structures that are likely to be subjected to impact loads. Following this, the low and high velocity impact response of FMLs will be discussed.

Table 2. Comparison of the tensile strengths of various FMLs with a 2024-T3 aluminum alloy. Data taken from Krishnakumar (1994).

Laminate	Ultimate tensile strength (MPa)	Tensile modulus (GPa)	Yield strength (MPa)
ARALL-1	800 (L)	68	545
	386 (LT)	48	333
ARALL-2	717 (L)	64	360
	317 (LT)	49	228
GLARE-1	1282 (L)	65	641
	352 (LT)	50	331
GLARE-2	1214 (L)	66	359
	317 (LT)	50	228
GLARE-3	717 (L)	58	587
	716 (LT)	58	317
AL 2024-T3	455	72	359

2- Rate Effects in Fiber-metal Laminates

As previously stated, FMLs are based on the stacked arrangement of composite and metal plies. Clearly, there are many components in this hybrid structure that could potentially exhibit a strain-rate sensitivity. It is therefore important to characterize and understand the rate-sensitivity of both the metal alloy and the composite material used to form the FML. It is also possible that the interface between the composite and the metal alloy may exhibit a rate-sensitive response and this also needs to be investigated. The following section will consider the influence of varying strain rate on the behavior of the metal and composite constituents, as well as the interface between them.

2.1- Strain Rate Effects at the Composite-metal Interface.

Very little work has been done to investigate the influence of strain rate on the fracture properties of FMLs. Reyes (2002) and Reyes and Cantwell (2002) investigated the effect of varying crosshead displacement rate on the degree of adhesion across the composite-metal interface in a glass fiber reinforced polypropylene (PP)/aluminum FML and a self-reinforced polypropylene steel FML (see Reyes and Gupta 2009). He used the single cantilever beam specimen shown in Figure 1 to conduct tests at crosshead displacement rates between 0.1 mm/minute and 3 m/s.

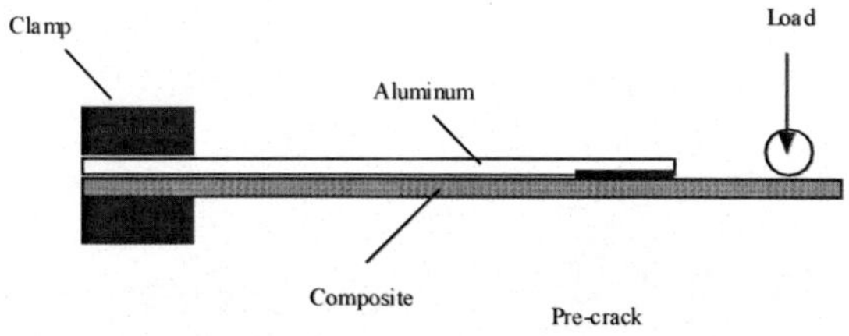

Figure 1. Single cantilever beam used to characterize the degree of adhesion in an FML.

In this test, a composite substrate is bonded to the metal alloy following the recommended manufacturing procedure for the composite. An aluminum starter defect is incorporated between the two materials to act as a starter defect. Load is then applied to the composite arm, forcing a crack to propagate along or close to the composite-metal interface. The interfacial fracture energy, G_c, can then be determined using a compliance calibration procedure (Reyes 2002). This procedure involves forcing the specimen compliance, C, to fit a curve of the form:

$$C=Co+ma^3 \quad (1)$$

Where a is the crack length and m is a constant.

The interfacial fracture energy, G_c, is then determined using the force, P, at each crack length using:

$$G_c=\frac{3P^2ma^2}{2B} \quad (2)$$

where B is the specimen width.

Figure 2 shows the variation of the interfacial fracture energy, G_c, with crosshead displacement rate for single cantilever beam specimens based on unidirectional and woven glass fiber reinforced polypropylene. The values of G_c average approximately 3000 J/m^2 at the lowest rate of loading for both material types. Increasing the crosshead displacement rate results in a steady increase in G_c until a maximum is reached at 100 mm/minute. Further increases in the rate of loading result in a significant drop in G_c, with the value at impact rates being only just above the quasi-static value. This pronounced rate sensitivity was explained by investigating the fracture surfaces of the laminates, Figure 3. At 0.1 mm/minute, a mixed cohesive/adhesive failure mode was observed, suggesting that the crack had propagated both along the interface and through the composite. Increasing the crosshead displacement rate resulted in a greater degree of adhesive failure, with large amounts of composite remaining on the fracture surface. Further increases in rate once again resulted in the mixed fracture appearance observed at the lowest rates (see Reyes 2002). This pronounced change in failure locus was associated with the relative rate-sensitivity of the composite and the adhesive material used to bond the FMLs (see Reyes and Cantwell 2000). The following sections will consider strain rate effects in the fiber direction of composites and FMLs.

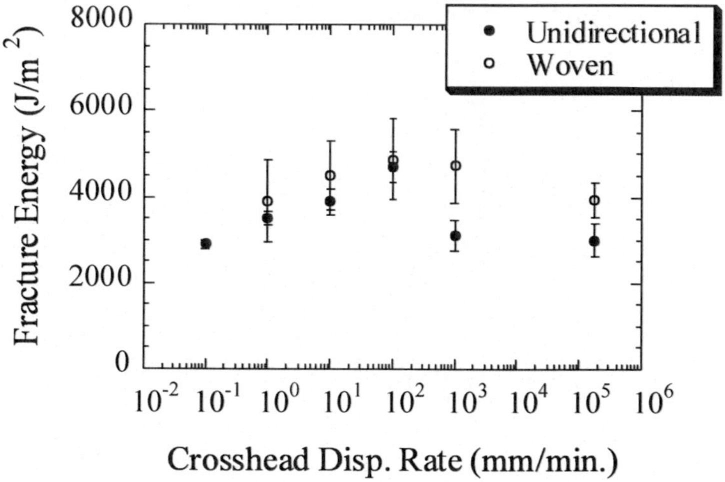

Figure 2. The influence of crosshead displacement rate on fracture energy for glass fiber reinforced/PP FMLs (Reyes, 2002).

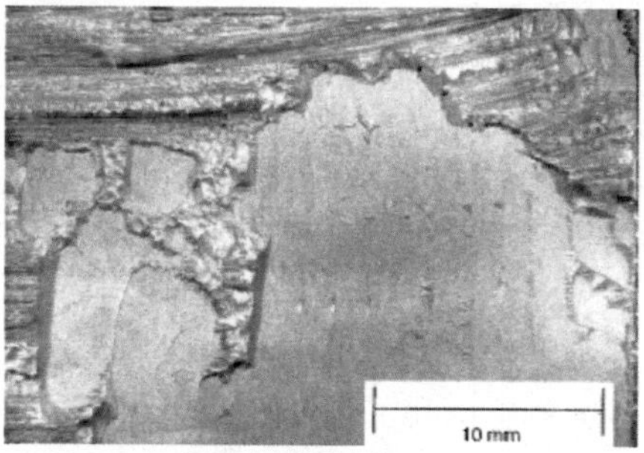

0.1 mm/minute

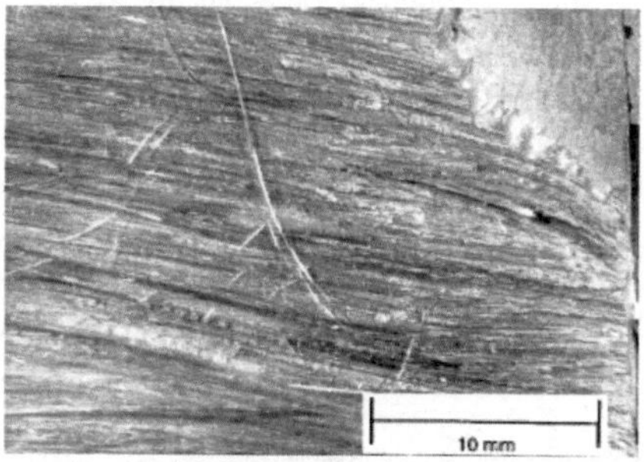

100 mm/minute

3 m/s

Figure 3. Fracture surfaces of SCB specimens at different crosshead displacements rates (Reyes, 2002).

2.2- Rate Effects in Glass Fiber-based FMLs

Vlot (1993) investigated the influence of strain rate on the tensile properties of FMLs. Tests on the FML were undertaken using dog-bone samples over a wide range of strain rates. He found that the tensile strength of GLARE increased by more than twenty percent over approximately five decades of strain rate, Figure 4, whereas ARALL only exhibited a small increase in strength at the highest strain rates. Tests on a plain 7075-T6 aluminum alloy suggested that the tensile strength of this alloy decreased by approximately ten percent over a similar range of strain rates. The rate-sensitivity of GLARE therefore appears to be determined by the rate-sensitive behavior of the glass fiber reinforced composite, since, previous work on such materials has shown the tensile strength of glass fiber reinforced composites increases with strain rate (see Haque and M. K. Hossain, 2003 and Harding, 1987).

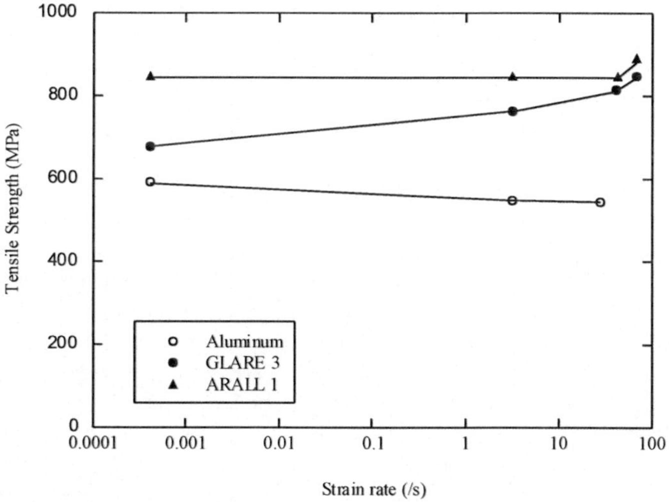

Figure 4. Strain-rate sensitivity of GLARE 3, ARALL 1 and an aluminum alloy. (Vlot, 1993).

The reason for the rate-sensitivity of glass fiber reinforced composites is not fully understood. Xia et al. (1994) conducted tensile tests on glass fiber bundles over seven orders of magnitude and found that their tensile strength increased by almost two hundred percent. Similarly, epoxy resins have been shown to exhibit an increase in tensile strength with increasing strain rate. Gilat et al. (2005) conducted tensile and shear tests on two epoxy resins at strain rates between 5×10^{-5} and 700 s^{-1} and showed that in tension, the failure mode changed from ductile at low rates to brittle at higher strain rates. The tensile strength increased by between 20% and 30% over the range of strain rates considered. More significant increases were observed in shear, with the maximum stress recorded during the test doubling in passing from low to high strain rates. It is therefore likely that both the matrix and the fibers in a glass fiber reinforced composite contribute to the rate-sensitivity of the FML.

Clearly, given the rate-sensitivity of GLARE, it is necessary to employ the appropriate material properties when modelling the dynamic response of FMLs. Indeed, prior to modelling the impact response of a FML leading edge subjected to bird strike impact, McCarthy et al. (2004a) conducted a series of tests in order to characterize and model the rate-sensitive behavior of GLARE-type FMLs based on the stacking sequence [A/0°/90°/A/90°/0°/A] where A represents a layer of 2024-T3 aluminum alloy. Tests were conducted on waisted samples with their fibers aligned at 0°/90° and +/-45° degrees to the loading axis. The length and width of the waisted region were 8 mm and 4 mm respectively. Strain rate effects were observed in both types of laminate, with an increase in strain rate from quasi-static to 3300 s^{-1} resulting in an increase in failure load from 4200 N to approximately 5000 N, Figure 5a. No difference was observed between the response at medium (nominally 100 s^{-1}) and quasi-static rates of loading, leading the authors to conclude that a threshold exists, above which the material begins to exhibit a rate-sensitive response. Tests on samples with fibers oriented at +/-45° showed that whereas the maximum extension increased significantly with strain rate, the tensile strength increased by only a small amount, Figure 5b. Since the load-deformation responses of samples tested at medium and high rates were similar, the authors argued that if a threshold exists in these +/-45° samples, it occurs at a strain rate below 75 s^{-1}.

Given that FMLs are likely be riveted when used as fuselage materials, it is important to characterize the rate-sensitive behavior of structures containing a notch or a similar form of stress concentration. Vlot et al. (1998) conducted tests on 20 mm wide plain-sided notched tensile samples and showed that GLARE laminates containing a sharp notch exhibited a greater rate sensitivity than both unnotched laminates and those containing a blunt notch. This greater degree of rate sensitivity in the samples with sharp notches was attributed to the fact that the strain rate at the root of the sharp notch is raised by presence of the stress concentration.

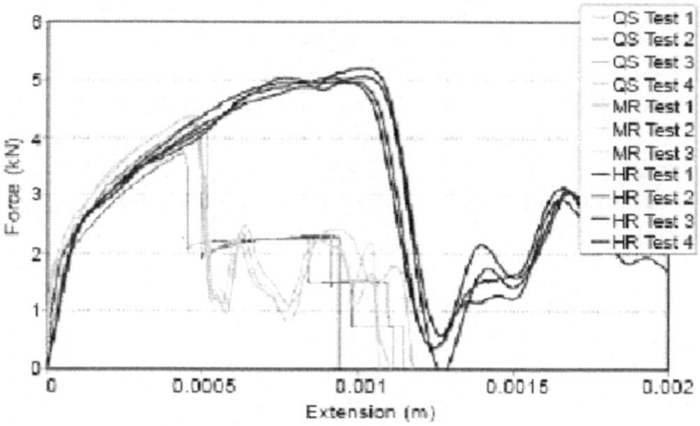

Figure 5a: Force deflection curves for a glass fiber epoxy FML loaded in the 0° direction. QS = quasi-static, MR = medium rate, HR = high rate (McCarthy, 2004a).

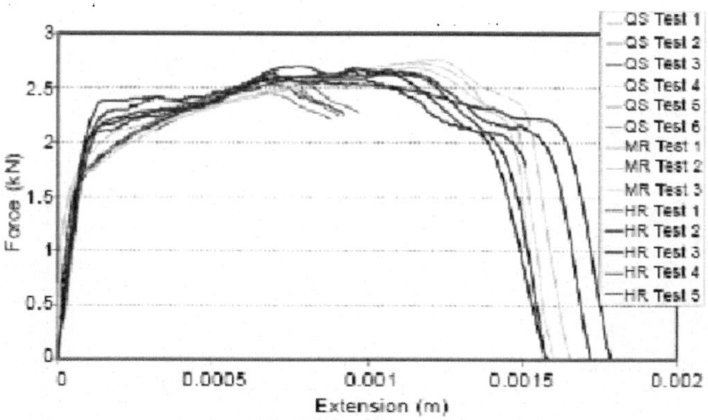

Figure 5b: Force deflection curves for a glass fiber epoxy FML loaded in the 45° direction. QS = quasi-static, MR = medium rate, HR = high rate (McCarthy, 2004a).

2.3- Rate Effects in Kevlar and Carbon-based FMLs

A number of workers have investigated the rate-sensitivity of FMLs based on other types of fiber, particularly Kevlar and carbon. Wu (1991) investigated the combined effect of temperature and strain rate on the tensile properties of ARALL-1 laminates based on a 7475-T61 aluminum alloy and two layers of unidirectional aramid fiber reinforced epoxy resin. Temperatures up to 121°C were considered. The findings of this study are summarized in Figure 6, where the ellipses in the figure enclose averages that were judged not to be statistically significantly different at the 95% level. Strain rate effects in the tensile strength were largely absent at room temperature and at 82°C, except at the very lowest rate. In contrast, at 121°C, the ultimate strength of the ARALL laminates increased by approximately thirty percent over seven decades of strain rate. Similar trends were observed in the modulus and strain data.

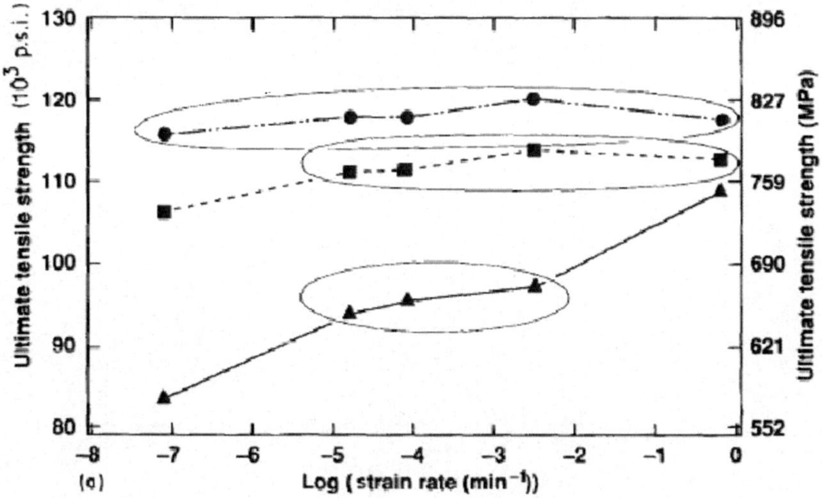

Figure 6. The effect of strain rate and temperature on the ultimate tensile strength of ARALL-1. ● 24°C ■ 82°C ▲ 121°C (Wu, 1991).

Zhou et al. (2004) conducted room temperature tensile tests on a Kevlar fiber reinforced aluminum laminate at strain rates between 100 and 1300 s^{-1} and found that the tensile strength increased steadily with strain rate over the relatively narrow range of rates investigated. They used a linear damage model combined with a Weibull distribution function to establish a constitutive equation for this FML and found that the strength obeyed a bimodal Weibull distribution. Zhou et al. (2004) also investigated the rate-sensitivity of pre-stressed Kevlar fiber FMLs and observed a similar response to that exhibited by the plain Kevlar-based FML laminate.

Xia et al. (2007) characterized the tensile properties of a carbon fiber reinforced epoxy/aluminum laminate at strain rates up to 1200 sec^{-1}. The experimental data indicated that this hybrid is also rate-sensitive, with the tensile strength and failure strain increasing with rate. The authors found that the aluminum alloy and the composite exhibited a strain-rate sensitive response, where the tensile strength doubled over a relatively narrow range of strain rates. Tests on a plain epoxy polymer resulted in a similar strain rate response, leading the authors to conclude that the rate sensitivity of this system is due to the polymer in the composite.

The majority of the published work investigating strain rate effects in the mechanical response of FMLs has shown that such hybrid materials do exhibit a clear strain-rate sensitivity. This appears to be true in both fiber and matrix-dominated modes of loading. Before attempting to model the impact behavior of FMLs it is clearly important, therefore, to ensure that their mechanical properties are correctly established and that any rate-sensitivity is accounted for in the analysis. Failure to do so could lead to grossly inaccurate predictions with potentially serious consequences.

3- Low Velocity Impact Response of FMLs

Impact on fiber metal laminates frequently involves dynamic loading of the structure by a heavy mass travelling at several meters per second. Under such conditions, where the contact time is relatively long, the target frequently exhibits a global response in which energy is absorbed in regions well away from the initial point of contact. Many fiber metal laminates have a greater capacity to undergo significant plastic deformation and should, in principle, offer a greater ability to absorb energy in this mechanism than do composite materials. A number of workers have investigated the low velocity impact response of FMLs and many have compared their performance to that exhibited by plain aluminum alloys and composites (see for example Atas, 2007, Johnson,

1986 and Alderliesten, 2003, Laliberte´ et al., 2000, Liaw et al. 2001).

Vlot et al. (1998) and Vlot (1993) conducted low velocity impact tests on a range of GLARE FMLs and compared their response to the dynamic behavior of a 2024-T3 aluminum alloy and various fiber-reinforced thermoplastics. The damage resistance of the laminates was characterized by measuring the dent depth, the width of the damaged area and the minimum cracking energy. Figure 7 shows the variation of dent depth and maximum displacement as a function of the specific impact energy (impact energy divided by areal density). The figure shows that the trends in the GLARE data are similar to those associated with the 2024-T3 aluminum alloy, with the permanent dent depth being approximately two thirds of the maximum displacement during impact. The dent depth in the fiber reinforced thermoplastic was very small up to the penetration threshold, making it more difficult to identify the impact location. First damage in the FMLs took the form of a crack in the lower surface aluminum sheet that extended parallel to the fiber direction. It was noted that fiber cracking in the composite ply did not occur without cracking in the aluminum layers.

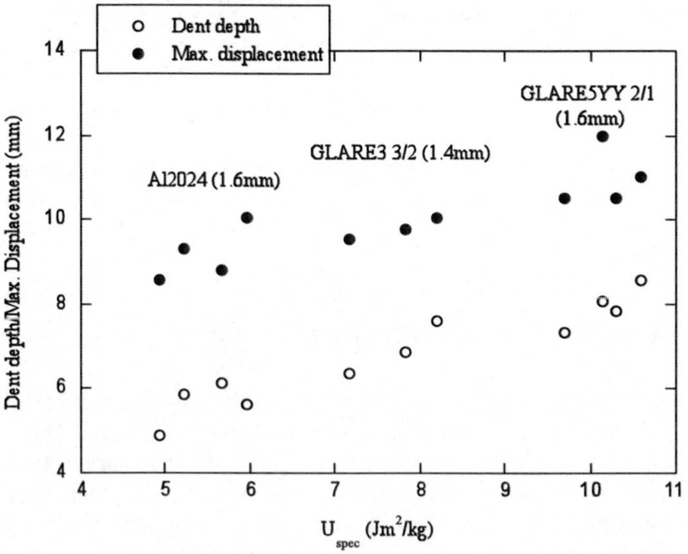

Figure 7. Variation of the maximum displacement and dent depth as a function of specific impact energy (Vlot et al, 1998).

Figure 8 shows the specific cracking energies (defined as the impact energy divided by the areal density) for a through-thickness crack for a range of FMLs, aluminum and composite panels. Interestingly, Figure 8 shows that a very thin plain aluminum offered the highest specific cracking energy, probably as a result of membrane stretching effects in this thin metallic sheet. It is clear that the GLARE laminates offer a similar resistance to that offered by the plain aluminum alloy. The FMLs do, however, exhibit a markedly superior response to that of the carbon fiber reinforced thermoplastic.

Laliberté (2005) et al. investigated the low velocity impact response of three types of GLARE based on a 2-1 configuration, these being GLARE 3, GLARE 4 and GLARE 5 with thicknesses of 0.85, 0.98 and 1.11 mm respectively. In order to avoid problems associated with the corners when using square or rectangular supports, the impact test samples were supported between two 292 mm square steel clamps with a 203 mm diameter cut-out.

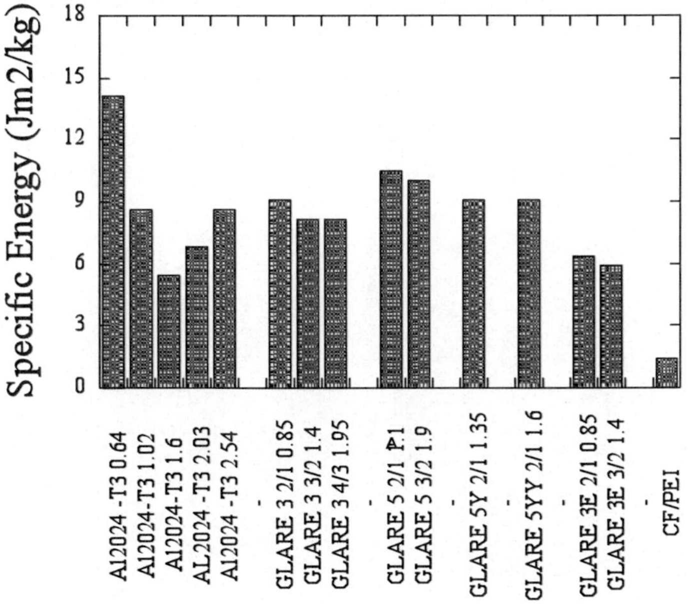

Figure 8. Specific energy for a through thickness cracking in GLARE, aluminum alloy and a carbon fiber PEI (Vlot et al., 1998).

The authors found that the absorbed energy increased with impact energy, Figure 9a, due to a number of mechanisms including plastic deformation of the panel and matrix cracking, delamination, fiber and aluminum fracture as well as a small amount in sound and heat.

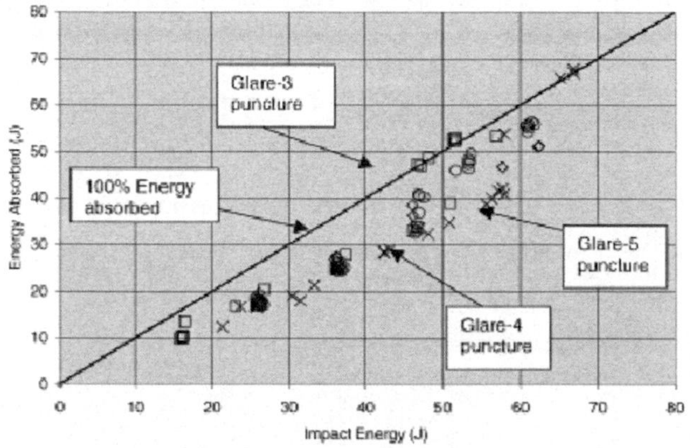

Figure 9a. Absorbed energy vs impact energy for impacts on aluminum 2024-T3 ◇ , GLARE 3□, GLARE 4 ○ and GLARE 5 ⨯ (Laliberté, 2005).

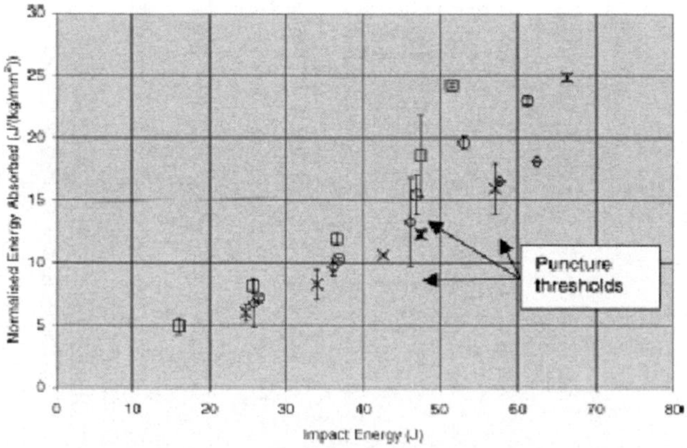

Figure 9b. Normalized absorbed energy vs. impact energy for tests on aluminum 2024-T3 ◇ , GLARE 3□, GLARE 4 ○ and GLARE 5 ⨯ (Laliberté, 2005).

The relative performance of the different panels was examined by normalizing the data in Figure 9a by the respective areal density. Here, it was noted that the GLARE 5 panels absorbed the least amount of energy in damage, Figure 9b, and therefore incurred the smallest amount of damage. It was stated that the additional composite plies in the GLARE 5 laminate increased its impact resistance whilst lowering the areal density of the laminate. Similar observations were made when the dent depth was measured, with the incorporation of additional composite layers in the GLARE 5 laminates resulting in smaller dent depths after impact. One of the advantages of FMLs such as GLARE, is that there is no hidden damage within the depth of the laminate and the plastically-deformed dent gives evidence of the impact event.

Wu et al. (2007) conducted low velocity impact tests on GLARE 4 and GLARE 5 laminates to investigate damage initiation and propagation in these composite metal hybrids. They showed that the specific energy for first cracking in GLARE 5 (with a 2/1 stacking configuration) was approximately ten percent higher than that of a 2024-T3 aluminum alloy. Cross-sections of damaged FML laminates highlighted considerable damage between the lower surface aluminum layer and the composite ply. It was suggested that such damage was due to interfacial shear stresses associated with bending of the laminate. It was also argued that delamination allows GLARE laminates to deform and fracture in a more efficient membrane action, contributing to the overall level of energy absorption.

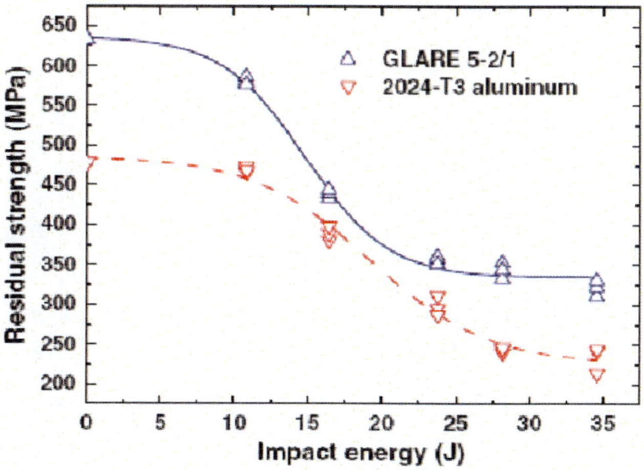

Figure 10. Post-impact residual strength of GLARE and 2024-T3 aluminum alloy (Wu at al., 2007).

Post-impact residual tensile strength tests were conducted on GLARE 5 and plain aluminum alloy laminates subjected to impacts up to approximately 35 Joules, Figure 10. At energies up to 10 Joules, only a local plastic dent was observed in the GLARE and aluminum alloy and the resulting strength reductions for both types of material were minor. At higher energies, the strengths of both laminates dropped rapidly, although the GLARE laminate offered superior residual properties at all impact energies.

In operational service, FMLs are likely to be subjected to some form of in-plane loading when impacted by a projectile, possibly leading to greater levels of damage. Vlot (1998) conducted a series of low velocity impact tests on FMLs and a plain aluminum alloy under a tensile pre-load. The test samples were clamped in the grips of a hydraulic testing machine and subjected to localized impact loading by an instrumented pendulum. An examination of the force-deflection traces showed that the specimens exhibited a higher stiffness and a higher maximum force as a result of the initial stress. Figure 11 shows the variation of residual strength as a function of the level of initial stress for ARALL laminates subjected to energies of 6.12 and 9.39 Joules. A significant reduction in residual strength was observed for preload stresses in excess of 200 MPa, presumably as a result of the combined initial stress and the impact-generated stresses exceeding the failure strength of the FML.

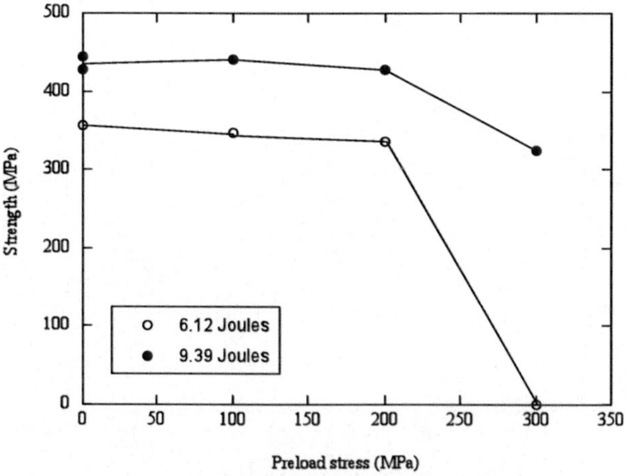

Figure 11. Strength of impact-loaded ARALL laminates as a function of initial stress. Legend gives impact energy (Vlot et al., 1998).

Caprino et al. (2004) investigated the low velocity impact response of fiberglass-aluminum hybrids and showed that these materials offer a superior resistance to perforation than plain composite materials. However, their perforation resistance was lower than that measured on a monolithic aluminum alloy of equivalent thickness. Caprino et al. (2007) modelled the impact behavior of these glass/aluminum hybrids using simple second order polynomials. The proposed model was capable of predicting the main impact parameters such as the force-displacement trace, the contact duration and the dissipated energy.

Other workers have also attempted to model the low velocity impact response of FMLs. Seo et al. (2010) developed a finite element model to predict the impact response of GLARE and compared their results to experimental data generated by Wu et al. (2007). The authors found that for different kinds of impact damage, agreement between the FE models and experimental data was good. Small discrepancies were noted in the contact durations with the numerical results giving smaller values. This was to the boundary conditions adopted in the model. Song et al. (2010) modelled a carbon fiber reinforced plastic/aluminum FML and observed reasonable agreement between the FE model and experimental data on hybrid plates.

Payenganeh et al. (2010) modelled the low velocity impact response of FMLs based on a glass fiber reinforced polyester. They showed that the layer stacking sequence is an important factor in determining the dynamic response of FMLs. They showed that the positioning of the aluminum layers determines the local and overall response of the structure. Placing the metal layers close to the impact surface had the most beneficial affect whereas placing they away from this surface reduces this effect or can even have adverse effects.

Kulkarni et al. (2008) investigated the low velocity impact response of a long fiber thermoplastic/aluminum alloy laminate. Long fiber reinforced thermoplastic (LFT) composites are based on a thermoplastic matrix such as polypropylene, polyamide or polyurethane and are reinforced with discontinuous fibers with lengths between 10 and 50 mm. LFTs offer a number of advantages, including their low cost and the fact that they can be processed using conventional molding procedures such as compression molding, extrusion and injection molding. Kulkarni et al. (2008) conducted drop-weight impact tests on an aluminum alloy/fiber reinforced Nylon 6,6 laminate and found that perforated panels showed failure in the form of shear fracture of the upper and lower aluminum plies, delamination between the composite and metal layers and fracture of the LFT composite. The perforation resistance of the fiber metal laminate was compared to that of the plain LFT by normalizing the perforation energy by the areal density of the laminate. The resulting specific perforation energy for the long fiber thermoplastic/metal laminate was 7.58 J/(kg/m^2) and that for the plain LFT was 1.72 J/(kg/m^2). Comparisons with other hybrid systems showed that the LML offered a superior perforation resistance to a thermosetting-based system.

Carrillo (2007) and Carrillo and Cantwell (2008) investigated scaling effects in the low velocity impact response of a polypropylene-based fiber-metal laminate (FML) structure. Two approaches were employed to investigate stacking sequence effects in scaling the FMLs. In the first approach, referred to as ply-level scaling, the laminate thickness was increased by scaling the thicknesses of the aluminum sheets and the number of 0°/90° layers in the inner composite core of the FML. This yielded four specimen thicknesses representing four scale sizes, n, equivalent to n = ¼, ½, ¾ and 1 as shown in Table 3. The second method, sublaminate-level scaling, involved increasing the specimen thickness by repeating the basic sublaminate block up to four times. This again yielded four scaled thickness and the resulting laminates are given in Table 4.

Table 3. Data for the ply level scaled FMLs. 2A, 3A and 4A refer to aluminum sheets that are nominally 2,3 and 4 times the thickness of the baseline aluminum with a thickness A (data from Carrillo, 2007).

n	*Stacking sequence*	*Thickness (mm)*
1/4	[A, 0/90]s	1.77
1/2	[2A 0/90, 0/90]s	3.51
3/4	[3A, 0/90, 0/90, 0/90,]s	5.45
1	[4A, 0/90, 0/90, 0/90, 0/90,]s	7.03

Carrillo (2007) generated load-displacement traces by combining the output of a laser-Doppler velocimeter with a piezo-electric load-cell. Wen and Jones (1993) conducted a similitude study to investigate scaling effects in the low velocity impact response of steel plates. They developed a replica scale law through the careful inspection of the pertinent input and output parameters.They showed that for the scaling laws to be satisfied, the panels dimensions should be scaled according to the scale size 'n', the impactor mass and associated energy should be scaled by the cube of the scale size i.e. 'n^3'. The resulting (measured) force and target displacement should scale with 'n^2' and 'n' respectively. Figure 12a shows the load-displacement responses from Carillo's (2007) tests on the four panel sizes at a constant impact energy of $302n^3$ Joules using an impact mass of $22.46n^3$ kg. From the figure, it is evident that the panels exhibit some recovery at the end of the test, indicative of an elastic response in the composite plies. It is worth nothing that none of these specimens showed external damage after the impact test. The residual displacement in the figures is therefore due to plastic deformation in the metal plies. Figure 12b shows the normalized responses, where it is evident that all four load-displacement traces collapse onto each other, suggesting that the scaling laws were satisfied.

Cross-sections of scaled samples at this energy level are presented in Figure 13. Following an examination of the ply-level scaled samples in Figure 13a, Carrillo (2007) stated that the level of gross plastic deformation is similar for the four scale sizes. All samples exhibit localized denting and some thinning around the point of impact as well as significant residual deformation.

Table 4. Data for the sublaminate level scaled FMLs (data taken from Carrillo, 2007).

n	*Stacking sequence*	*Thickness (mm)*
1/4	[A, 0/90]s	1.77
1/2	[A, 0/90, A, 0/90]s	3.62
3/4	[A, 0/90, A, 0/90, A, 0/90]s	5.44
1	[A, 0/90, A, 0/90, A, 0/90, A, 0/90]s	7.24

The aluminum-composite interface exhibited a small amount of debonding and delamination for the largest scaled samples. In general, damage tended to become more severe as the scale size increased. A closer examination of the full-scale sample indicates that the composite core has failed directly under the point of impact, this being attributed to the fact that the structure of the weave was not scaled in the study. Figure 13b shows the cross-sections of four samples scaled at a sublaminate-level.

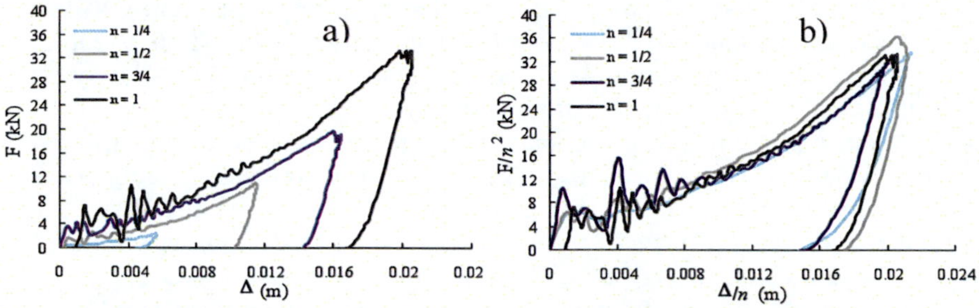

Figure 12. Impact force vs. displacement traces for scaled samples and b) Normalized traces. Energy = $302n^3$ Joules (Carrillo, 2007).

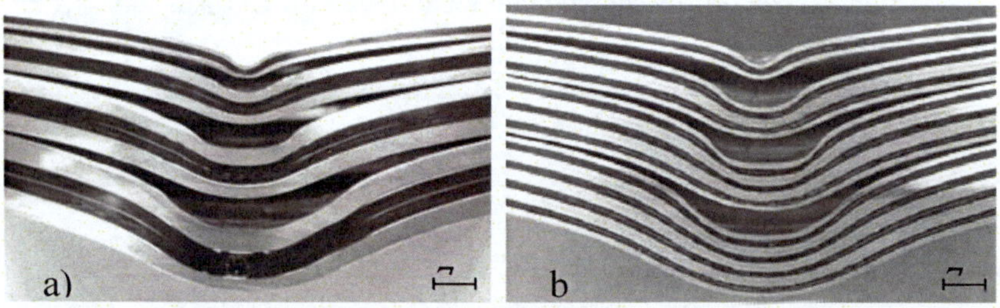

Figure 13. Scaled samples at an impact energy of $302n^3$ Joules a) Ply-level scaling and b) Sublaminate-level scaling (Carrillo, 2007).

Here, the plates exhibit significant plastic deformation without incurring fracture in the outermost aluminum layers. No delamination was observed between the composite plies in any scaling sizes.

The low velocity impact response of FML plates was further investigated using the non-dimensional approach developed by Wen and Jones (1993). They showed that graphs of the non-dimensional residual target deflection, δ, against the non-dimensional impact energy, χ, should yield a common curve over the range of scale sizes considered.

The non-dimensional energy was defined as:

$$\chi = \frac{E}{\sigma_y d^3} \tag{3}$$

where E is the impact energy, σ_y is the yield stress of the material (steel in the case of Wen and Jones' work) for the given scale size n and d is a characteristic scale length.

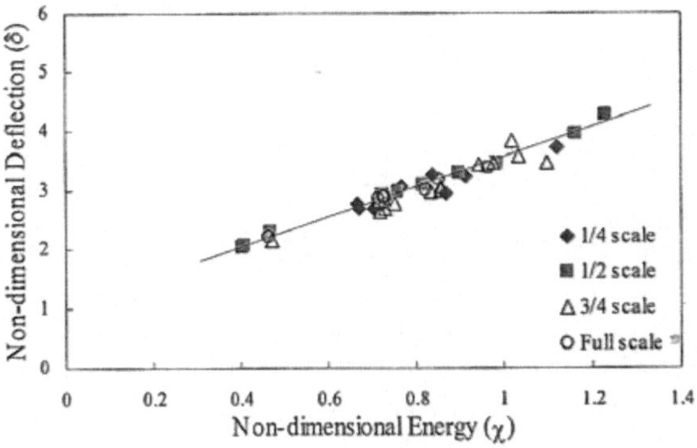

(a)

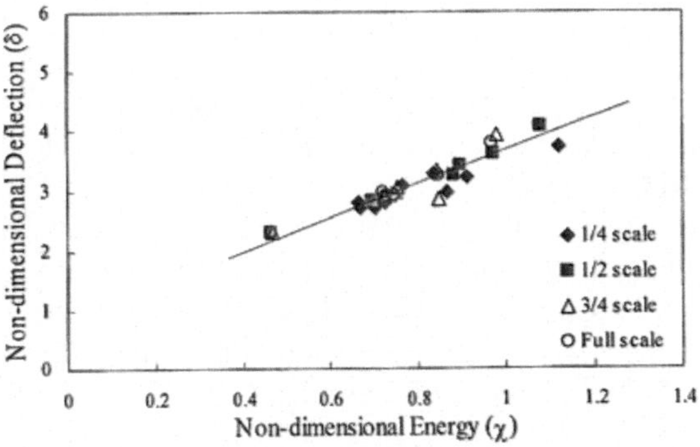

(b)

Figure 14. Non-dimensional permanent deflection vs. non-dimensional impact energy following tests on FMLs scaled at a) ply-level and b) sublaminate-level (Carrillo, 2007).

Figure 14 shows plots of non-dimensional permanent deflection, δ, against non-dimensional impact energy, χ, for specimens scaled using both approaches outlined above. From the figure, it is clear that both sets of data appear to exhibit similar trends with the data collapsing onto straight lines. It was noted that the slopes of both traces were similar, suggesting that both sets of laminates exhibit a similar impact response.

A number of workers have investigated using metals other than aluminum in FMLs (see for example McVay and Johnson 2009). Khalili et al. (2005) tested FMLs manufactured from combinations of steel, aluminum and glass fiber reinforced epoxy resin. They showed that a steel GFRP hybrid offered a Charpy fracture energy that was more than double that of the plain glass fiber composite. Bernhardt et al. (2007) conducted low velocity impact tests on a titanium/carbon fiber reinforced thermoplastic FML and showed that this hybrid out-performed a carbon fiber reinforced epoxy composite exhibiting a smaller damage zone after impact.

Cortes and Cantwell (2007) and Cortes (2005) investigated the low velocity impact of FMLs based on titanium sheets and both carbon fiber reinforced poly-ether-ether-ketone (PEEK) and poly-ether-imide (PEI). Testing showed that both unidirectional and cross-ply PEEK FMLs offered a superior impact resistance to the plain carbon fiber composite. However, when the data were normalized by the areal density of the laminates, the cross-ply FML offered a similar response to the plain composite and the unidirectional FML offered a poorer impact resistance, Figure 15. Similar tests on the PEI-based system showed that the titanium FML exhibited a poorer impact resistance to that the plain composite. An examination of the samples after testing highlighted the presence of longitudinal splits in the metal layers as well as the absence of any significant ductility in the metal plies. The values of specific perforation measured on these systems were lower than those measured on thermosetting-based FMLs (see Reyes, 2002).

This evidence suggests that, in terms of low velocity impact resistance, there is no benefit from combining titanium and tough thermoplastic matrix composites to form fiber metal laminates. In contrast, however, tests by Cortes and Cantwell (2006) on FMLs based on magnesium alloys showed that a glass fiber reinforced polypropylene/ magnesium FML offered a specific perforation energy that was more than double that offered by a glass fiber reinforced epoxy/aluminum FML. This suggests that magnesium-based FMLs do offer some potential in terms of their specific impact resistance, although the added cost of magnesium alloys may limit their wider use.

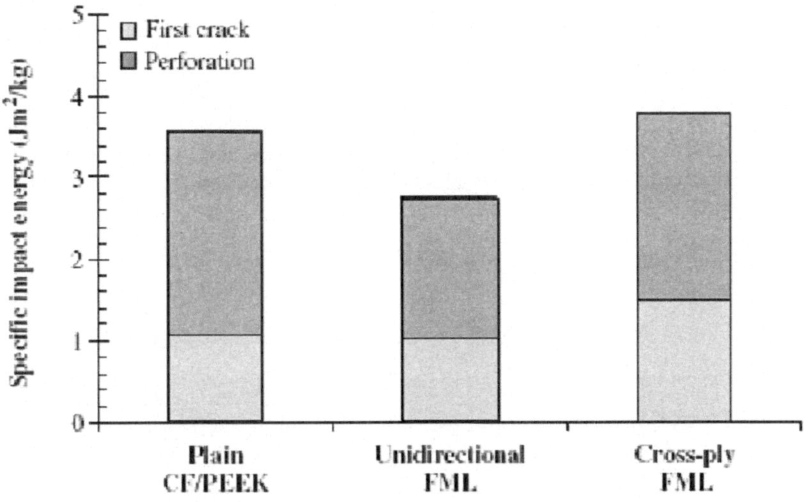

Figure 15. Specific impact energies of the plain carbon fiber/PEEK and FMLs based on CF/PEEK under low velocity impact (Cortes, 2005).

Few workers have investigated the long-term fatigue response of impact-damaged FMLs. Laliberté et al. (2002) conducted a series of tension-tension fatigue tests on several types of GLARE and an aluminum alloy following low velocity impact loading at energies up to 65 Joules. The aluminum samples failed suddenly through the impact dent after approximately 25,000 cycles. The GLARE FMLs offered superior fatigue lifetimes and never failed from the dent. Instead, crack propagation often initiated at the radius of the dogbone sample before propagating across the samples away from the point of impact. One sample that had been subjected to an impact energy of 52 Joules failed at the grips of the test machine after 127,000 cycles. The authors argue that the stress state around the dent in FMLs is complex with a non-symmetrical stress state developing in which high tensile stresses are developed on the impacted surface and compressive stresses develop on the lower face. This evidence suggests that superior fatigue response of notched FMLs is reflected in a similarly impressive post-impact fatigue response.

Few workers have studied the impact response of FML sandwich structures. Reyes (2010) investigated the low velocity impact response of aluminum form sandwich structures with FML skins. A number of failure mechanisms were observed in the sandwich structures including buckling and interfacial delamination in the FML skins as well as crushing and densification of the aluminum foam. An energy-balance analysis indicated that the greatest amount of energy was absorbed in contact effects within the metal core.

Summary

A number of detailed studies have been undertaken to characterize the low velocity impact response of FMLs. Testing has shown that when compared to plain composite materials, such as a relatively tough carbon fiber reinforced thermoplastic, damage in FMLs is more localized for a given impact energy. Cross-sections of damaged FMLs have highlighted a number of failure mechanisms including delamination between the composite layers, interfacial failure between the composite and metal plies, fiber fracture, membrane stretching, extensive yielding and ductile fracture of the metal sheets. With respect to perforation resistance, FMLs do not appear to offer any significant advantage over conventional aluminum alloys, with values of specific perforation energy for both types of material being similar for a given set of impact conditions. Initial scaling studies have shown that FMLs obey a simple scaling law, suggesting that small-scale samples can be used to predict the impact behavior of larger, more representative structures. Attempts to hybridize composites with other metals other than aluminum alloy have not yielded any significant benefit in terms of perforation resistance, with FMLs based on titanium alloys offering a relatively poor impact response. Few attempts have been made to model the low velocity impact response of FMLs and more work is needed in this area to understand and optimize the low velocity impact response of FMLs.

4- High Velocity Impact Response of FMLs

Fiber metal laminates are being considered for use in structures that can be subjected to impact by a lightweight projectile travelling at high velocity. Impact scenarios of this nature frequently result in a very localized mode of loading that is very different from the global response observed under conditions of low velocity impact by a heavy mass. This can lead to a change the energy-absorbing mechanisms involved in the impact process and material properties such as the transverse fracture properties and delamination resistance of the composite may become important.

Hoo Fatt et al. (2003) presented experimental data obtained from a series of ballistic tests showed that the ballistic limit data for GLARE and bonded GLARE laminates were not significantly different from those measured on a

2024-T3 aluminum alloy. In thinner FMLs, the ballistic limit of the FMLs was up to fifteen percent greater than that of the aluminum alloy, whilst in thicker laminates the ballistic response was similar to that of the plain aluminum alloy. Hoo Fatt et al. (2003) stated that GLARE could be used to raise the ballistic limit in certain instances. A cross-section of an impacted FML showed that there was extensive delamination between the glass/epoxy layers without debonding between the composite and metal plies. Hoo Fatt et al. (2003) developed an analytical solution to explain energy absorption in GLARE and to explain the disparity in the test data. Here, an equivalent mass-spring model was employed to establish the dynamic response of GLARE when impacted by a blunt titanium cylinder. Expressions for the inertial and stiffness responses of the panels were developed by considering static indentation and these were then expressed at time-dependent functions by equating the panel dimensions to the distance travelled by shear waves propagating the impact location. The ballistic limit was determined by solving the non-linear differential equation of motion and using energy conservation.

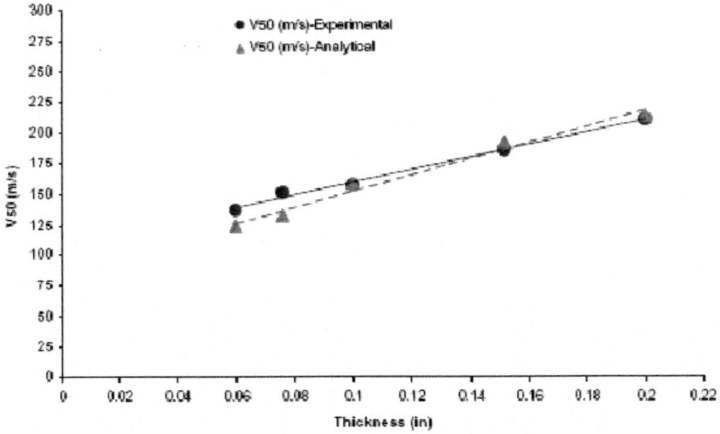

Figure 16. Variation of V_{50} of GLARE with target thickness. (Hoo Fatt et al., 2003).

The analytical model was found to be within thirteen percent of the perforation threshold data as shown in Figure 16. The model was then used to conduct an energy partition to highlight the relative contribution of the deformation and failure mechanisms. The partition suggested that most of the energy was dissipated in global panel deformation with bending and membrane deformations accounting for between 84 and 92% of the total energy. The authors showed that thinner panels absorbed a higher percentage of deformation energy than their thicker counterparts, since it was easier for then to deform in bending and stretching prior to failure. In contrast, delamination accounted for only 2 to 9% of the total energy, with thinner laminates absorbing a lower percentage of the overall energy due to their reduced number of interfaces. The remaining 7% of energy was absorbed in tensile fracture of both the composite and the metal plies.

Vlot et al. (1998) determined the minimum cracking energy for a range of FMLs under high velocity impact loading and compared their response to that observed under drop-weight impact conditions. Table 5 summarizes the data for a number of 2/1 FMLs and compares their behavior to that offered by an aluminum alloy and a carbon fiber reinforced PEI. It is clear that the minimum cracking energy is greater at high velocities that at low velocities. This is most pronounced in the GLARE 5Y system (an FML with extra layers of glass fiber composite). The data also suggests that the rate-sensitivity, as defined by the ratio of the high velocity threshold energy normalized by its low velocity equivalent, increases with fiber content, an effect that can be attributed to the rate-sensitivity of the fiber reinforcement. It is interesting to note that the aluminum alloy does not exhibit any rate-sensitivity whereas the carbon/PEI suffers a decrease. When the data were normalized with respect to areal density, the GLARE 5 laminates, with additional composite plies, again offered the most impressive impact resistance. The authors also argued that the impressive impact resistance of the FMLs is due to membrane deformation in the thin aluminum layers, particularly after the metal sheets have delaminated from the composites plies.

Figure 17 shows the variation of the width of damage with impact energy for two GLARE FMLs and the carbon fiber PEI composite (see Vlot et al. 1998). Included in the figure are data following low velocity impact tests on the same materials. From the figure, it is clear that the FMLs offer a superior impact resistance relative to the plain composite. Once again, the FML with the higher volume fraction of composite plies (GLARE 5YY with a 2/1 stacking sequence) offered a superior impact resistance than its counterpart with a lower composite volume fracture (the GLARE 3 laminates with a 3/2 stacking sequence).

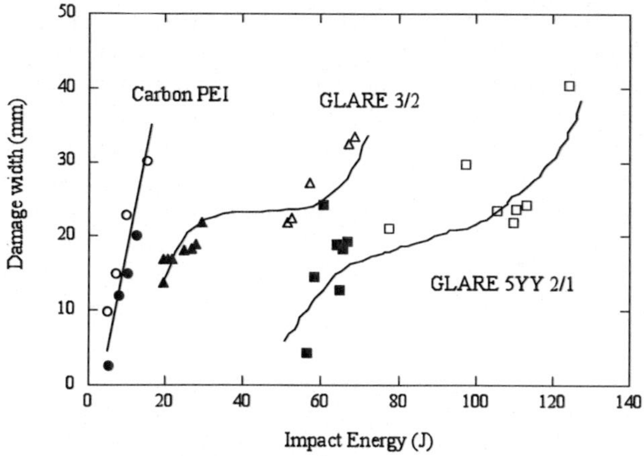

Figure 17. Variation of damage width with impact energy. Open symbols: high velocity impact, closed symbols: low velocity impact . (Vlot et al., 1998).

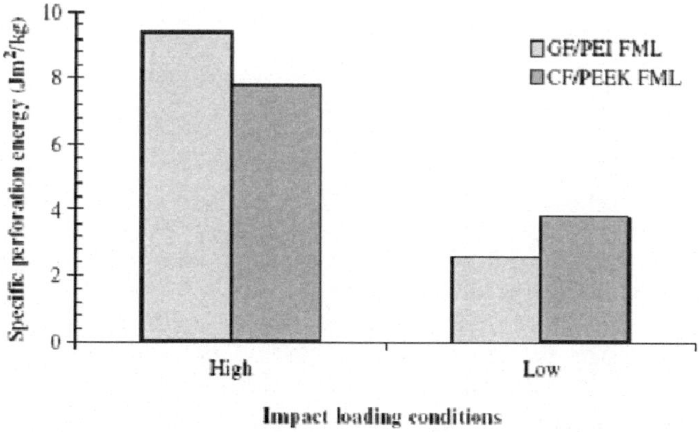

Figure 18. Specific perforation energies of FMLs based on CF/PEEK and GF/PEI after low and high velocity impact (Cortes, 2005).

Cortes and Cantwell (2007) conducted high velocity impact tests on titanium FMLs based on both carbon fiber reinforced PEEK and glass fiber reinforced PEI. The impact resistance of the FMLs were compared by determining their perforation energies and Figure 18 compares the high velocity specific perforation energies of these systems with those measured under low velocity impact conditions. From the figure it is evident that these thermoplastic-based systems exhibit a pronounced rate sensitivity, with the specific perforation energy increasing by up to three hundred percent in passing to from low to high velocity impact loading. In spite of this, these values are significantly lower than those based on other thermoplastic matrix systems (see Reyes, 2002).

Table 5. Minimum cracking energies lfor GLARE, aluminum alloy and a carbon fiber reinforced PEI (data taken from Vlot, 1991).

Material	Lay-up	t (mm)	Areal density (kg/m^2)	Min. cracking energy (J) Low velocity	Min. cracking energy (J) High velocity
Al 2024-T3	-	0.64	1.67	31.8	31.0
	-	1.02	2.78	28.1	48.7
	-	1.60	4.45	23.5	43.2
	-	2.03	5.56	22.4	49.8
	-	2.54	6.95	35.6	78.6
	-	3.18	8.62	46.7	
GLARE 3*	2/1	0.85	2.16	21.4	43.3
	3/2	1.4	3.49	26.2	69.5
	4/3	1.95	4.82	37.0	110.5
GLARE 5*	2/1	1.10	2.66	28.4	70.9
	3/2	1.90	4.46	44.1	99.0
	4/3	2.70	6.31	56.9	
GLARE 5Y*	2/1	1.35	3.15	28.6	110.5
	3/2	2.40	5.47	52.2	
GLARE 5YY*	2/1	1.60	3.65	34.0	126.7
	3/2	2.90	6.46	57.7	
GLARE 3E+	2/1	0.85	2.18	14.0	27.7
	3/2	1.40	3.52	13.7	47.5
	4/3	1.95	4.86	24.0	
Carbon/PEI	$[0/90]_7$	2.00	3.46	3.8	3.0

* = Based on S2 glass fibers + = Based on E glass fibers

Sun et al. modeled the contact behavior in an ARALL laminate and investigated the residual tensile properties of laminates in the longitudinal (fiber) and transverse directions (see Sun et al., 1993 and Wu and Sun, 1993). The loading portion of the indentation trace was successfully modeled using a generalized contact law of the form:

$$P = k\alpha^{n} \tag{4}$$

Where 'α' is the indentation, 'k' is contact rigidity and 'n' is a power index, Tests showed that a value for 'n' of 1.5 yielded a good fit to the data.
The unloading path was modeled using:

$$P=P_m\left(\frac{\alpha-\alpha_o}{\alpha_m-\alpha_o}\right)^q \tag{5}$$

where P_m is the contact force at which unloading begins, α_m is the corresponding indentation and α_o is the permanent indentation. The power index, q, was found to equal 2.5 for ARALL laminates. Sun et al. (1993) observed a linear relationship between local indentation and impact velocity and showed that the residual tensile strength is related to the depth of the dent in the FML following impact. This is shown in Figure 19 where the residual strength of the 0° FML decreases rapidly with increasing dent depth. It was suggested that this is a convenient tool for estimating the residual strength of damaged ARALL laminates. Clearly, by measuring the depth of the dent and using the data in Figure 19, it should be possible to estimate the residual load-bearing capacity of the laminate.

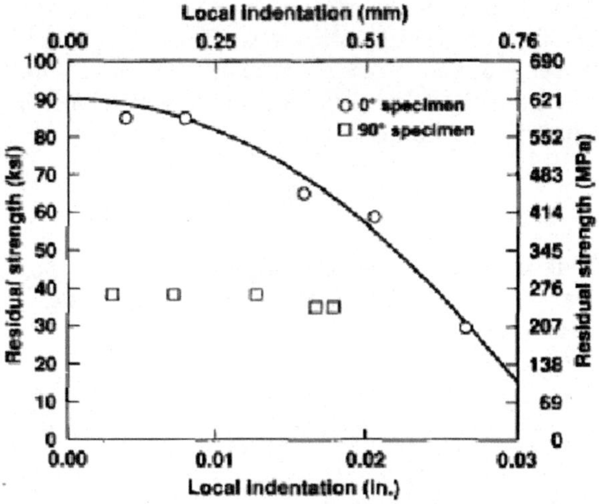

Figure 19. Residual tensile strength vs. local indentation for ARALL laminates with fibers at 0° and 90° (Sun et al., 1993).

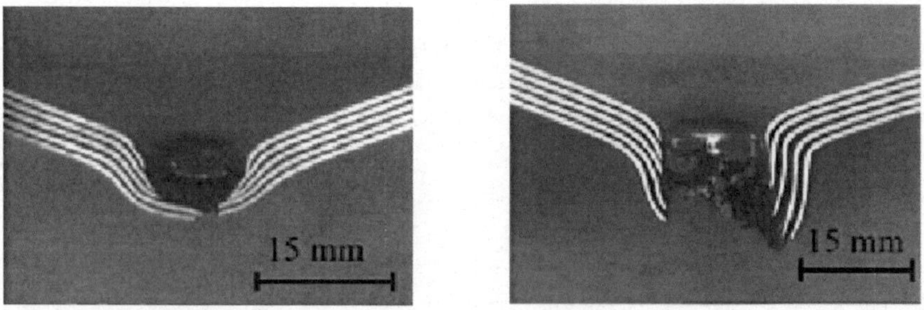

(a) (b)

Figure 20. Cross-sections of 5/4 aluminum alloy/self-reinforced PP. High velocity impact (a) 228 Joules (b) 244 Joules (Abdullah, 2006).

Abdullah (2006) Abdullah and Cantwell (2006a and 2006b) investigated the high velocity impact response of FMLs based on glass fiber reinforced polypropylene and a self-reinforced polypropylene. Figure 20 shows cross-sections of a self-reinforced PP fiber-metal laminate with a 5/4 stacking configuration, where the composite and aluminum layers have been fractured and some thinning of the metal layers is apparent. It is also evident that the FML has absorbed a considerable amount of energy in plastic deformation, with the residual deflection being more than double the initial thickness of the laminate. Impact tests on FMLs based on 2024-T3 and 2024-O aluminum alloys showed that those structures based on the T3 alloy offered a superior perforation resistance, due to the superior energy-absorbing capability of this particular alloy, Figure 21. The perforation resistance of the FMLs was modeled using Wen and Reid's impact model that gives the perforation velocity as:

$$V_b = \frac{\pi \Gamma \sqrt{\rho_t \sigma_e} D^2 T}{4m} \left[1 + \sqrt{1 + \frac{8m}{\pi \Gamma^2 \rho_t D^2 T}} \right] \tag{6}$$

where ρ_t is the density and T is the thickness of the target, m and D are the mass and diameter of the projectile, σ_e is the static elastic compression limit of the FML and Γ is a constant having a value of 1.5 for a hemispherical impactor. The solid lines in Figure 21 correspond to the predictions of this model, where it is clear that the above equation predicts the impact response with some success. The perforation resistances of the various self-reinforced polypropylene FMLs were subsequently compared by dividing the perforation energy by the areal density. Figure 22 compares the specific perforation energies of a range of self-reinforced polypropylene FMLs based on 2024-O and 2024-T3 aluminum alloys (see Abdullah and Cantwell, 2006a and Abdullah, 2006). Similar observations were made by Abdullah and Cantwell (2006a) and Abdullah (2006) following high velocity impact tests on a glass fiber reinforced PP/FML. Increasing the thickness of the composite core of these laminate resulted in an increase in the specific perforation resistance. The figure shows that 2/1 FMLs with a thick composite core out-perform multi-layer systems such as 3/2 and 4/3 laminates. A comparison of the perforation resistances of a number of different FMLs showed that those based on a self-reinforced polypropylene offered a specific perforation resistance that was more than double that of an aramid fiber/aluminum alloy hybrid.

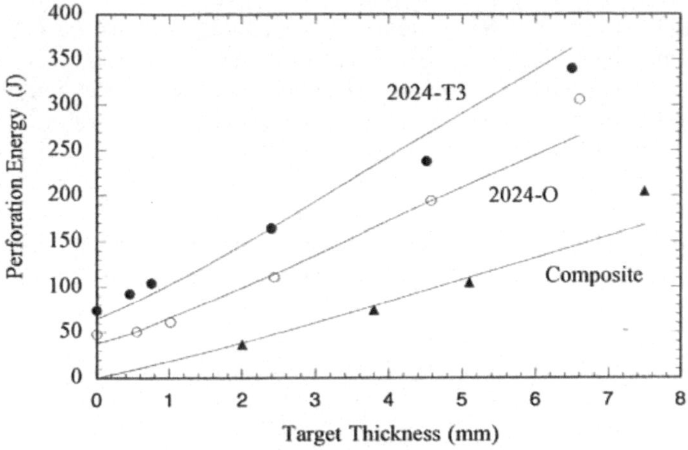

Figure 21. Predicted and experimental perforation data for high velocity impact on SRPP FMLs. (Abdullah, 2006).

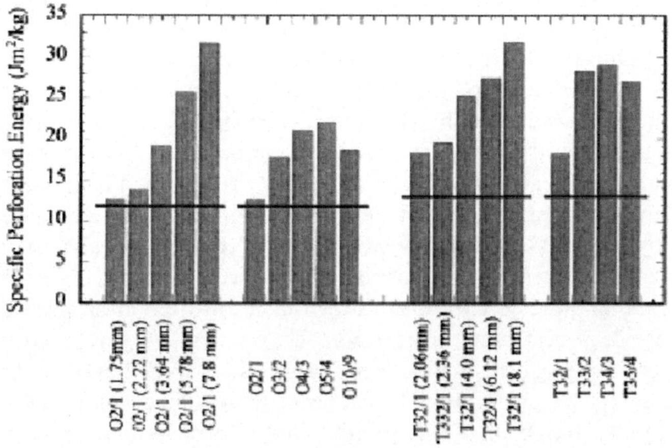

Figure 22. Specific perforation energies of SRPP FMLs based on 2024-O and 2024-T3 aluminum alloys (Abdullah, 2006).

Guan et al. (2009) modeled the high velocity impact response of FMLs based on a self-reinforced PP/aluminum alloy composite. The composite was modeled as an isotropic material with a specified tensile cut-off stress to allow for the automatic removal of failed elements. The aluminum was modeled as an elasto-plastic material with a specified shear failure strain and a tensile failure cut-off stress. Figure 23 shows deformed samples after impact and the corresponding predictions from the FE analysis. Agreement between the deformation modes and failure mechanisms was found to be very good for a range of laminates and impact conditions.

The accuracy of the FE model was also investigated by comparing the variation of the predicted and measured non-dimensional residual displacement (permanent displacement normalized by the laminate thickness) with the normalized impact energy (energy divided by perforation energy). Figure 24 shows a comparison of the numerical and experimental data for FMLs based on a T3 aluminum alloy. Agreement between the model and the experimental data is very good over the range of energies considered. The residual deflections were greater in the lower strength 2024-O alloy than it in its stronger T3 counterpart.

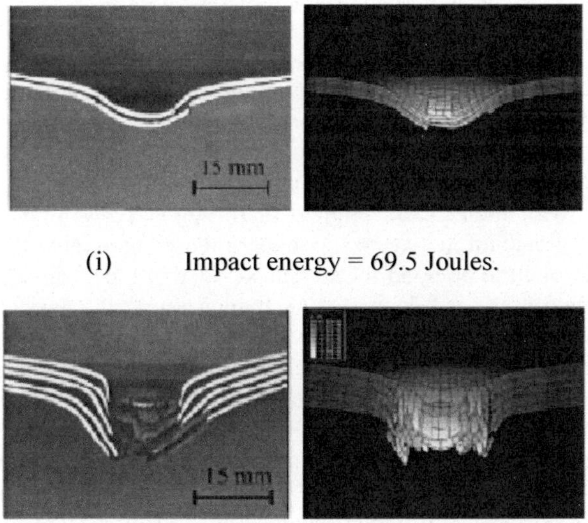

(i) Impact energy = 69.5 Joules.

(ii) Impact energy = 353 Joules.

Figure 23. Comparison of the cross-sections of impact-damaged with numerical predictions. SRPP/2024-T3 FMLs: (i) 2/1 laminate (ii) 5/4 laminate (Guan et al., 2009).

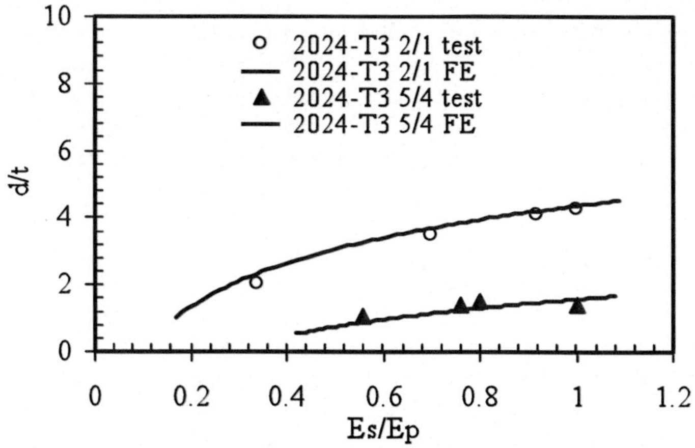

Figure 24. Non-dimensional maximum permanent displacement vs.normalized impact energy for 2024-T3 FMLs. Solid lines are predictions of numerical model (Guan et al., 2009).

McCarthy et al. (2004a, 2004b) modeled the case of a bird strike impact on a GLARE-type leading edge. Initially, attempts were made to model the strain-rate sensitive response of the FMLs. Here, the material model was based on a continuum damage model for composites introduced into the finite element code PAM-CRASH/SHOCK. The model was compared with experimental data from tests on FMLs with their fibers aligned at 0°/90° and +/-45° degrees to the loading axis. Agreement between the experimental data and the numerical predictions was found to be good for samples loaded in the 0° direction at quasi-static and medium rates of loading. At higher rates, the model accurately predicted the extension at failure but overestimated the maximum force. The authors argue that a more complex model for fiber fracture, which allows a bi-linear rather than a linear damage development, was needed at high rates. The model accurately predicted the failure strength of the +/-45° samples but over-estimated the maximum extension at quasi-static and high rates. Smooth particle hydrodynamics was used to model the bird, with the bird parameters being obtained from a system identification analysis of impacts on flat plates. Figure 25 compares two video sequences with the numerical predictions for a 4/3 FML at two separate times during the impact process. In general, the model successfully predicted the overall deformation of the structures, although there was a greater level of deformation in the test and a different folding pattern. The model predicted a greater degree of structural rebound due to the fact that plasticity was not incorporated in the material model because of a lack of experimental data. Pre-test numerical simulations correctly predicted that the bird would not penetrate the structure and the predicted loads transferred to the support structure were in good agreement with those measured experimentally. More recent work by Guida et al. (2009) showed that a leading edge based on a 3/2 stacking sequence offered a superior resistance to impact loading than a 2/1 system. This enhanced performance was attributed to the greater number of aluminum layers in the 3/2 laminate, resulting in increased energy absorption and reduced deformation relative to its 2/1 counterpart.

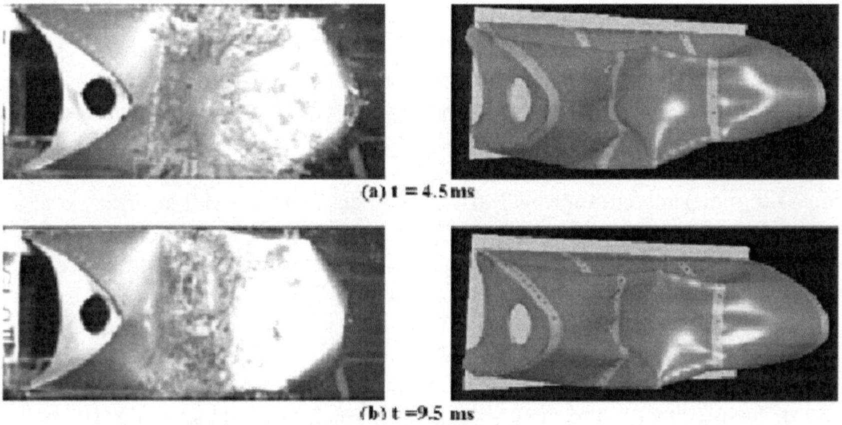

Figure 25. Video sequences from bird strike on an FML leading edge. Left hand images show video clips and right hand images show predictions (McCarthy et al., 2004b).

Summary

Fiber metal laminates offer significant potential for use in structures subjected to high velocity impact loading. Testing has shown that aluminum-based FMLs, such as GLARE, offer specific perforation resistances significantly higher than those associated with plain metal alloys and composites. Examinations of failed samples have highlighted many of the failure mechanisms observed under low velocity impact loading conditions. Glass-based FMLs profit from the rate-sensitivity of the glass fibers, with the strength and energy-absorbing capacity of the composite, and therefore the FML, increasing rapidly with strain rate. Successful attempts have been made to model the high velocity impact response of FMLs and this has resulted in a greater understanding of the dynamic behavior of these materials. A number of studies have shown that fiber-reinforced thermoplastics offer considerable potential for incorporation into FMLs. Thermoplastic-based systems can be readily moulded, shaped and repaired. They also offer the advantage that they are tougher, offering higher interlaminar fracture toughnesess and impact resistances than most thermosetting-based composite materials.

5- Conclusions

Fiber-metal laminates represent a represent a new and exciting group of hybrid materials that are currently finding use in a range of high-performance engineering application. By combining the relatively high stiffness and ductility of a metal alloy with the elevated fracture toughness and fatigue resistance of composite materials, a range of FMLs have been developed that offer excellent short and long-term mechanical properties. A number of researchers have shown that FMLs offer attractive impact properties, particularly at very high rates, such as those associated with high velocity impact loading.

At both low and high impact velocities, damage in high performance GLARE FMLs is often significantly less than in relatively tough carbon fiber reinforced thermoplastics. FMLs also offer the advantage that damage is often easy to locate due to the fact that impact loading frequently results in the formation of a visible dent at the point of impact.

Further work is required in order to fully establish the potential offered by these materials. Very little work has been conducted at velocities above 300 m/s and this should be addressed in order to establish if the advantages offered in the velocity range of 50 to 200 m/s are maintained under more extreme loading conditions. Few studies have been undertaken with a view to model the impact results of FMLs and work is required in order to predict the response of larger, more representative structures. Further work is also required to assess the potential of thermoplastic-based FMLs, which combine the potential for very short manufacturing cycles with high values of interlaminar fracture toughness and impact resistance.

References

Abdullah, M.R (2006). *The high velocity impact responseof novel fibre-metal laminates.* PhD Dissertation, University of Liverpool.

Abdullah, M.R., and Cantwell, W.J. (2006a). The impact resistance of polypropylene-based fiber-metal laminates. *Composites Science and Technology* 66:1682-1693.

Abdullah, M.R., and Cantwell, W.J. (2006b). The impact resistance of fiber-metal laminates based on a glass fiber reinforced polypropylene. *Polymer Composites* 27:700-708.

Alderliesten, R.C., Hagenbeek, M., Homan, J., Hooijmeijer, P.A., De vries, T.J., and Vermeeren, C. (2003). Fatigue and Damage Tolerance of Glare. *Applied Composite Materials* 10:223–242.

Atas, C. (2007). An Experimental Investigation on the Impact Response of Fiberglass/Aluminum Composites. *Journal of Reinforced Plastics and Composites* 26:1479-1491.

Bagnoli, F., Bernabei, M., Fugueroa-Gordon, D., and Irving, P.E. The response of aluminium/GLARE hybrid materials to impact and to in-plane fatigue, (2009). *Materials Science and Engineering A* 523:118-124.

Bernhardt, S., Ramulu, M., and Kobayashi, A.S. (2007). Low velocity impact response characterization of a hybrid titanium composite laminate, *Journal of Engineering Materials and Technology* 129:220-226.

Caprino, G., Spataro, G., and Del Luongo, S. (2004). Low velocity impact behaviour of fiberglass-aluminium laminates, *Composites Part A* 35:604-616.

Caprino, G., and Lopresto, V., and Iaccarino, P. (2007). A simple mechanistic model to predict the macroscopic response of fiberglass-aluminium laminates under low velocity impact, *Composites Part A* 38:290-300.

Carrillo, J.G. (2007). *A study of the mechanical properties and scaling effects in a thermoplastic fibre-metal laminate.* PhD Dissertation, University of Liverpool.

Carrillo, J.G., and Cantwell, W.J. (2008). Scaling Effects in the Low Velocity Impact Response of Fiber-Metal Laminates, *Journal of Reinforced Plastics and Composites* 27:893-907.

Cortes, P. (2005). *The mechanical properties of high temperature fibre-metal laminates.* PhD Dissertation, University of Liverpool.

Cortes, P., Cantwell, W.J. (2006). The fracture properties of a fibre–metal laminate based on magnesium alloy. *Composites: Part B* 37:163–170.

Cortes, P., and Cantwell, W.J. (2007). The Impact Properties of High-temperature Fiber-Metal Laminates. *Journal of Composite Materials*

41:613-632.

Gilat, A., Goldberg, R.K., and Roberts, G.D. (2005). Strain rate Sensitivity in Epoxy Resin in Tensile and Shear Loading. NASA/TM 2005-213595.

Guan, Z., Cantwell, W.J., and Abdullah, M.R. (2009). Numerical modeling of the impact response of fiber-metal laminates. *Polymer Composites* 30:601-611.

Guida, M., Marulo, F., Polito, T., Meo, M., and Riccio, M. (2009). Design and testing of a fiber-metal-laminate bird-strike-resistant leading edge. *Journal of Aircraft* 46:2121-2129.

Haque, A., and Hossain, M.K. (2003) Effects of Moisture and Temperature on High Strain Rate Behavior of S2-Glass–Vinyl Ester Woven Composites. *Journal of Composite Materials* 37:627-647.

Harding, J. (1987). in: Blazynski, T.Z. (ed.), Materials at High Strain Rates. p. 133, Elsevier Applied Science, London and New York.

Hoo Fatt, M.S., Lin, C., Revilock, D.M., and Hopkins, D.A. (2003). Ballistic impact of GLARE fiber-metal laminates. *Composite Structures* 61:73-88.

Johnson, W.S. (1986) Impact and Residual Fatigue Behavior of ARALL and AS6/5245 Composite Materials. NASA Technical Memorandum 89013.

Khalili, S.M.R., Mittal, R.K., Gharibi Kalibar, S. (2005). A study of the mechanical properties of steel/aluminum/GRP laminates. *Materials Science and Engineering A* 412:137–140.

Krishnakumar, S. (1994). Fiber metal laminates – The synthesis of metals and composites. *Materials. and Manufacturing Processes* 9:295-354.

Kulkarni, R.R., Chawla, K.K., Vaidya, U.K., Koopman, M.C., and Eberhardt, A.W. (2008). Characterization of long fiber thermoplastic/metal laminates. *Journal of Materials Science* 43:4391-4398.

Laliberte', J.F., Poon, C., Straznicky, P.V., and Fahr, A. (2000). Applications of fiber-metal laminates. *Polymer Composites* 21:558-567.

Laliberte', J.F., Poon, C., Straznicky, P.V., and Fahr, A. (2002). Post-impact fatigue damage growth in fiber–metal laminates. *International Journal of Fatigue* 24:249–256.

Laliberté, J.F., Straznicky, P.V., and Poon, C. (2005). Impact damage in fiber metal laminates, Part 1: Experiment. *AIAA Journal* 43:2445-2453.

Liaw, B.M., Liu, Y.X., and Villars, E.A. (2001) Impact damage mechanisms in fiber-metal laminates. *Proceedings of the SEM Annual Conference on Experimental and Applied Mechanics* 536-539.

McCarthy, M.A., Xiao, J., Petrinic,, N., Kamoulakos, A., and Melito, V. (2004a). Modelling of Bird Strike on an Aircraft Wing Leading Edge Made from Fiber Metal Laminates – Part 1: Material Modelling. *Applied Composite Materials* 11:295-315.

McCarthy, M.A., Xiao, J.R., McCarthy, C.T., Kamoulakos, A., Ramos, J., Gallard, J.P., and Melito, V. (2004b). Modelling of a bird strike on an aircraft leading edge made from fiber metal laminates - Part 2: Modelling of impact with SPH bird model. *Applied Composite Materials* 11:317-340.

McVay, A.C., and Johnson, W.S. (2009). Permeability of various hybrid composites subjected to extreme thermal cycling and low velocity impacts. *Journal of Composite Materials* 22:2507-2520.

Payenganeh, G.H., Ashenai Ghasemi, F., and Malekzadeh, K. (2010). Dynamic response of fiber-metal laminates (FMLs) subjected to low velocity impact. *Thin Walled Structures* 48:62-70.

Reyes, G. (2002). *Processing and characterization of the mechanical properties of novel fibre-metal laminates*. PhD Dissertation, University of Liverpool.

Reyes, G. (2010). Mechanical behavior of thermoplastic FML-reinforced sandwich panels using an aluminum forma core: Experiments and Modeling. *Journal of Sandwich Structures and Materials* 12:81-96.

Reyes, G., and Cantwell, W.J. (2000). The mechanical properties of fiber-metal laminates based on glass fiber reinforced polypropylene. *Composites Science and Technology* 2000:1085-1094.

Reyes, G., and Gupta, S. (2009) Manufacturing and mechanical properties of thermoplastic hybrid laminates based on DP500 steel. *Composites Part A*:176-183.

Seo, H., Hundley, J., Hahn, H.T., and Yang, J.-M. (2010). Numerical simulation of glass fiber-reinforced aluminium laminates with diverse impact damage. *AIAA Journal* 48:676-687.

Song, S.H., Byun, Y.S., Ku, T.W., Song, W.J., Kim, J., and Kang, B.S. (2010) Experimental and numerical investigation on impact performance of carbon reinforced aluminum laminates. *Journal of Materials Science and Technology*, 26:327-332.

Sun, C.T., Dicken, A., and Wu H.F. (1993). Characterization of impact damage in ARALL laminates. Composites *Science and Technology* 49:139-144.

Vlot, A. (1991*). Low velocity impact loading on fiber reinforced aluminum laminates (ARALL and GLARE) and other aircraft sheet materials.* PhD Dissertation, TU Delft.

Vlot, A. (1993). Impact properties of fiber metal laminates. *Composites Engineering* 3:911-927.

Vlot, A. (2001) *Glare—History of the Development of a New Aircraft Material.* Kluwer Academic Publishers, Dordrecht, Netherlands.

Vlot, A., Kroon, E., and LaRocca, G. (1998). *Impact Resistance of Fiber Metal Laminates.* Key Engineering Materials 141-143:235-276.

Vogelesang, L.B., and Vlot, A. (2000). Development of fiber metal laminates for advanced aerospace structures, *Journal of. Materials and Processing Technology* 103:1-5.
Wen, H.M., and Jones, N. (1993). Experimental investigation of the scaling laws for metal plates struck by large masses. *International Journal of Impact Engineering* 13:485-505.
Wu, G., Yang, J.-M., and Hahn, H.T. (2007). The impact properties and damage tolerance of bi-directionally reinforced fiber metal laminates. *Journal of Materials Science* 42:948-957.
Wu, H.F. (1991) Effect of temperature and strain rate in tensile mechanical properties of ARALL-1 laminates. *Journal of Materials Science* 26:3721-3729.
Wu, H.F., and Sun, C.T. (1993). Impact damage characterization of aramid aluminum laminates. Volume 4, *Proceedings of the Ninth International Conference on Composite Materials,* Woodhead Publishing, 157-165.
Xia, Y., Wang, Y., Zhou, Y., and Jeelani, S. (2007). Effect of strain rate on tensile behavior of carbon fiber reinforced aluminum laminates. *Materials Letters* 61:213-215.
Xia, Y.M., Uan, J., and Yang, B. (1994). A statistical model and experimental study of the strain-rate dependence of the strength of fibers. *Composites Science and Technology* 52:499-504.
Zhou, Y., Wang, Y., and Mallick, P.K. (2004). An experimental study on the tensile behavior of Kevlar fiber reinforced aluminum laminates at high strain rates. *Materials Science and Engineering* A381:355-362.

Ballistic Impacts on Polymer Matrix Composites, Composite Armor, Personal Armor

R. Zaera

Department of Continuum Mechanics and Structural Analysis

University Carlos III of Madrid

1 Introduction

Throughout recorded history, humans have used various types of materials to protect themselves from injury in combat or other dangerous situations. In recent decades, the appearance of high-performance fibers and ceramics has allowed the development of protection against ballistic impact, with some exceptionally lightweight and protective characteristics. This chapter, divided in two parts, provides a review of the design and use of lightweight composite amours. The first part contains preliminary information of interest for the understanding of the second part, in which the characteristics of the various composite armours are described, as well as the methodologies for analysis and simulation.

In the first part, section 2 describes the use of light protection against high-velocity impact, together with a summary of the history of its technological evolution since the twentieth century. In section 3 the projectiles launched by handguns, rifles, or heavy guns are described, along with the fragments, threats against which composite protections show their greatest efficiency. Section 4 presents the fundamental characteristics of the ballistic impact tests and covers the main standards and specifications used by the North American, European, Asian, and Oceania countries to test ballistic materials.

In the second part, section 5 focuses on high-performance fiber armors and section 6 on ceramic-faced armors. Section 7 shows the various methodologies for simulating the impact on composite armors, used for analysis and design. Chapter 8 describes the specific features of oblique impact, chapter 9 presents new developments in lightweight ballistic armor, and chapter 10 offers some concluding remarks about the challenges currently being faced in the field of ballistic protection using composite materials.

2 The Importance of Lightness in Ballistic Protection: a Historical Perspective

2.1 The Interest of Lightness in Ballistic Protection

Methods for protecting people, structures, weapons and vehicles from the effects of projectile impacts have been studied carefully for many years. The usefulness of certain systems depends, essentially, on their ability to resist impulsive loads. Although methods to protect people and objects have been especially important in the military sphere, there are also other branches of engineering where devices are at a serious risk of being damaged by high-velocity impacts from objects which form part of the system itself or are unrelated to it. The inside of a centrifugal separator, the interior casing of an aviation turbine or the exterior panels on an aerospace vehicle are all clear examples of structural components which must provide protection from impulsive loads so that any damage caused by an impact does not affect or break the rest of the system. The characteristics of the system which needs to be protected are important factors which must be taken into account when designing armor. When it comes to designing armor to protect fortifications or buildings, weight is not a key factor. In these cases, it is quite common for cheap materials to be used, as thicker layers of these materials can compensate for their poor or average resistance. Sandbags in military fortifications, or walls, panels and heavy duty concrete slabs in military shelters, guard shacks and roofs over nuclear reactors, are all examples of cases where the use of heavy armor does not have a negative effect upon the efficiency of the systems that they are designed to protect.

However, this is not the most common scenario faced by engineers when designing armor. Weight is one of the most important factors when it comes to designing impact armor, as most systems which require protection from impulsive loads (military or civilian vehicles, boats and airplanes, and defense and security personnel) are mobile. In the field of aeronautics, the weight of each system component must be kept to a minimum, and the armor used to protect vulnerable parts from impulsive loads must therefore be made of low-density materials. For armored vehicles, using light protections makes it possible to use engines with a lower cylinder capacity without reducing their maximum speed and maneuverability. In armored military vehicles, the advantages of using light protections are even greater. As fuel is of great strategic importance in times of war, given that it is very expensive and only limited amounts of it can be stored in reserve for use during crises, it is obviously very important to be able to reduce the weight of motorized vehicles. Reducing the weight of these vehicles makes them more useful, as they are easier to drive on poor quality roads, tracks and bridges. In addition, lighter vehicles can be transported in airplanes or used as amphibious vehicles.

The development of light armor is therefore a very interesting topic, and one which has become more and more important over the years. As its relevance has increased, so has the range of materials used to make this type of armor and the variety of designs created to provide greater protection.

2.2 Metallic Materials for Lightweight Ballistic Protections

What we call light armor today first came about at the beginning of the 20th century. When the first tanks were used in 1917, the link between mobility and protection became increasingly relevant. In order to reduce their vulnerability as much as possible, vehicles had to be maneuverable and well-protected from enemy fire. Initially, Rolled Homogeneous Armor (RHA) was used. This armor consisted of a rolled steel plate of a hardness between 35 and 40 Rockwell C. In order to stop normal impacts from rifle fire, the basic infantry weapon used at the time, RHA had to be quite thick, and it therefore added far too much excess weight to the vehicle. The first improvements were made through the use of hardened steel, with a hardness of between 50 and 60 Rockwell C. The ratio between the areal density of RHA and that of high-hardness steel armor, which provided the same protection from the threats in question, was up to 1.15. This reduced weight meant that this type of armor became very popular. The major problem with these sheets of metal was that they were very brittle. Plates broke quite often, even during storage operations, and this armor type began to be used less and less often in the 1930s.

2.3 High Performance Fiber Armors

In the early stages of World War II, the United States designed manganese-steel body armor for infantrymen. However, these vests were heavy, restricted mobility and led to overheating, and were therefore rejected by the troops. The next major improvement in the field of impact protection took place at a later stage of the Second World War. During this period, the company Wilkinson Sword developed the flak jacket or flak vest to protect British Royal Air Force pilots from fragments of exploding shells launched by German anti-aircraft guns. These vests were made using 1 mm-thick sheets of manganese steel sewn into multi-layered nylon, a fiber developed in 1935 by the DuPont company, that would catch low velocity fragments. Sample suits were received in the United States in 1943, and the Army Ordnance Department took over the task of improving the design and producing the vests in bulk for the United States Army

Air Corps. US flyers wore this type of vest and this led to a great decrease in the number of casualties. At the end of the Second World War, high-strength steel was substituted by aluminum, which provided greater protection and a decrease in the total weight of the armor.

At the end of the War, efforts began to focus on the development of non-metallic armor (Tobin and Iremonger 2006). New developments led to the use of an all-nylon cloth, where metal plates were completely removed from the vest, as well as to the use of Doron, a glass-fiber reinforced plastic manufactured in 1943 by the Dow Chemical Company. The initial product consisted of layers of glass fibers bonded with an ethylcellulose resin under high pressure. Later modifications made it possible to increase ballistic performance over a wider temperature range by introducing variations in the resin. This composite material also made it possible to make curved plates instead of the flat plates used previously, and this meant that the plates were a better fit around the curvature of the torso.

Fiberglass vests became standard equipment during the Korean war, where US soldiers' vests were made with 90% Doron to protect the body, and 10% nylon to protect the shoulders and the upper thorax. These vests were a vast improvement in terms of lightness. An evaluation of the US Army concluded that they served to stop more than 75% of fragment impacts (main cause of military ballistic casualties in the Korean war) and 25% of pistol bullets, however these vests were inefficient against rifle threats (Tobin and Iremonger 2006). The vests used by US troops in Vietnam were updated versions of the Korean models and were still not able to protect wearers against standard rifle projectiles. In 1965, this problem was solved by a DuPont polymer research group, which developed an aramid fiber of a high stiffness and strength, and low density. Eight years later, the final product, called Kevlar™, reached the marketplace and proved its efficiency as a ballistic protective material. Since the 1970s, several new high-performance fibers (Twaron™, Dyneema™, Spectra™, Zylon™) and manufacturing methods for armor panels have been developed. These newer materials provide outstanding protection against fragments and bullets and they have since then been used to protect combat vehicles and aircraft.

2.4 Dual-Hardness Armor

One of the greatest advances in the field of ballistic protection took place as a result of a change in armor design. Up until the beginning of the 20th century, metal armor had just consisted of a monolithic plate of a single metal, and the materials used were selected because of their hardness and resistance so that they could decrease the impact energy of projectiles. However, the brittleness of these steels meant that the kinetic energy of the projectiles was not effectively dissipated through plastic deformation of the armor plate. As a result, in the middle of the 20th century, armor manufacturers came up with the idea of using two different materials to manufacture one piece of armor. The first material was very hard and used on the outside of the plate to deform the nose of the projectile. The second material was highly ductile and located behind the first in order to absorb the kinetic energy of the projectile through plastic deformation. The first results of this new design concept were seen through the creation of steel *dual-hardness* armors, which consisted of plates made of two sheets of different steels layered together. The sheet of the harder material was carburized so that it would have a hardness of about 60 Rockwell C. This made it possible to reduce the weight of the armor in comparison with RHA. As these armors were difficult to manufacture, their use did not become very widespread, although subsequent technological improvements in the field have made it possible to use this design concept again.

When aluminum began to be used for ballistic protection, significant improvements were made in the design of armors. The USA was the first to use aluminum as an armor material for armored vehicles; in late 1950s they started production of M113 APC with a 5083 aluminum hull. This protected the crew and passengers against small arms fire and shrapnel impacts. However, because it was so vulnerable to impacts from projectiles with a hard core (steel or tungsten), armor manufacturers began to think about protecting the external armor using layers of high-hardness steel, also following the dual-hardness design philosophy. These steel/aluminium armors were about 17% lighter than RHA, and this weight reduction was greater than that achieved through the use of 60 Rockwell C steel plates. Adding a high-hardness steel plate to a primary aluminum armor became common practice at the beginning of the 1960s and is one of the most effective systems of ballistic protection used today.

Although the use of steel/aluminum armors led to great improvements in terms of reduced weight, the real breakthrough came when ceramic materials began to be used for ballistic protection. In 1963, the Goodyear Aerospace Corporation developed an armor which matched the performance of RHA but weighed half as much. This armor was designed based on the same concept as that of the dual-hardness armors described above. The exterior high-hardness layer consisted in an array of alumina tiles, and a fiberglass reinforced composite served as backing layer. Both materials were joined with an adhesive layer (Figure 2.1 left). As technological advancements were made in the field of ceramics, new armor types were developed: boron carbide, silicon carbide, aluminum nitride or titanium diboride were used as a hard layer in ceramic armors. Furthermore, metals (steel, aluminum, titanium, aluminum foam) and more efficient fiber-reinforced materials (aramid, polyethylene and others) were used as backing panels, increasing the efficiency of ballistic protection, and always at a higher cost (Figure 2.1 right). Flexible backing panels can only be used in "appliqué" or "add-on" armors, which can be easily applied and removed to a main armor, thus helping to reduce the kinetic energy of the projectile to a percentage of its initial value. Rigid backing plates are more convenient for "integral" or "structural" armors which are used as part of a structure to perform load-carrying functions as well as for impact protection.

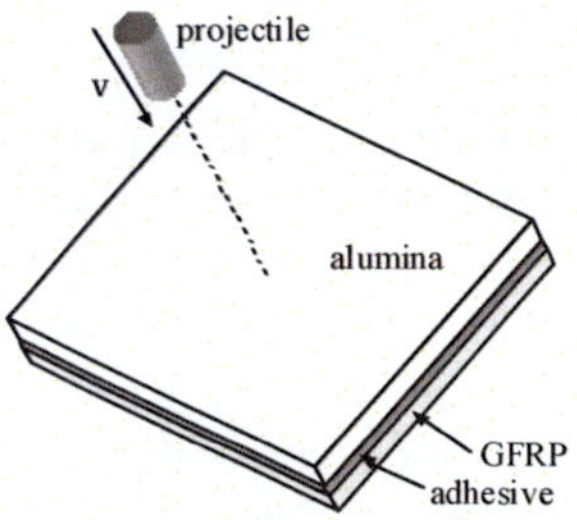

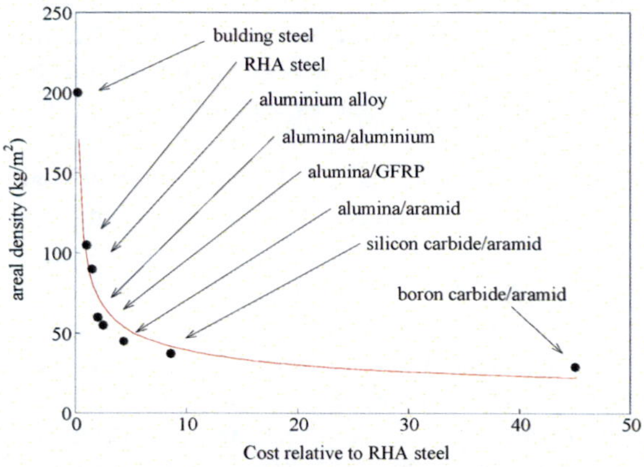

Figure 2.1. Left: Alumina/Glass fiber reinforced panel armor, as developed by Goodyear Aerospace Corporation. Right: Areal density of the armor needed to arrest a 7.62 armor piercing projectile at 850 m/s (Roberson, 1995).

3 The Threat: Handgun Bullets, Rifle Bullets, Heavy Gun Projectiles and Fragments

3.1 Introduction

Armors used to protect military aircraft, vehicles and personnel have been designed to resist the impact of a number of different threats, typically projectiles launched by direct-fire weapons such as handguns and rifles, or by indirect-fire weapons such as grenades and mortar-shells. Armor must be designed taking into account the threats which may affect that armor, and it is therefore essential to know about the basic characteristics of the fragments projected by the explosion of grenades or mortars, and of the projectiles launched by handguns, rifles or heavy guns. These firearms cover a broad range of calibers, defined as the measure of the inside diameter of the barrel: from small in handguns and rifles (below 3.5 inches) to medium and large (above 7 inches) in heavy guns. Lightweight composite amours are effective defeating this range of projectiles. Warheads such as shaped charges do not fall within the scope of this review, as reactive armors are required to stop them.

3.2 Handgun and Rifle Bullets

Handguns and rifles are firearms designed to be held by the shooter when used. The handgun is held in one hand, whereas the rifle is held in one hand to support the barrel and in the other arm and shoulder to support the stock. The cartridge which is inserted into a handgun or rifle consists of the bullet, the case, the primer and the propellant. The primer starts the detonation process when the head of the case is hit, burning the propellant that generates the gases expanding inside the case of the cartridge and inside the gun bore. The bullet is then pushed down the barrel. The bullet is the part of the cartridge that impacts the target, constituting the threat that has to be stopped by the armor.

Traditionally, bullets have been made of lead. This material's high density gives the projectiles more energy, and its deformability means that it can cause greater injury to the body. Since the lead is easily deformable, the nose of the bullet is flattened during impact and its energy is spread over a larger area. However, The Hague Convention (The Hague Convention, Declaration III, 1899) established that the lead used for bullets had to be covered with a copper full metal jacket to reduce mushrooming during the penetration process. The copper jacket also protects against melting of the lead core due to friction with the bore and heat inside the barrel, and this allows much higher muzzle velocities.

Mushrooming of a lead bullet helps to transfer the kinetic energy of the projectile to the target, causing more damage in soft materials, like human body tissues. However if the target is made of a hard material, lead projectiles have reduced piercing capabilities. Thereby hard core or "armor piercing" projectiles have gradually been designed, using materials such as steel, tungsten, tungsten carbide or uranium (Figure 3.1 left). These projectiles are more effective when it comes to perforating armor, and are therefore used as "anti-material" weapons. The lead projectiles, on the other hand, are effective if they impact directly with the human body, and are therefore used as "anti-personnel" weapons. As such, the bullets which are usually taken into account when designing armor are lead core and Armor Piercing (AP) bullets, both of which are jacketed with a copper layer.

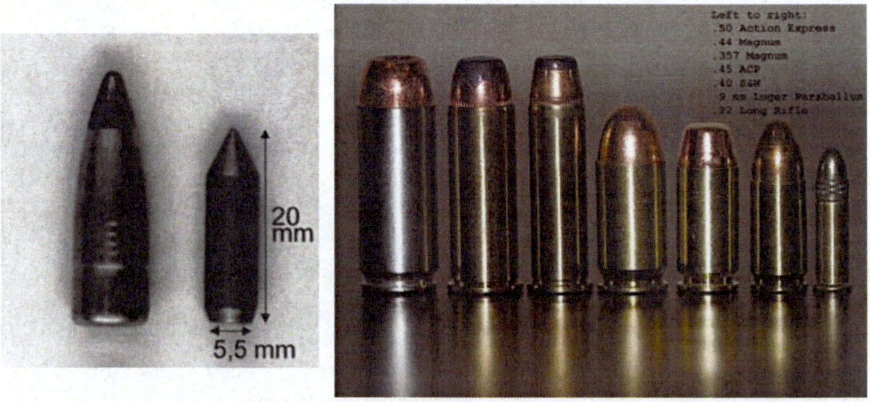

Figure 3.1. Left: 7.62 Armor Piercing projectile with its tungsten carbide core (Zaera et al. 2000). Right: different handgun cartridges (Wikimedia Commons, Paularized, public domain).

The bullets launched by handguns tend to have a muzzle velocity slightly higher than the speed of sound, between 350 and 450 m/s. Their short barrels make it impossible for the bullets to move any faster. However, they tend to have a higher mass than rifle bullets (Table 3.1), and this can partially compensate for their lower kinetic energy due to the lower speed. In almost all cases, the projectile's core is made of lead, which is easily deformable, and their cross-sectional area is higher than that of rifle bullets (lower energy per unit of cross section area). This means that they can be more easily stopped by armor. The most common handgun bullets are the 9 mm, used in a variety of automatic handguns, the .357 Magnum, mounted on a larger cartridge with more gunpowder, and the .44 Magnum, a higher powered cartridge (Figure 3.1 right).

Table 3.1. Characteristics of different threats (* approximate value, depending on the cartridge).

Threat	Projectile mass (g) *	Muzzle velocity (m/s) *	Sub-calibrated	Gun type	Core material
9 mm Parabellum	7.5	350	No	handgun	lead
0.357 Magnum	10.2	385	No	handgun	lead
0.44 Magnum	15.6	440	No	handgun	lead
7.62 mm NATO Ball	9.4	830	No	rifle	lead
7.62 mm AP	9.7	850	No	rifle	steel, tungsten, tungsten carbide
5.56x45 mm AP	4	940	No	rifle	steel/lead
12.70 mm AP	46.7	820	No	heavy gun	tungsten, tungsten carbide
14.50 mm	60	1000	No	heavy gun	tungsten, tungsten carbide
20 mm APDS	72	1250	Yes	heavy gun	tungsten
25 mm APDS	122	1250	Yes	heavy gun	tungsten
30 mm APDS	219	1250	Yes	heavy gun	tungsten
Fragment Simulating Projectile (FSP)	1.1, 2.8, 5.3, 13.4, 53.9	-	-	-	steel

Bullets fired by rifles have a muzzle velocity of between 650 and 950 m/s. These bullets' cores can be made of lead or high strength metals, and the cross-sectional area is smaller compared to that of handgun bullets. Due to their higher kinetic energy, higher energy density and core hardness, it is more difficult to stop them using armor. The most commonly used bullets include the 7.62 mm, the standard NATO rifle round until it was replaced by the 5.56×45 mm, which has lower kinetic energy and lower recoil, lower ammunition weight and better accuracy.

3.3 Heavy Gun Projectiles

A heavy gun refers to a larger-caliber, high-power firearm fired from heavy mounts or static positions. Heavy gun projectiles were developed to perforate armored systems such as tanks or infantry fighting vehicles. Rifle armor piercing bullets are unable to defeat a thick steel plate, and increased penetration capacity is achieved by increasing the mass and the velocity, and using harder core materials. If we examine the impulsion system inside the gun, two different categories of medium caliber perforating projectiles can be defined:

i) The projectile's maximum diameter is the same as the interior diameter of the bore. In this case, the pressure created inside the canon is applied to the rear face of the projectile, which usually has a metal jacket. The most common projectiles in this category are the 12.70 mm AP (Figure 3.2 left), used by NATO countries, and the 14.50 mm AP, used by armies from eastern countries (Table 3.1).

ii) The projectile diameter is smaller than the bore diameter, and pressure is applied both to the back face of the projectile and the back face of the sabot, a lightweight solid surrounding the projectile and divided into several pieces that are pulled away by the air once it reaches the muzzle of the barrel (Figure 3.3). The projectile is then called "sub-calibrated". This projectile category includes the 20 mm APDS, 25 mm APDS and 30 mm APDS (Figure 3.2 and Table 3.1), where APDS means Armor Piercing Discarding Sabot. All of them are made with tungsten, a high-density and high-strength material. The density of tungsten is 1.6 greater than that of lead and 2.3 times greater than that of steel; the yield stress of tungsten is, typically, 80 and 2 times greater than that of lead and steel used in ballistic applications, respectively. This means that a projectile made of tungsten will decelerate slower during impact and will be more difficult to deform, thus increasing its piercing ability. For instance, a steel core projectile hardly penetrates a tile of alumina (used for armor applications) during the first phase of impact, whereas a tungsten projectile starts to penetrate it just after the contact [Zaera and Sánchez-Gálvez 1998a].

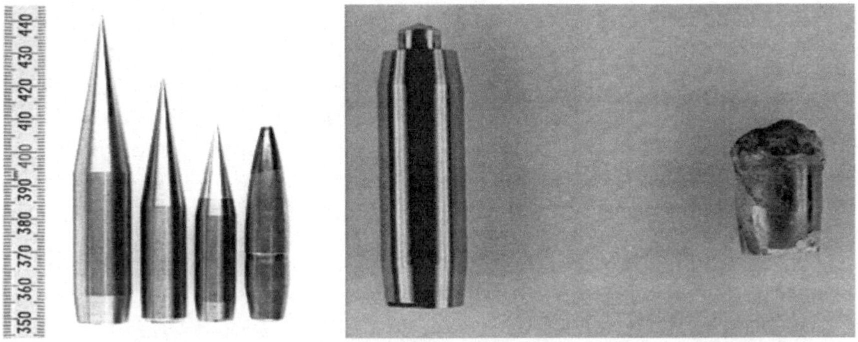

Figure 3.2. Left: From left to right, bullet of the 30 mm APDS, 25 mm APDS, 20 mm APDS and 12.7 mm AP cartridges [Zaera 1997]. Length units in cm. Right: Picture of the tungsten core of a 20 mm APDS before and after impacting an alumina/aluminum armor. [Zaera et al. 1998a]

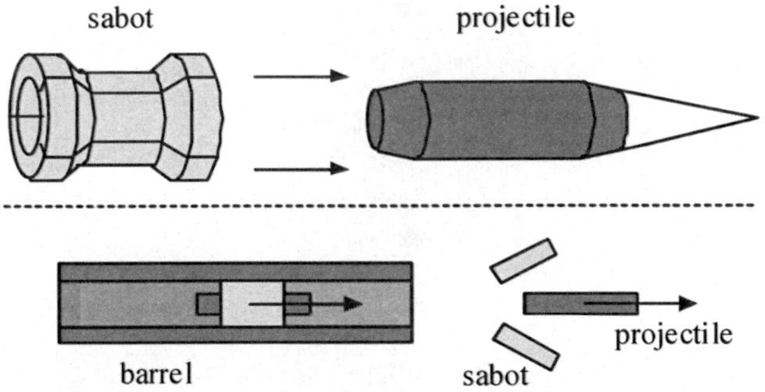

Figure 3.3. Top: Sabot and sub-calibrated projectile (drawn separately). Bottom: Separation of the sabot at the end of the barrel.

3.4 Fragments

Mortar ammunition (Figure 3.4a) and grenades (Figure 3.4b) are designed to produce and propel a large number of small fragments, also called shrapnels, to injure troops. Usually, mortar rounds are launched by muzzle-loading indirect fire weapons, whereas grenades are designed to be thrown by hand. Combustible material is ignited to produce an explosion that breaks up the case into small pieces and propells them at high velocity. In general a fragment from a modern shell will travel at about 1000 m/s (Moss et al., 1995), but they lose their velocity quickly and are effective up to 15 or 20 meters. The shape and size of fragments varies a great deal and they have a mass of between 0.1 and 3 g. This wide range of different fragment types makes it difficult to characterize them in order to carry out impact tests. To draw up a standard geometry, Fragment Simulating Projectiles (FSP) have been defined. FSPs are cylinders with three flat faces at the tip (Figure 3.4c) to simulate the shape of an irregular fragment impacting on the target. Conforming to military specification MIL-P-46593 (1964), FSPs should be manufactured with a steel of 30 Rockwell hardness and diameters of 0.22 inches, 0.30 inches, 0.50 inches or 20 mm (Table 3.1). Steel right circular cylinders or spheres, cheaper and readily available, may be also used to simulate a fragment.

(a) (b) (c)

Figure 3.4. (a): Polish artillery mortars (Wikimedia Commons, Tomasz Szulc, public domain). (b): MK2 grenade (Wikimedia Commons, US Army, public domain). (c): Shape of a 1.1 g Fragment Simulation Projectile.

4 Testing Armors: Experimental Methodologies and Standards

4.1 Introduction

The design of armors was traditionally carried out empirically, relying on real impact tests using the given projectile-target in each test. This method is very reliable but it is also a costly procedure, as testing in these conditions requires sophisticated facilities and equipment: shooting galleries, pistols, rifles, guns. It becomes even more costly if the aim of the tests goes beyond the discovery of the resistance quality of a target, or the characteristics of the deformation of a projectile capable of perforating an armor or destroying a structure; to obtain information on the deformation of the target or on the position of the projectile during penetration, sophisticated data acquisition systems are required since the briefness of the process requires very high sample rates.

This section describes the essential features of impact tests on armors, which are destructive in nature, and presents general aspects of the tests standards and specifications used for testing ballistic amours. It also includes some particularities of testing personal armors (ballistic vests and helmets) where an excessive deformation of the surface of the armor in contact with the body during impact may produce severe damage in chest, abdomen or head.

4.2 Experimental Methodologies

Among the variables that should be measured during the impact test, an accurate determination of the velocity of the projectile is obviously of great importance, since its square is related to the kinetic energy applied to the specimen. The impact velocity may be measured by *chronographs* (two sensors which detect the passage of the projectile and calculate the velocity dividing the base length by the difference of the times when the projectile arrives at these locations), *Doppler radar* (an antenna is aligned to the projectile path which sends a continuous radio signal reflected by the projectile with a frequency shift due to the Doppler effect) or *ultra high-speed digital* or *film cameras*. The residual velocity after perforation can be measured only with radar or photographic systems because the trajectory of the projectile may change, and the projectile may be followed by a number of fragments of the protection and the projectile itself, making it difficult to detect the passage.

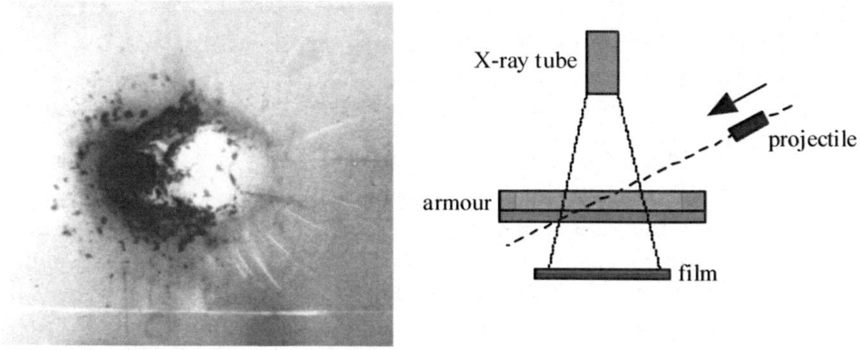

Figure 4.1. Left: X-Ray image of the oblique penetration process of a 200 mm APDS projectile into an alumina/aluminum target (Performed at Empresa Nacional Santa Bárbara). Right: Sketch of the experimental set-up. (Zaera et al. 1998a).

The use of a high speed camera may provide additional and valuable information about the position of the projectile tail and about the deformation and failure mode of the armor. When the position of the projectile tip has to be determined during perforation –e.g. to analyze the erosion- high-speed X-Ray cameras are needed (Figure 4.1), increasing the cost of the information acquisition system even more.

Generally, ballistic armor is designed so that it is not perforated when it is subjected to the impact of the projectile which represents the characteristic threat. As we are examining the problem from the perspective of defense, the fundamental task consists of determining the armor thickness and the material to be used to stop the projectile. To complement this, we could also determine the ballistic limit, defined as "the minimum velocity at which a particular projectile is expected to completely penetrate an armor of given thickness and physical properties at a specified angle of obliquity". The ballistic limit may also be defined as "the maximum velocity at which a particular projectile is expected to fail to penetrate an armor of given thickness and physical properties at a specified angle of obliquity" (according to US MIL-STD-662F standard, see Table 4.1). Once the criteria to be used to distinguish between a case of perforation and one of projectile arrest has been established, the experimental tests can provide a probability of perforation curve (Figure 4.2) of which each region is related to one of the situations depicted: (A) corresponds to projectile arrest, (C) to target perforation and (B) to the range of impact velocities at which perforation is probable. The critical impact velocity at which the probability of perforation is P % is known as the ballistic limit v_P. The most common probability used to define the ballistic limit is 50 % thus leading to v_{50}.

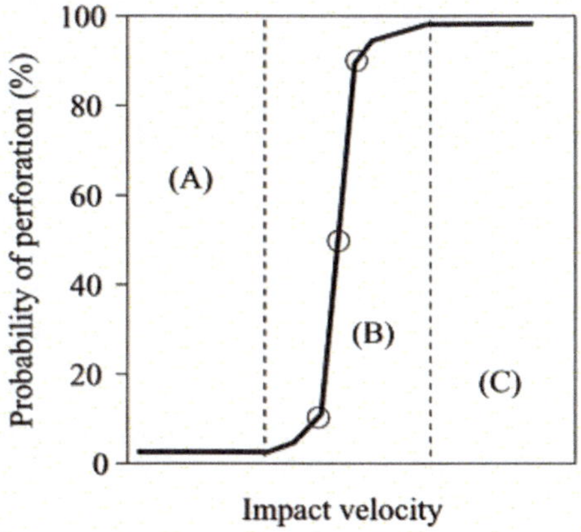

Figure 4.2. Probability of perforation vs. impact velocity.

When testing personal armor, a key parameter to be measured is the Behind Armor Blunt Trauma, defined as the non-penetrating injury resulting from the rapid deformation of personal body armors covering the body. The protection may impede the perforation by a projectile, but some of the kinetic energy is transferred to the body, causing effects such as sudden disturbance of heart rhythm, pulmonary contusion, apnea or even death even though the projectile has not perforated the armor. The maximum deformation of the rear face of a body armor during ballistic impact is determined using a deformable block of a plasticine-like substance, placed behind the protection during testing. The deformation of the armor causes an indentation in the plasticine block. After removal of the armor, the indentation depth resulting from a shot is measured to provide an indication of the level of trauma inflicted on the wearer.

4.3 Standards

Standards describe the guidelines for ballistic threads, testing equipment, physical conditions and procedures to test amours under the same specific conditions, so that results obtained in different laboratories can be compared. Most standards provide acceptance conditions for helmets, vests, armored seats, armored aircrafts or armored vehicles resisting against the impact of a given projectile or fragment.

Table 4.1. List of the most utilized ballistic standards.

Standard name	Agency	Scope	Threats
NIJ 0108.01; Ballistic Resistant Protective Materials	US National Institute of Justice	Applicable to all ballistic resistant materials intended to provide protection against gunfire, with the exception of police body armor and ballistic helmets	Small caliber handguns to high-powered or armor-piercing rifles
NIJ 0101.04; Ballistic Resistance of Personal Body Armor	US National Institute of Justice	To establish minimum performance requirements and test methods for the ballistic resistance of personal body armor intended to protect the torso against gunfire	Small caliber handguns to high-powered or armor-piercing rifles
NIJ 0106.01; Standard for Ballistic Helmets	US National Institute of Justice	To establish performance requirements and test methods for helmets intended to protect the wearer against gunfire. Requirements for face shields are not included in this standard	Handguns
EN 1063; Security Glazing. Testing and Classification of Resistance Against Bullet Attack	European Committee for Standardization	Specifies performance requirements and test methods for the classification of the bullet-resistance of glass	Small caliber handguns to high-powered or armor-piercing rifles
CEN prEN ISO 14876; Bullet Resistance and Knife Stab Resistance of Body Armor	European Committee for Standardization	Specifies performance requirements for bullet resistance and knife stab resistance of body armor	Handguns, rifles and shotguns
AS/NZS 2343 1997; Bullet Resistance Panels and Elements	Joint Technical Committee MS/43 (Australia and New Zealand)	Specifies requirements for bullet-resistant panels and elements according to their performance in preventing penetration by projectiles	Handguns, rifles and shotguns
Rusia-GOST R 50744-95; Armor Clothes. Classification and General Technical Requirements	State Standardization Committee of Russian Federation	Specifies performance requirements and testing methods for the ballistic resistance of armor clothes	Small caliber handguns to high-powered or armor-piercing rifles
NATO STANAG 4569; Protection Levels for Logistic and Light Armored Vehicles	North Atlantic Treaty Organization	Describes the system qualification and acceptance procedure for determining the protection level of logistic and light armored vehicles for KE and artillery threats	Rifles and heavy guns, from 5.56 to 25 mm ADPS
MIL-STD-662F; V_{50} Ballistic Test for armor	US Department of Defense	Provides general guidelines to determine the V50 ballistic limit of an armor. Applicable to any type of armor	Handguns, rifles and heavy guns

Others define the methodology to determine the ballistic limit or, in case of personal armors, to measure the behind armor blunt trauma through the back face signature produced by the impact in a block of plasticine placed behind the armor.

There are a number of standards, each one proposed by a different agency. However, the standards used by different countries have many similarities. Table 4.1 shows the most common standards in American, European, Euro-Asian, South Asian and Pacific Rim countries. All standards used by governments for testing ballistic materials specify diverse levels of protection. For each threat, the geometrical characteristics, material and speed of the projectile vary considerably, so it would be impossible to protect each system with an armor capable of stopping all possible types of projectiles. Then, each level is defined by a test ammunition type, its bullet mass, the impact velocity range, maximum back face deformation, barrel length, distance from the barrel, number of shots required or shot pattern for multi-hit performance. Standards consider the cartridges of weapons commonly used in the region of influence. This ensures that materials are tested with the most probable threats to be protected against. Ballistics specifications do not specify materials, any material able to arrest a specific projectile is rated according to that standard. Table 4.2 shows a summary of the NIJ 0108.01 Standard Performance Test.

Table 4.2. Ballistic armor levels defined by the NIJ 0108.01 Standard.

Protection Level	Weapon	Cartridge	Mass (g)	Projectile velocity (m/s)
I	long rifle handgun	0.22 long rifle high velocity, lead	2.6	320 ± 12
	Handgun	0.38 special round nose, lead	10.2	259 ± 15
II-A	Handgun	0.357 Magnum	10.2	381 ± 15
	Handgun	9 mm full metal jacket	8.0	332 ± 12
II	Handgun	0.357 Magnum	10.2	425 ± 15
	Handgun	9 mm full metal jacket	8.0	358 ± 12
III-A	Handgun	0.44 Magnum lead	15.55	426 ± 15
	Handgun	9 mm full metal jacket	8.0	426 ± 15
III	Rifle	7.62	9.7	838 ± 15
	Rifle	0.308 Winchester full metal jacket	9.7	838 ± 15
IV	Rifle	30-06 Armor Piercing	10.8	868 ± 15

5 Fiber Armors

5.1 Fibers

The most important element of armors made using reinforced composites is the *fiber*. Each individual fiber has a very small diameter and must therefore be spun into a yarn of parallel fibers, which are combined to produce non-woven or woven fabrics. Although glass fiber is also used for ballistic protective applications, the most effective fibers in this field are para-aramids, High Molecular Weight Polyethylene (HMPE), Poly-phenyleneBenzobisOxazole (PBO) and poly-diimidazo-pyridinylene-dihydroxy-phenylene (M5), which are all organic fibers. The main features of all of them will be presented next. Fibers suitable for use in the field of ballistic protection must have a low density, a high Young modulus and a high strength. These fibers can be produced using raw materials and production processes which guarantee a low mass density and a high molecular weight, with a marked orientation of these molecules in the direction of the fiber. Good temperature resistance, good stability after light exposure and good adhesion to matrix (for composite usage) are also desirable qualities.

5.1.1 Glass fiber

The basic component of glass fibers is silica (SiO_2) blended with other types of oxides. Unlike other fibers used in ballistic protection, it is an amorphous solid. Glass fibers are known for their high strength, good temperature and corrosion resistance, and they are mainly used in the naval, automotive and sports equipment industries. They are manufactured by spinning; small filaments of around 10 μm are pulled out through a nozzle from molten glass at a very high production rate. Its main drawbacks are its relatively low fatigue resistance and, compared to other ballistic fibers, higher density, lower stiffness and strength. However, its low price could make them an optimal solution for armors in which weight is not a primary concern. There are two main types of glass fibers, E-glass and S-glass. E-glass represents most of the fiberglass production in the world. S-glass is a formulation with higher mechanical performance. Both types are used in the manufacture of armors.

5.1.2 Para-aramids

Para-aramid (or simply aramid in the field of ballistic protection) fiber consists of long molecular chains of poly-para-phenylene terephthalamide. Developed at DuPont in 1965, it was first used commercially in the early 1970s under the trade name *Kevlar*™. A few years later, the Dutch company Akzo presented its high performance aramid with the name *Twaron*™ (owned by Teijin Group since 2000). Its use for impact protection constituted a significant technological advancement, and from the early 1970s it was commonly used in helmets and vests for security and defense corps. In addition to its excellent mechanical properties (Table 5.1) has low flammability and good integrity at both high and low temperatures. Aramid fibers are chemically stable under a wide variety of exposure conditions, and only certain chemicals can cause degradation over long periods. The tensile properties of aramid fibers are unaffected by moisture. The main drawback of aramid is its sensitivity to ultraviolet radiation. Extended exposure to unprotected fibers can cause a loss of mechanical properties.

Typical para-aramid fibers manufactured by DuPont include Kevlar™ 29 (all purpose), Kevlar™ 49 (high modulus) and Kevlar™ 129 (high tenacity providing enhanced bullet resistance). Teijin produces the CT microfilaments for helmets, soft and hard ballistics applications.

5.1.3 HMPE

HMPE stands for High Molecular Weight Polyethylene, a type of organic fiber which contains extremely large and highly oriented molecules. These molecules make the fiber very stiff and strong, which, when combined with its low density (less than that of water), means that it has excellent ballistic qualities. It was originally developed by Ruhrchemie AG in the 1950s, and in the late 1970s the Dutch company DSM marketed fibers for ballistic protection and defense applications under the trade name *Dyneema*™. Honeywell developed a product identical in chemical structure, which is sold under the brand name *Spectra*™. It is highly resistant to corrosive chemicals, has a very low coefficient of friction, is highly resistant to abrasion and is extremely durable and resistant to moisture and UV light. According to DSM, it is not advisable to use HMPE fibers at temperatures exceeding 80 to 100°C over prolonged periods of time. It becomes brittle at temperatures below -150°C.

DSM supplies SK25, SK60 and SK65 yarns (multi-purpose) and SK75 yarn (high-tenacity), as well as layers of unidirectional fibers UD-SB (soft ballistic to protect against handgun ammunition) and UD-HB (hard ballistic to protect against rifle fire). Honeywell develops Spectra™ 900 (multi purpose) and Spectra™ 1000 (high tenacity).

5.1.4 PBO

PBO stands for poly-phenylenebenzobisoxazole. It is a high-performance fiber developed by Stanford Research Institute International in the 1980s, and marketed by Toboyo with the trade name *Zylon*. PBO has a high tensile modulus and is very strong, and this means that it is a useful fiber for reinforcing materials used for items such as ropes, protective clothing and sporting goods.

When a police officer who was wearing a vest made with Zylon died in late 2003, the National Institute of Justice ordered that an extensive analysis of this material be carried out. Zylon fiber was found to be susceptible to hydrolytic (moisture) and photolytic (light) degradation. However, if PBO is isolated from external sources of moisture, there is no significant change in its mechanical properties.

5.1.5 M5

M5 is a high-performance organic fiber developed at Akzo Nobel and is currently produced by Magellan Systems and DuPont Advanced Fiber Systems. M5 fiber is based on the rigid-rod polymer poly-diimidazo-pyridinylene-dihydroxy-phenylene. M5 has potential as an ultra-high strength, ultra-high thermal and flame resistant alternative to products currently available in the advanced fibers market. This new fiber technology has the potential to represent a dramatic improvement over materials currently employed in critical-use applications. M5 processing technologies are still under development. However, ballistic impact tests of composite materials prepared from low-strength M5 fiber have shown that the performance of these systems is almost as good as that of the best composite materials. It is estimated that armor manufactured with M5 will reduce the areal density by approximately 50% compared to Kevlar™ systems at the same level of protection.

5.2 Yarns

Once the fiber has been produced and it needs to be woven to manufacture ballistic protection devices, the next step is to make the yarn by twisting or entangling a set of fibers. The yarn is the basic component of any ballistic protection material. Yarns are assembled together to form woven fabrics. Each yarn is defined according to the material used to make it (fiber type) and its mass per unit length. In the field of ballistic protection, the mass per unit length is expressed using two different units: Tex and Denier. Tex is the preferred unit among the scientific community, and it expresses the mass in grams per 1000 meters of yarn. Denier is a unit of measurement which expresses the mass in grams per 9000 meters of yarn.

The fundamentals of the behavior of yarns when undergoing ballistic impact are well understood. When a yarn is transversely impacted, a transverse deformation propagates away from the impact point at a finite velocity c_t. In addition, a longitudinal stress wave propagates outwards at a higher velocity (Figure 5.1). In a seminal work, Smith et al. (1958) developed a theoretical model to treat an infinitely long flexible filament after transverse impact. According to their theory, based on the differential equations that govern the motion of a differential portion of the filament, the outermost longitudinal wave propagates at a constant velocity c_l

$$c_l = \sqrt{\frac{1}{\rho}\left(\frac{d\sigma}{d\varepsilon}\right)_{\varepsilon=0}} \tag{5.1}$$

ρ being the mass density and the term inside the brackets the slope of the stress-strain curve evaluated at zero strain, that is, the Young modulus (see Table 5.1 for values of c_l for different fibers). This wave is followed by a plastic wave front propagating at a slower velocity c_p

$$c_p = \sqrt{\frac{1}{\rho}\left(\frac{d\sigma}{d\varepsilon}\right)_{\varepsilon=\varepsilon_p}} \tag{5.2}$$

Material points of the fiber are accelerated towards the impact point at a velocity that increases to the value

$$U = \int_0^{\varepsilon_p} c(\varepsilon)\,d\varepsilon = \int_0^{\varepsilon_p} \sqrt{\frac{1}{\rho}\left(\frac{d\sigma}{d\varepsilon}\right)}\,d\varepsilon \tag{5.3}$$

holding a constant strain ε_p. Afterwards a transverse wave is formed, shaped like an inverted V with the impact point at the vertex. The transverse wave travels at a velocity

$$c_t = \sqrt{\frac{1}{\rho}\frac{\sigma_p}{1+\varepsilon_p}} \tag{5.4}$$

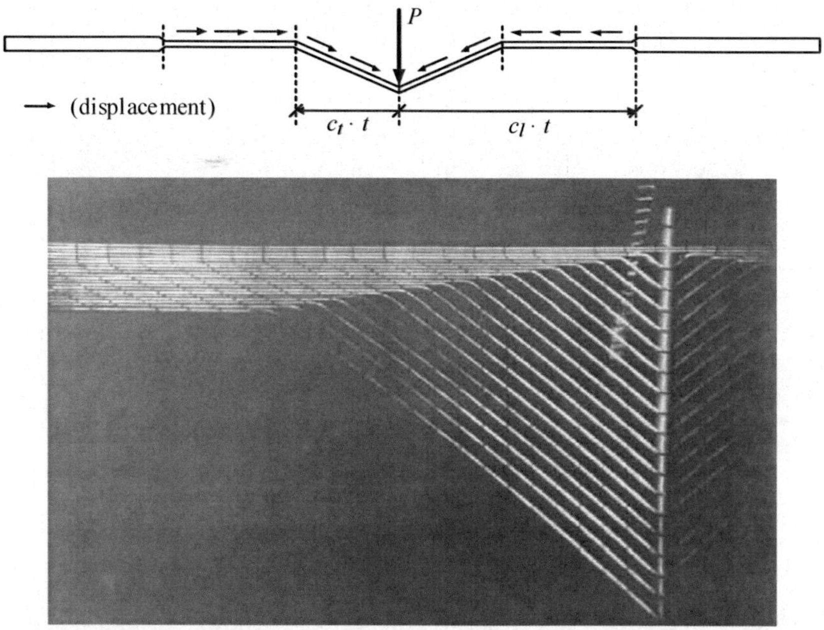

Figure 5.1. Top: Deformation of a filament subjected to transverse impact. Bottom: Superposed and slightly displaced images of a polymeric fiber impacted by a projectile. Marks on the fiber make it possible to see the propagation of longitudinal and transverse waves (Jameson et al. 1962).

Increased propagation velocities (for higher Young modulus and low density) lead to a greater dissipation of the energy of the projectile. The longitudinal wave contributes to it mainly through strain energy, and the transverse wave mainly through kinetic energy. These waves propagate until the yarn breaks. Thus the Young modulus of the fiber, its density and its tensile strength are the main parameters in the ballistic performance of the yarn. Table 5.1 provides characteristic values of these three parameters for different fibers used in ballistic protections, and also compares values with those corresponding to armor steel. Additionally, a high ductility of the fiber contributes to dissipate more energy. According to Cunniff (1999), for a fixed armor areal density and a specific threat, the ratio of the ballistic limits may be computed by the ratio of the number $\Phi^{1/3}$, Φ being the product of the specific toughness –which characterizes the energy absorbed per unit mass up to failure- and the longitudinal elastic wave velocity of the fiber

$$\Phi = \frac{\sigma_{failure}\varepsilon_{failure}}{2\rho}\sqrt{\frac{E}{\rho}} \tag{5.5}$$

This number includes the ultimate tensile strain, as a measure of the ductility of the fiber. Tabulated values for this number are shown in Table 5.1, showing the higher performance of M5 fiber, and the drawback of using glass fibers.

Table 5.1. Mechanical properties of glass, aramid, HMPE, PBO and M5 fibers (Cunniff and Auerbach, 2005), compared to steel.

Material	Density (kg/m^3)	Young modulus (GPa)	Tensile strength (GPa)	Ultimate tensile strain (%)	$\Phi^{1/3}$ (m/s)	c_l (m/s)
Glass fiber E	2550	75	2.4	4.5	480	5400
Glass fiber S	2500	90	4.2	5.0	631	6000
Aramid (200 den. Kevlar™ 29)	1440	91	2.9	2.95	618	7950
HMPE (Spectra™ 1000)	970	120	2.6	3.5	805	11100
PBO (Zylon™)	1540	180	5.8	3.5	893	10800
M5 (goal properties)	1700	450	9.5	2.5	1043	16300
Armor Steel (Armotec 500)	7850	210	1.6	8	-	5200

The viscous behavior of the yarns, inherited from the mechanical properties of the constituent material, must also be examined. The mechanical properties of the yarn may suffer important variations when deformed at high strain rates. The higher the strain rate level the higher the stiffness and tensile strength and the lower the failure strain. This results in a rate-dependent behavior of the composite laminate (Figure 5.2). The incorporation of a rate-dependent material model in simulation of fiber reinforced armors results in a better prediction of the impact behavior of the armor compared with a rate-independent material model

(Silva et al. 2005, Gower et al. 2008, van Hoof et al. 1999). This strain-rate effect is included in the model through the constitutive equation describing the stress-strain response of the yarn (Lim et al. 2003, Tan et al. 2005, Tan and Ching 2006).

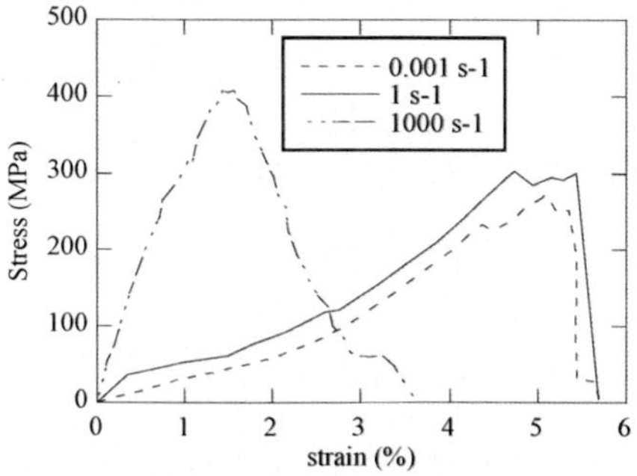

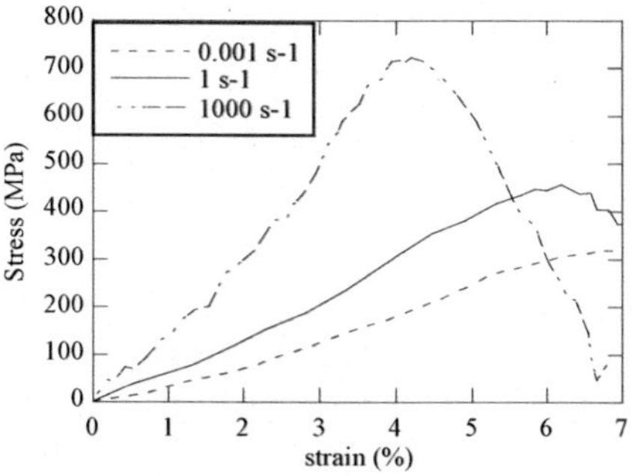

Figure 5.2. Tensile tests of different woven fiber composite laminates at three different strain rates. Left: Woven Aramid/Polyethylene terephthalate (PET). Right: Woven Polyethylene/Polyethylene. Chocrón-Benloulo et al. (1997).

5.3 Assembling Fibers and Yarns: Woven, Non-Woven and Prepreg Fabrics

The backbone of a fiber composite armor is the high performance fiber. But fibers placed in the path of the projectile cannot arrest a high-speed projectile by themselves. In order to engage a higher number of fibers, they are combined into a woven fabric, into a felt material or into a non-woven cross-plied unidirectional set of fibers (Bhatnagar et al. 2006).

5.3.1 Woven fabrics

Woven fabrics are built through interlacing *yarns* in 2D or 3D regular geometric arrangements. Two-dimensional woven fabrics created by positioning yarns at 0° and 90° are the most common for armor applications. The yarns running parallel to the direction of weaving are termed *warp* and transverse yarns are termed *weft*. The three basic weaves are *plain*, *twill* and *satin*.

Plain weave is the most basic of the 2D types. The warp and weft form a simple criss-cross pattern. Each weft thread crosses the warp threads by going over one, then under the next. Adjacent weft threads go under the warp threads that its neighbor went over (Figure 5.3 left). Twill weave is made by passing weft threads over one warp thread and then under two or more warp threads, with an offset between rows. This creates a pattern of diagonal ribs (Figure 5.3 centre). Satin is a more flexible type of weave, and it is therefore used to manufacture complex shapes. In this case, one warp runs over three or more weft yarn and under one weft yarn (Figure 5.3 right).

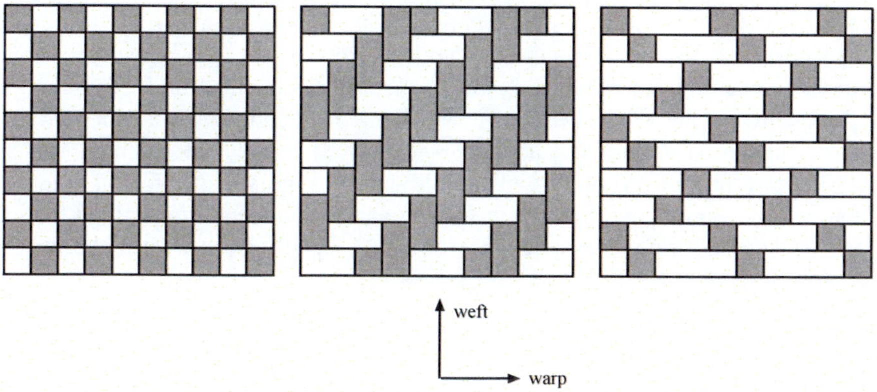

Figure 5.3. Left: Plain weave fabric. Centre: 2/2 twill weave fabric. Right: 4-harness satin weave.

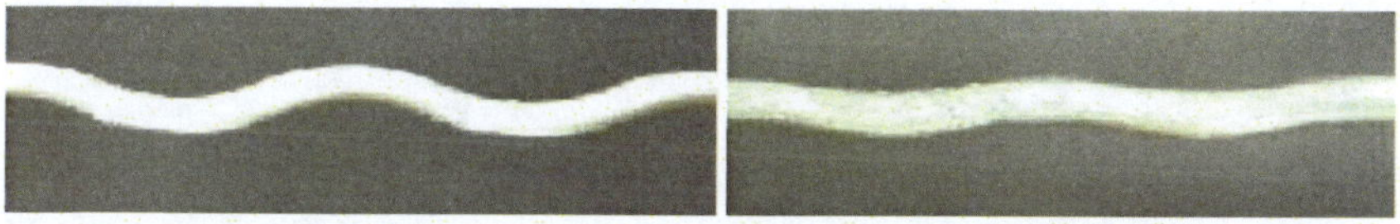

Figure 5.4. Crimp in warp (left) and weft (right) yarns from Twaron™ CT716 fabric (Tan et al, 2005).

Balanced weaves are fabrics in which the warp and weft are made of threads of the same weight and the same number of threads per inch. An unbalanced fabric may be used to obtain different mechanical properties along the warp and weft directions. It is well accepted that balanced fabrics dissipate more energy than unbalanced ones. However, ballistic fabrics normally have different levels of crimp in warp and weft yarns because of the weaving process, resulting in weft yarns having lower levels of crimp than warp yarns (Figure 5.4). This is believed to cause weft yarns to break preferentially to warp yarns during ballistic impacts. New generation fabrics for ballistics applications are now manufactured with equal crimp in both yarns so that they are loaded equally during ballistic impacts (Tan et al., 2005). Moreover, the process of weaving degrades the mechanical properties of the yarns. According to Lee et al. (2002), the tensile stiffness of E-Glass dry yarns is reduced by 7% and the strength is reduced by 30% after processing (Figure 5.5).

Fabrics (and fiber laminates in general) exhibit high mechanical in-plane properties, but they usually have poor interlaminar strength and energy release rate. Various techniques are considered to improve their out-of-plane properties, creating 3D fabrics. These basically consist of in-plane warp and weft yarns and out-of-plane yarns for through-thickness reinforcement. The most common process to manufacture 3D fabrics, due to the high production rates, is stitching, a three-dimensional reinforcement in which stitch threads are embedded along through-thickness direction. Karahan et al. (2008) found that stitching has a significant effect on properties of aramid fabric armors: reduction of trauma depth without significant increase in weight, and lower energy transmitted to the back of the panel (higher protection efficiency), but it also increases the panel stiffness.

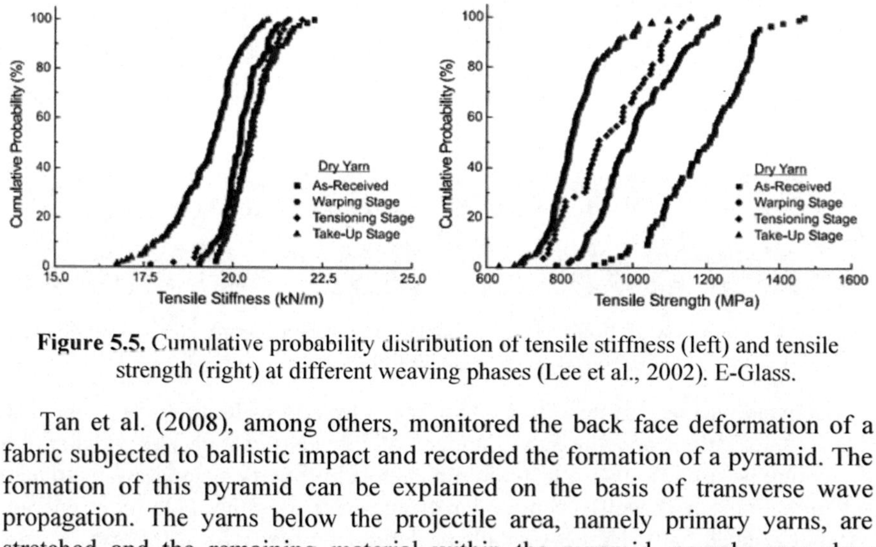

Figure 5.5. Cumulative probability distribution of tensile stiffness (left) and tensile strength (right) at different weaving phases (Lee et al., 2002). E-Glass.

Tan et al. (2008), among others, monitored the back face deformation of a fabric subjected to ballistic impact and recorded the formation of a pyramid. The formation of this pyramid can be explained on the basis of transverse wave propagation. The yarns below the projectile area, namely primary yarns, are stretched and the remaining material within the pyramid, namely secondary yarns, are also engaged in the process resulting in a larger volume being deformed and absorbance of energy (Figure 5.6 left). Figure 5.6 (right) shows two photographs from experiments of fabric perforation by a projectile at 251 m/s. The images were taken from the back as well as from the side of the fabric. When the spherical projectile impacts the armor, the fibers are pushed out of the fabric plane. Primary yarns, in direct contact with the projectile, are stretched and tensile waves travel down the primary yarns at the elastic wave speed, followed by a transverse displacement wave, engaging secondary yarns. Due to the crosswave of yarns, the elastic waves and the transverse deformation doesn't travel radially away from the impact point, but along the orthogonal directions of the yarns. This gives rise to the observed pyramidal deformation. The base of the pyramid is longer towards the clamped edges yarns; according to the model of Smith et al. (1965) deflection propagates faster in this direction since stress in clamped yarns is higher (Tan et al. 2005, Tan et al. 2008).

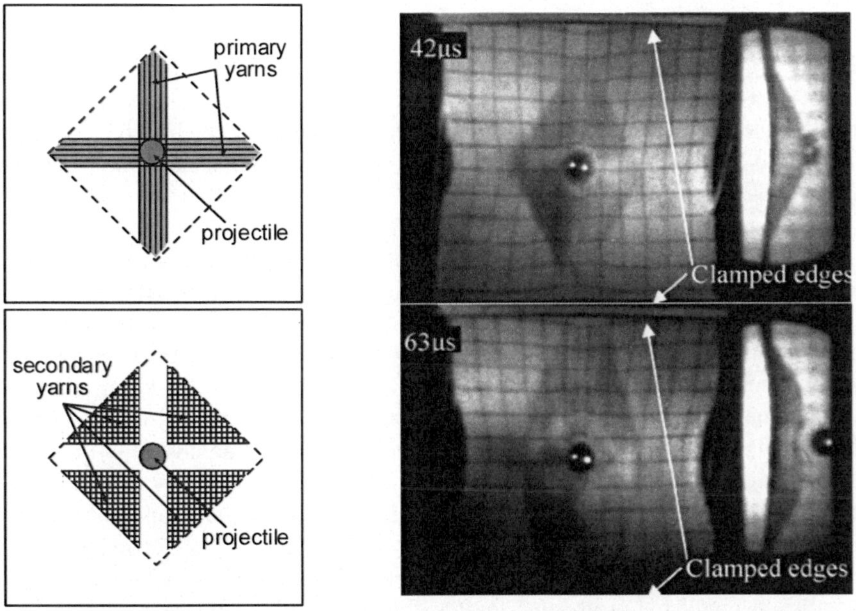

Figure 5.6. Left: Primary and secondary yarns affected during ballistic impact of a fabric. Right: Development of perforation in Twaron™ CT716 fabric clamped along two edges, corresponding to projectile impact at 251 m/s. Tan et al. (2008).

Three major components are again identified as contributing to the energy lost by the projectile, the energy converted into elastic deformation of the fabric, the energy converted into kinetic energy of the moving portion of the pyramid and the energy absorbed in tensile failure. Friction between fibers also influences the ballistic performance of the fabric, stabilizing the structure of the fabric in the impacting zone, delaying fabric failure and dissipating energy by rubbing and engaging adjacent yarns (Duan et al. 2005). Friction with the projectile contributes to strip the jacket of the bullet and to deform ductile projectiles. Thus it is usual to increase the friction coefficient between fibers using coating materials. However, if friction between fibers is too high, the contact between them may result in yarn tearing and this reduces the performance of the fabric. Another aspect which should be considered when designing a fabric armor is the weight per unit length of the yarns. Low denier yarns provide a higher ballistic efficiency but, at the same time, they are more expensive since weaving needs a higher number of operations per unit surface. Aramid fibers have a higher friction coefficient than HMPE fibers, 0.31 versus 0.20, and this could be a significant advantage since weavers can utilize higher denier aramid yarns achieving acceptable properties at a lower cost (Bhatnagar et al. 2006).

5.3.2 Non-woven fabrics

Yarn is a unit of woven fabrics made by weaving. Another method for creating armors is to combine fibers to form a non-woven layer. In this case, *staple fibers* or *continuous filaments* are used.

Staple fibers are a set of shorter, randomly oriented and unconsolidated fibers. The production process consists of cutting continuous fibers into pieces of a few centimeters long that are mixed to create a randomly oriented felt. The felt is needle-punched to entangle the fibers. Felt stops small fragments travelling at a low velocity very efficiently: a felt with an areal density of 1.2 kg/m² stops a 1.1 g fragment at 450 m/s (Chocrón et al. 2008). The most significant drawback of these felts is that they are not very effective at protecting against high-velocity projectiles, especially perforating projectiles. The most commonly used felts in the field of ballistic protection are Dyneema™ Fraglight, a felt manufactured by DSM from high density polyethylene, and ProTechtor, which is made from aramid fibers by National Nonwovens. Both of these non-woven felts were specially designed for protection against fragments of exploding landmines or artillery shells, although they do also protect against non-perforating projectiles travelling at speeds less than 450 m/s. Chocrón et al. (2008) performed quasi-static tensile tests and drop weight tests on Fraglight felts, showing a very non-linear, non-isotropic inelastic behavior, greatly deformable (more than 100% strain before failure, Figure 5.7 left) and softening after failure. The anisotropy is a result of the alignment of the fibers along the cross direction rather than the roll direction, indicating that this fabric is stiffer along the cross direction. Moreover, the non-woven felt properties were found to be strongly dependent on size and temperature (Chocrón et al. 2008). Drawing up models of how these materials react when they are subjected to impacts is still a complicated task which involves some major challenges.

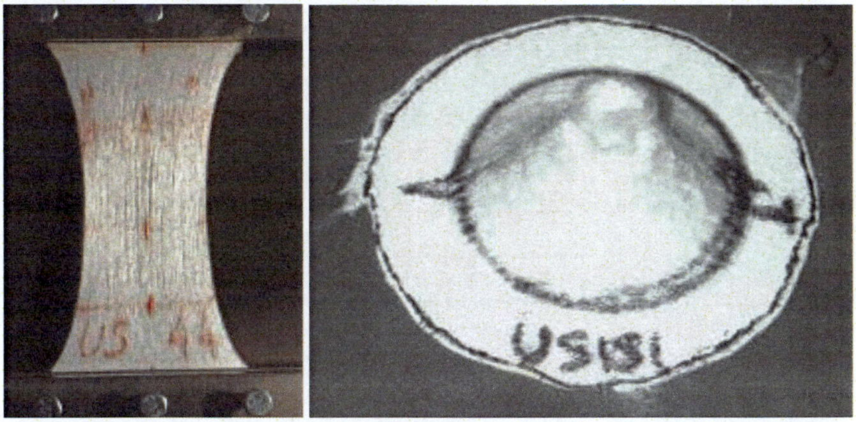

Figure 5.7. Left: Snapshot of an originally square specimen of Fraglight during a tensile test. Right: A close view of a Fraglight specimen tested with a drop weight tower. (Chocrón et al. 2008).

Felt deformation on impact is usually high and this can be a drawback when it comes to using these materials to make body armors (Figure 5.7 right). According to Jearanaisilawong (2004), macroscopic deformation of the felt is primarily accommodated by rotation of the constituent fibers. During a uniaxial tensile test the strain in the felt increases initially by alignment of the fibers with the loading direction, resulting in a contraction of the fibers in the perpendicular direction as a Poisson-like effect. Fiber realignment continues until fibers are uncurled and compacted, when further deformation can only be accommodated by fiber stretching, resulting in an increase in stiffness and a decrease in the Poisson-like effect (Figure 5.8). If strain still increases, softening starts due to disentanglement or breakage of the fibers until failure.

Table 5.2. Effect of thickness upon the ballistic resistance of nylon needle-punched felts (Liable and Henry 1969). Impact tests performed with a 1.1 g fragment simulating projectile.

Felt thickness (mm)	Areal density (kg/m^2)	Ballistic limit (m/s)
34	2.68	454
20	2.68	368
16	2.68	423

According to Liable and Henry (1969), the manufacturing process greatly influences the behavior of non-woven fabrics. The thickness of the felt for a given weight is one of the most important parameters in determining the ballistic efficiency. The thicker the felt, for a given areal density, the higher the ballistic resistance (Table 5.2). Thus, a pressing operation conducted to attain a thickness of felt amenable to tailoring results in reduced ballistic performance. Moreover, the behavior of the felt is mainly dependent upon fiber-fiber interaction rather than fiber strength. Anything which interferes with this interaction tends to lower ballistic resistance (Table 5.3).

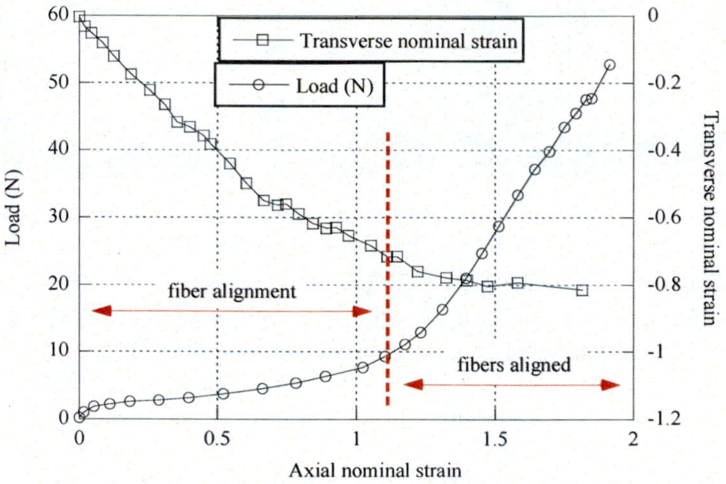

Figure 5.8. Experimental results of the uniaxial monotonic loading test of Dyneema Fraglight (Jearanaisilawong 2004).

Table 5.3. Influences of polymeric treatment upon ballistic resistance of nylon needle-punched felts (Liable and Henry 1969). Impact tests performed with a 1.1 g fragment simulating projectile.

Copolymer	Add-on weight (%)	Ballistic limit (m/s)
Control (no Add-on)	0.0	323
50-50 2-ethylhexyl acrylate and butyl methacrylate	2.5	269
50-50 2-ethylhexyl acrylate and butyl methacrylate	5.0	244
30-70 2-ethylhexyl acrylate and butyl methacrylate	2.5	279
30-70 2-ethylhexyl acrylate and butyl methacrylate	5.0	274

Continuous filaments may be defined as a single, flexible, small-diameter fiber of indefinite length. They are often presented rolled up on spools. This material can be used to manufacture a continuous resin-fixed layer of parallel fibers, which is then piled up in several layers to create the armor, usually at 0/90 orientations (Figure 5.9). The armors made using unidirectional layers usually have a lower areal density than those made with woven fabrics, but provide the same level of protection. This is because the energy is distributed more quickly between the fibers and the deformation concentration is reduced in those which receive the impact directly. In fabrics, the armor response is slower, as the fibers' crimp must be eliminated before they can start to absorb load. Furthermore, it is well accepted (Roylance 1973, Cunniff 1999) that at yarn intersections in woven layers, some energy is reflected back towards the impact point. Transverse displacement waves are generated in the yarns as a result of a bullet impact. Since cross-over points between yarns act as an obstacle to this displacement, the waves are partially reflected, increasing the strain in the fibers close to the projectile/armor contact. This may lead to a rapid failure of the yarns before the projectile has been significantly slowed down. Unidirectional composites have no such barriers, which means that much more material can be involved in stopping the bullet. In addition, this type of non-woven composite makes it possible to align a greater density of fibers, and the resin helps to keep the fibers in place.

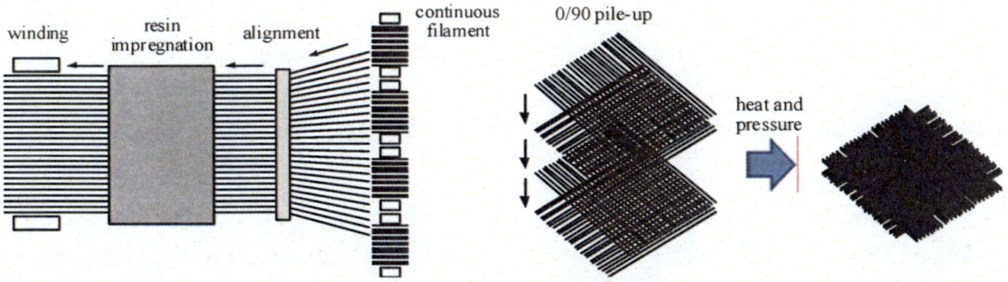

Figure 5.9. Left: Manufacturing a unidirectional layer. Right: Stacking of unidirectional layers to obtain a ballistic laminate.

The two most commonly used types of unidirectional non-woven fabrics in the field of ballistic protection are Dyneema™ UD (manufactured by DSM) and Spectra™ Shield (manufactured by Honeywell). Both of these fabrics are made using HMPE fibers. However, some unidirectional laminates are manufactured with Twaron™ or Kevlar™ fibers. All these products consist of a roll product of two or four unidirectional plies, crossplied at 0/90 and sandwiched or not in a thermoplastic film. They provide excellent protection against high velocity rifle bullets and fragments.

5.3.3 Prepregs fibers

The term "prepreg" refers to "preimpregnated" fibers, in other words, resin-impregnated woven, unidirectional or felt fibers formed into sheets. So, this definition applies to all the former fiber assembling architectures. Prepregs for ballistic applications tend to differ from those used in structural applications (Bhatnagar et al. 2006). The most important difference is the resin content; for armor purposes it should be much lower, from 10-20% by weight, as compared to 40-60% for structural composites. Also resins are more flexible, do not wet each fiber and limited bonding between fibers and resin is desirable for ballistic applications. Structural prepregs should usually be stored below 0 ºC, whereas ballistic prepregs can be stored at room temperature for a long time and the defined quality specifications remain acceptable.

The resin in the prepreg significantly improves the engagement between fibers and the shear strength between plies, and also helps, to some extent, to dissipate the impact energy by matrix damage. However, a high resin content may decrease the flexibility of the armor and consequently its efficiency. Besides the improvement of the ballistic properties, the addition of resin provides beneficial functions during manufacturing by keeping the fibers in the proper orientations. Both thermoplastic and thermoset resins are used. Thermoplastics, like polyolefins (polyethylene or polypropylene), polyurethane or acrylic resins, provide low production costs and a long shelf life. Thermosets, like vinylester, epoxy or phenolic resins, present high strength and toughness.

5.4 Influence of the Areal Density on the Ballistic Performance of Fiber Armors

One of the most important issues in the design of armors is the performance-to-weight ratio of the ballistic package. Adequate determination of the number of woven, non-woven, or prepregs layers needed to stop a given projectile permits adjustment of the weight of the protection to the considered threat, avoiding excess weight. Jacobs and Dingenen (2001) determined the ballistic limit of non-deformable Fragment Simulating Projectiles of different sizes impacting against Dyneema (fabric-based) soft and hard amours. These tests showed a linear relation between the energy absorption of the armor E_{abs} –normalized by the strike face area S - and the areal density of the ballistic package AD, that may be written as

$$E_{abs}/S = AD \cdot C \tag{5.6}$$

C being a ballistic material-related constant, the slope of the curve in Figure 5.10 left. This expression provides a phenomenological instrument to predict protection levels against rigid projectiles (Jacobs and Dingenen 2001). However, the same authors pointed out that when deformable bullets are considered, an area larger than that corresponding to the initial shape of the projectile should be considered. In that case, the energy of the bullet is assumed to be absorbed through three mechanisms: first, penetration as a rigid projectile with a strike face area S_1; second, deformation of the bullet consuming kinetic energy; and third, penetration of the bullet with a large strike face area S_2. The relation between absorbed energies associated with the first and third mechanisms are form-identical to equation [5.6], but with the respective face areas

$$E_{abs1}/S_1 = AD_1 \cdot C \tag{5.7}$$

$$E_{abs2}/S_2 = AD_2 \cdot C \tag{5.8}$$

where the sum of the areal densities AD_1 and AD_2 is the total areal density of the package. Cork and Foster (2007) also reported that energy absorbed increases linearly with the number of layers in the laminate (Figure 5.10 right).

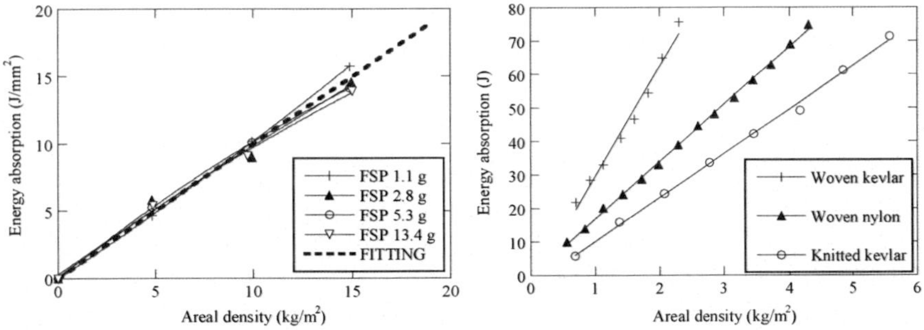

Figure 5.10. Left: normalised energy absorption versus areal density for fabric-based Dyneema™ hard ballistic packages impacted by Fragment Simulating Projectiles (Jacobs and Dingenen 2001). Right: energy absorption versus areal density for fiber armors impacted by 1 g cylinder at 520 m/s (Cork and Foster 2007).

This relation between energy absorption and the areal density of the fiber armor was previously observed by Song and Egglestone (1987). They proposed a relationship between the ballistic limit v_{50} and the areal density of a fiber armor AD in the following equation

$$v_{50} = \lambda(AD)^{\varphi} \tag{5.9}$$

Where λ and φ are constants related to the armor material properties. The values of these two constants are listed in Table 5.4 for various fiber reinforcements and resin systems (Bhatnagar et al. 2006). For most armor, the exponent is close to 0.5, showing again the roughly linear relation between absorbed energy and areal density.

Table 5.4. Values of constants λ and φ for different fiber reinforcements and resin systems (Bhatnagar et al. 2006). PH-PVB: Phenolic Polyvinyl Butyral. LLDPE: Linear Low Density Polyethylene. PSU: Polysulfone.

	S2 Glass/ PH-PVB	Kevlar-29/ PH-PVB	Kevlar-KM2/ PH-PVB	Kevlar-KM2/ LLDPE or PSU	Spectra-900	Spectra-Shield
λ	99	158	216	234	223	200
φ	0.64	0.56	0.50	0.49	0.58	0.67

6 Ceramic-Faced Armors

6.1 Introduction

The use of ceramic materials in the field of ballistic protection was a major breakthrough, and numerous studies have been carried out as a result: the seminal works of Wilkins (1967-69) and Florence (1969) helped to understand why these materials are so effective in the field of ballistics. Since those studies were published, a large number of authors have continued to carry out research in the field, and this has led to improvements in the performance of these armors. Whereas metallic armors absorb the energy of the thread mainly through plastic deformation, and fiber-reinforced armors accomplish this mainly by spreading the load to a large volume of fabric target which deforms elastically (friction may also play a principal role in both types of armor, Duan et al. 2005, Arias et al. 2008), ceramics are efficient as ballistic materials due to their capacity to blunt and fragment the tip of the projectile.

An initial analysis of the armor impact process reveals three distinct phases: fragmentation of the ceramic material, penetration of the ceramic tile and penetration of the backing panel (Figure 6.1).

i) *Fragmentation of the ceramic material*: one of the most important stages in the response of a ceramic tile under impact is the initial phase. Immediately upon contact with the projectile, stress waves are generated in the ceramic tile and a conical cracking front propagates from the border of the projectile/ceramic contact area. When the compressive wave reaches the adhesive layer, with lower mechanical impedance, hoop tensile stresses develop at the rear face of the tile and radial cracks propagate back to the projectile. The consequence is a general break-up of the ceramic from the coalescence of the cracks. This fragmentation process has been described in a number of studies (e.g. Wilkins 1978, Mayseless 1987, den Reijer 1991, Zaera and Sánchez-Gálvez 1998a). The projectile can advance only if the pulverized ceramic material in its path is pushed ahead or to the sides. But this flow is impeded by the heavy confinement of the crushed and fragmented ceramic due to the intact surrounding material and the adhesive and metallic plate, so the ceramic powder can only flow back against the advance of the projectile, along the cavity it has produced, eroding the projectile tip (Figure 6.1 a). Fragmentation continues during the whole process of penetration, but that occurring in the first microseconds after impact is an important factor in decreasing the ceramic strength. Before the end of this fragmentation stage, the ceramic along the path of the projectile deforms essentially in a uniaxial strain condition, thus the material shows its highest strength to be penetrated and the tip of the projectile is strongly eroded. Soft-core projectiles (lead or steels with low yield stress) can hardly penetrate into the tile. The time necessary to complete this stage is that required for the different crack fronts to pass through the ceramic tile. Den Reijer (1991) assumed that the ceramic break-up time was the sum of the time it takes for the compression wave to reach the adhesive/ceramic interface and the time required for the radial fracture front to traverse back the ceramic tile. The latter depends on the crack propagation speed whose maximum velocity is that of the Rayleigh waves in the material. Senf et al. (1994) and Strassburger et al. (1994) measured the propagation velocity of cracks in glasses and ceramics, and found values in between 1/3 and 1/6 of the elastic wave speed. Den Reijer (1991) proposed a value of 1/5 of the elastic wave speed to fit the numerical simulations made by Wilkins (1978) for the development of the cracked conoid.

ii) Penetration in the ceramic tile: After fragmentation, the projectile penetrates a conoid of comminuted, pulverized and fragmented ceramic (damaged ceramic) whose mechanical properties are lower than those of the undamaged tile and are dependent on the degree of fragmentation at each point (Figure 6.1 b). Crack generation as described above allows the displacement of small fragments, making projectile penetration easier. However, the erosion of the projectile tip may continue during this phase. The volume of fragmented ceramic ahead of the projectile, physically separated from the remaining intact ceramic tile, has a conoid shape. The conoid is accelerated in the impact direction and distributes the load over a large area of the backing panel which undergoes extensive deformation and contributes remarkably to increasing the ballistic performance of the ceramic-faced armor.

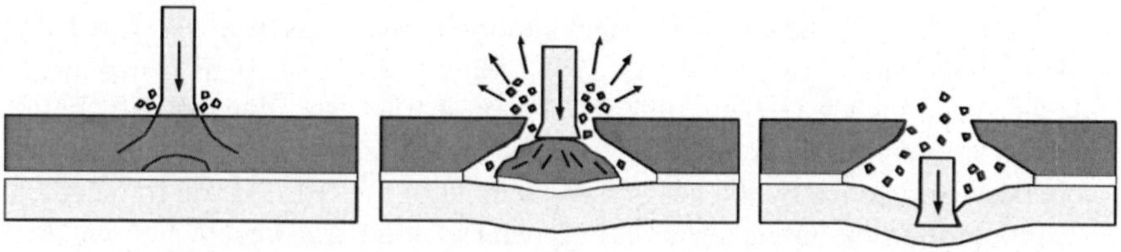

(a) (b) (c)

Figure 6.1. Phases of the penetration of a projectile into a ceramic-faced armor. (a) fragmentation, (b) penetration in the ceramic tile, (c) penetration in the backing panel.

iii) Penetration of the backing plate: Once the thread has completely penetrated into the ceramic conoid, its tip contacts the backing plate (metallic or fiber-reinforced). If the energy is still high it continues penetrating the target until arrest or perforation (Figure 6.1 c).

6.2 Efficiency of Ceramics as Protective Materials

Ceramics are the most important materials available when it comes to the effectiveness of light armor. When the projectile interacts with the ceramic, the erosion process takes place, and this plays a major role in terms of the behavior of an armor when subjected to impact. As such, it would be useful to have a method which would allow us to evaluate the capacity of a ceramic material to withstand the penetration of a projectile. Ballistic limit tests are not the most useful tests to determine how effectively a ceramic material can withstand an impact, as the ceramic's behavior is also influenced by the material and the thickness of the backing panel, and by the joint between the two. The most suitable experiment which could be used to determine how a ceramic material responds, regardless of the geometry of the backing panel, would be to use a ceramic block which is thick enough to be considered semi-infinite. This block would be subjected to an impact and the penetration of the projectile could be measured. The resulting value would be used to represent the effectiveness of the ceramic material. As this experiment would be impossible to carry out for technical and economic reasons, Bless et al. (1987) proposed an alternative method which makes it possible to eliminate the premature stages of tensile stresses in the ceramic and the consequences of it. This method, known as the Depth of Penetration Test (DOP), consists of backing up the ceramic tile using a metal block which is thick enough to stop the projectile before affecting the surface of the backing panel and can therefore be considered semi-infinite (Figure 6.2).

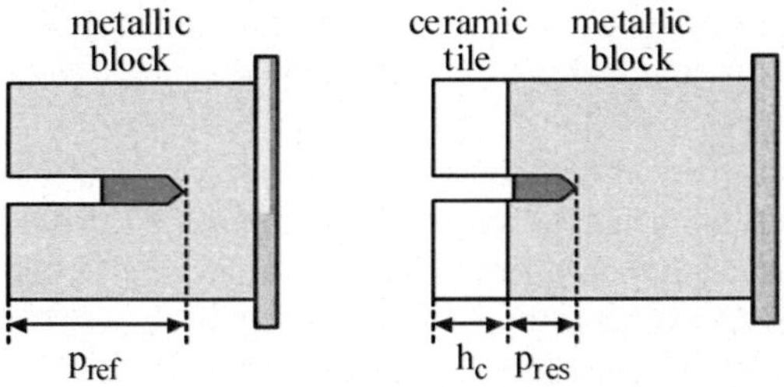

Figure 6.2. Armor configuration required to carry out the Depth of Penetration Test.

A number of different factors can be used to characterize the protective capacity of ceramic materials, based on the data obtained directly through the Depth of Penetration Test. The simplest of these is obtained based on the ratio between the penetration of the block without the protection of the ceramic tile and the penetration of the block when it is protected by the ceramic tile, as expressed in the following equation

$$F = \frac{p_{ref}}{h_c + p_{res}} \tag{6.1}$$

where h_c is the thickness of the ceramic tile, p_{ref} is the residual penetration of the block without the ceramic tile and p_{res} is the residual penetration with the ceramic tile (Figure 6.2). The drawback with this factor is that it does not take the density of the two materials into account. As such, other types of factors which take into account the reduction in weight as a result of using the ceramic tile are used, such as the Mass Efficiency Factor (MEF) (Hohler et al. 1995b)

$$MEF = \frac{\rho_b p_{ref}}{\rho_c h_c + \rho_b p_{res}} \tag{6.2}$$

where ρ_b is the density of the metal and ρ_c is the density of the ceramic tile. The most widely used factor for determining how effective ceramic materials are at protecting against impacts is the Differential Efficiency Factor (DEF) defined by Yaziv et al. (1986), as expressed in the following equation

$$DEF = \frac{\rho_b \left(p_{ref} - p_{res} \right)}{\rho_c h_c} \tag{6.3}$$

The validity of a test carried out to characterize the intrinsic properties of a ceramic material depends on how constant the factor chosen remains when different tile thicknesses, impacts velocities, projectiles and armor configurations are involved. In terms of the first of these variables, most authors have established an inverse linear relationship between p_{res} and the thickness of the ceramic tile in front of the metal block for a number of different projectiles at different impact speeds and on different types of material (Rosenberg et al. 1988, Mellgard et al. 1989, Anderson and Morris 1992, Woolsey 1992, Hohler et al. 1995b). This is shown in Figure 6.3, which has been drawn up based on tests carried out on different ceramic materials.

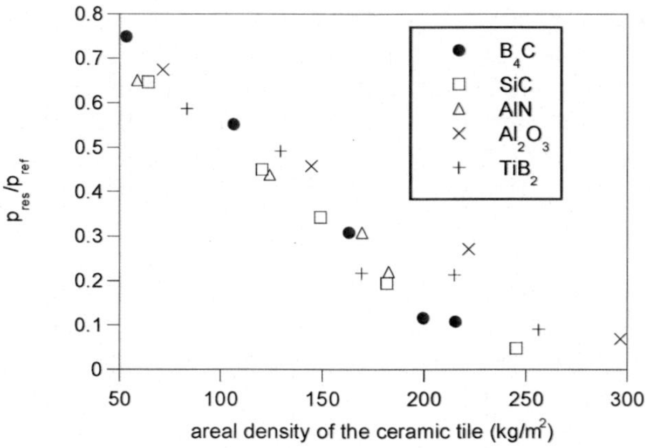

Figure 6.3. Residual penetration in the DOP Test in relation to the thickness of the ceramic for different ceramic materials. Sintered Tungsten projectile (L/D=12.5) at 1700 m/s. RHA semi-infinite backing plate. (Hohler et al. 1995b).

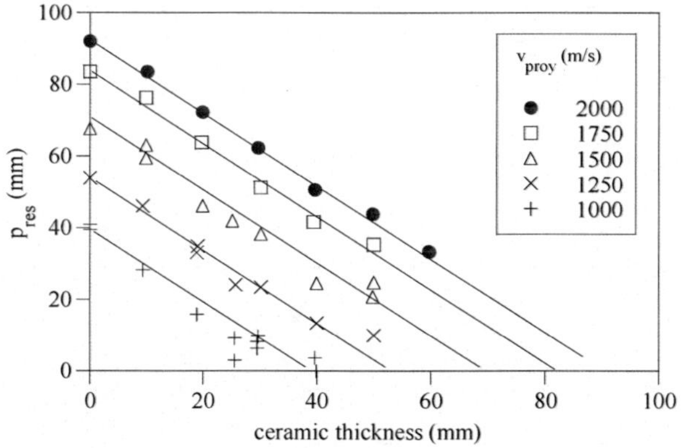

Figure 6.4. Residual penetration in the DOP Test in relation to the thickness of the ceramic for different impact speeds. Sintered Tungsten projectile (L/D=3.2). 99.5% alumina (Senf et al. 1995).

Researchers have failed to reach a consensus with regard to variations in DEF at different speeds. While Senf et al. (1995) obtained straight lines at the same angle in their graphs showing penetration in relation to ceramic thickness at different speeds (Figure 6.4), Hohler et al. (1995a) found that the DEF increases as the speed increases. Hauver et al. (1992) and Sternberg (1989) have explained why the second case is true. When the projectile hits the ceramic layer, a cracking front propagates through the ceramic at a finite speed, fissuring the material as the resulting fragments of ceramic slide between each other, creating gaps. The projectile therefore comes into contact with a material other than the intact ceramic, with inferior resistant properties. The faster the projectile is travelling, the less time it takes for the ceramic material to break up into fragments and lose its resistance to penetration, resulting in a higher ballistic effectiveness.

The shape of the projectile also seems to affect the value of the DEF. Woodward et al. (1994) carried out the DOP test using high-hardness projectiles in two different shapes: one with a blunt nose and one with a conical nose. The resulting DEF values were completely different (Figure 6.5). A conical nose leads to greater penetration in the aluminum block, giving rise to higher p_{ref} values. However, this greater penetration does not occur when a ceramic tile is used, as the projectile's nose is eroded quickly as a result of the hardness of the ceramic. Higher efficiency values are therefore obtained when using sharp projectiles.

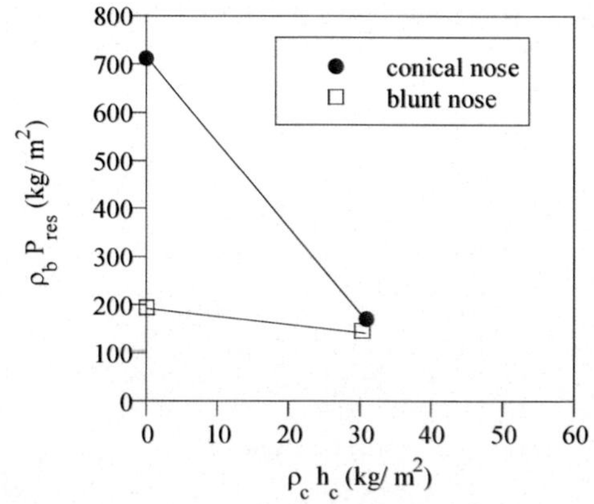

Figure 6.5. Residual penetration in the DOP test in relation to the thickness of the ceramic for tungsten projectiles with a conical nose and with a blunt nose. Semi-infinite aluminum block. Impact speed = 1200 m/s. 99.5% purity alumina. (Woodward et al. 1994).

With regard to the size of the projectile, Rosenberg et al. (1988) calculated the DEF for 12.70 AP and 14.50 AP projectiles. Both projectiles are very similar in terms of speed, core material and length/diameter ratio, and the main difference between them is their size. As Table 6.1 shows, the DEF obtained for four different types of ceramic materials are very similar, and this suggests that the size of the projectile does not affect the results.

Table 6.1. Differential Efficiency Factors for 12.70 AP and 14.50 AP projectiles, obtained for several ceramics (Rosenberg et al. 1988).

Ceramic	DEF 12.70 AP	DEF 14.50 AP
B_4C	7.8	8.3
SiC	6.9	7.0
TiB_2	5.0	5.2
Al_2O_3 85% purity	4.2	3.6

The configuration of the DOP specimen, on the other hand, does seem to have an effect on the efficiency of the ceramic. A number of authors (Hauver et al. 1992, Ernst et al. 1994, Bless et al. 1995, James 1995) have found that both radial and axial confinement of the ceramic leads to increased resistant properties of the material which is subjected to an impact in the DOP test. The DOP test can therefore only be used to compare the effectiveness of different ceramics under similar impact conditions.

6.3 Penetration Strength of the Ceramic

In addition to examining the DEF, many authors have tried to define another parameter which makes it possible to compare the efficiency of ceramic materials when subjected to impacts. To do this, they have used a model proposed independently by both Tate (1967) and Alekseevskii (1966) to study penetration in high-thickness metal targets. This model has become the most commonly used reference for the analytical simulation of penetration in semi-infinite targets. Projectiles are considered to behave as rigid-perfectly plastic, thus the projectile remains undeformed except for a small zone near the projectile tip which is assumed to be plastic, and erodes immediately afterwards. Such a hypothesis is reasonable because the elastic energy stored in the projectile is negligible compared to the energy dissipated in plastic deformation and erosion. The erosion process at the interface is controlled by the variable Σ, which is obtained from the modified Bernouilli equation

$$\Sigma = Y_p + \frac{1}{2}\rho_p (v-u)^2 = R_t + \frac{1}{2}\rho_t u^2 \tag{6.4}$$

Σ represents the normal stress at the interface and is the sum of two terms, one related to the material strength and the other related to the hydrodynamic pressure. ρ_p and ρ_t are the densities of the projectile and target respectively, v the projectile velocity and u the penetration speed. Y_p is the dynamic strength of the projectile material and R_t is the penetration strength of the target. The projectile is decelerated by the yield stress

$$\frac{dv}{dt} = -\frac{Y_p}{\rho_p L} \tag{6.5}$$

L being the actual length of the projectile. On the other hand, the decrease of the projectile length is expressed by

$$\frac{dL}{dt} = -(v-u) \tag{6.6}$$

This set of equations can be utilized to determine whether the projectile or the ceramic or both solids erode, depending on their dynamic strengths. Actually, there is a threshold in v

$$v_{\text{lim}} = \sqrt{\frac{2(R_t - Y_p)}{\rho_p}} \tag{6.7}$$

below which no penetration occurs.

6.4 Applying the Tate-Alekseevskii Model to the DOP Test: Determining R_t for the Ceramic Material

The Tate-Alekseevskii model was originally developed and validated for penetration into metallic materials. Bless et al. (1987) were the first workers who used this model to determine R_t for ceramic materials. Performing impacts with aluminum, steel and tungsten projectiles onto 85% and 90% purity aluminas, they delimited the impact velocity at which no perforation occurs. The values of R_t were found to be close to the Hugoniot Elastic Limit (HEL), defined as the yielding stress under uniaxial-strain dynamic conditions. The same procedure was used by Rosenberg and Tsaliah (1990) and they obtained similar results (Table 6.2).

Table 6.2. Critical velocities and R_t values obtained with the Tate-Alekseevskii Model (Rosenberg and Tsaliah 1990).

Projectile material	Ceramic type	Ceramic HEL (GPa)	v_{lim} (m/s)	R_t (GPa)
Copper	Al_2O_3 85%	6.0	1115±20	6.06
Steel	Al_2O_3 85%	6.0	990±20	5.75
Tungsten	Al_2O_3 90%	7.0	630±20	5.82
Copper	Al_2O_3 90%	7.0	1210±20	7.05

Rosenberg and Tsaliah (1990) also performed DOP tests with the same ceramic materials and evaluated the penetration in the metallic block with the Tate-Alekseevskii model using the values of R_t previously determined by the critical velocity method. The results (Table 6.3) demonstrate a good agreement. Nevertheless, other authors have suggested the need to consider a value of R_t which varies during the penetration process: an experimental determination of the position of the projectile tip, obtained with X-Ray high-speed shadowgraphs (Hauver et al. 1992) and its comparison with the predictions of the Tate-Alekseevskii model showed a very high value of R_t at the beginning of the impact, decreasing as the projectile penetrates into the ceramic block (Figure 6.6).

Table 6.3. Penetration depths in DOP tests. Critical velocities and values obtained with the Tate-Alekseevskii Model (Rosenberg and Tsaliah 1990).

Ceramic type	Ceramic thickness (mm)	Block steel	Impact velocity (m/s)	Measured penetration (mm)	Calculated penetration (mm)
Al_2O_3 85%	19.1	RHA	1330	40	38
Al_2O_3 85%	12.7	1020	1122	51	49
Al_2O_3 85%	12.7	1020	1400	82	73
Al_2O_3 85%	19.1	1020	1135	41	40
Al_2O_3 90%	19.1	1020	1350	57.5	54
Al_2O_3 90%	19.1	RHA	1412	44	46

Summarizing, most authors consider the Tate-Alekseevskii model sufficiently accurate assuming an average value of R_t. This value can be taken as the Hugoniot Elastic Limit (Bless et al. 1987, Anderson and Morris 1992,

Hohler et al. 1995b). Other authors (Rosenberg et al. 1988, Ernst et al. 1992) found a better correlation with the following factor $\overline{Y}$

$$\overline{Y} = \frac{HEL + \dfrac{(1-\nu)Y_s}{(1-2\nu)}}{2} \tag{6.8}$$

HEL being the Hugoniot Elastic Limit, ν the Poisson ratio and Y_s the elastic limit.

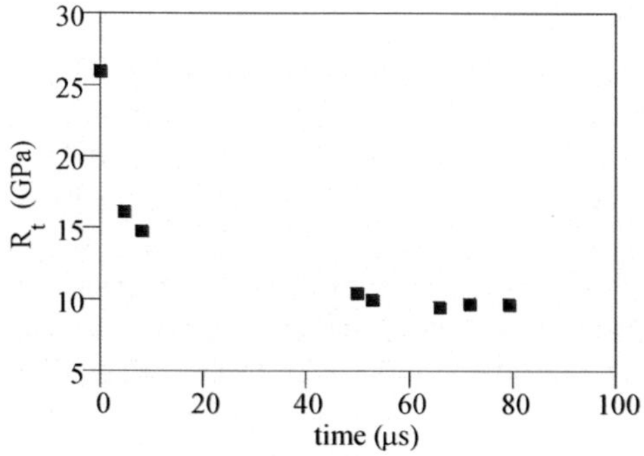

Figure 6.6. Variation with time of the penetration resistance R_t for a 99% purity alumina (Hauver et al. 1992).

6.5 Performance of Ceramic/Faced Armors Against Ballistic Impact

The ballistic limit test is the most useful technique when it comes to validating a protection system for a specific threat. A number of classic experiments carried out by different authors (Wilkins 1978, Mayseless et al. 1987, den Reijer 1991) provided results which can be used to analyze the response characteristics of the complete ceramic/backing system in more detail.

Wilkins (1978) carried out a series of firearm tests using a 7.62 AP projectile with an 8.1 g steel core, fired at 85% purity alumina tiles with a 6061-T6 alumina backing. All impacts hit the target at normal incidence and ballistic limits were determined for ceramics of thicknesses of 4.06, 6.35, 7.85 and 8.64 mm and aluminum of thicknesses of 3.05 and 9.40 mm. The results of this experiment are shown in Figure 6.7. The sudden jump in the ballistic limit for aluminum of a thickness of around 5.5 mm indicates an increase in the effectiveness of the ceramic, because the metal backing is thick enough to prevent fragmentation when the tile is deformed. According to Woodward (1990) and den Reijer (1991), for aluminum plate thickness up to 5.5 mm the deformation mode is governed by membrane deflection, whereas above 5.5 mm bending and shear dominates the deformation process. This change in the structural behavior causes a stiffer response of the back-up plate which restrains fragmentation of the ceramic tile.

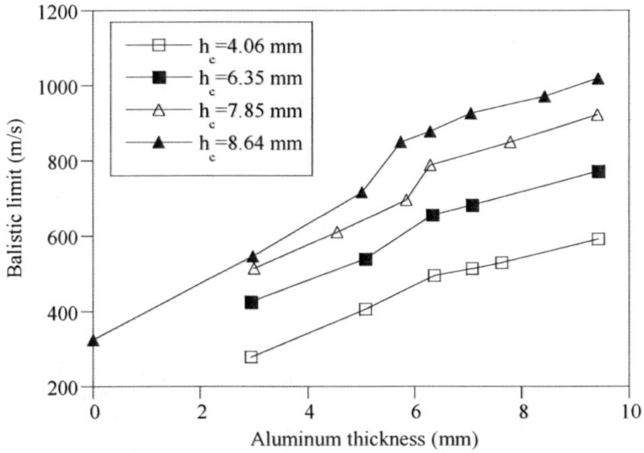

Figure 6.7. Ballistic limit for 85% purity alumina tiles backed by aluminum 6061-T6, impacted by a 7.62 AP projectile (Wilkins 1978).

In order to analyze the behavior of ceramic/faced armors when impacted by projectiles of a higher caliber, Mayseless et al. (1987) used a 12.70 AP projectile with a 25.3 g steel core fired at armors consisting of a 6.35 mm-thick 85% purity alumina tile, with backing plates made of 6061-T6 and 2024-0 aluminum alloys and SAE 4130 and SAE 1020 steels. The results, which are shown in Figure 6.8, indicate that steel is more efficient than aluminum of the same thickness. However, in terms of areal density, the aluminum offers a greater protection capacity.

The set of experiments carried out by Den Reijer (1991) can be used as a basis to carry out a complete study of the penetration process for small caliber projectiles in ceramic-metal armors. Den Reijer designed an alternative system which used two radiation emission devices. The first was positioned in line with the armor and was used to obtain an image of the position of the projectile's tail and the plate deformation. Another X-ray image was taken at the same time by a second device positioned at an angle to the armor plate so that the beam only passed through a narrower section. These complementary images make it possible to determine the length of the projectile (Figure 6.9).

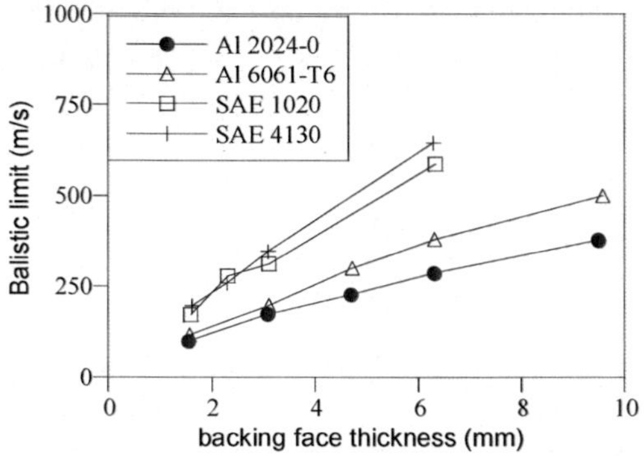

Figure 6.8. Ballistic limit for 85% purity alumina tiles (6.35 mm) backed by aluminum or steel impacted by a 12.70 AP projectile (Mayseless et al. 1987).

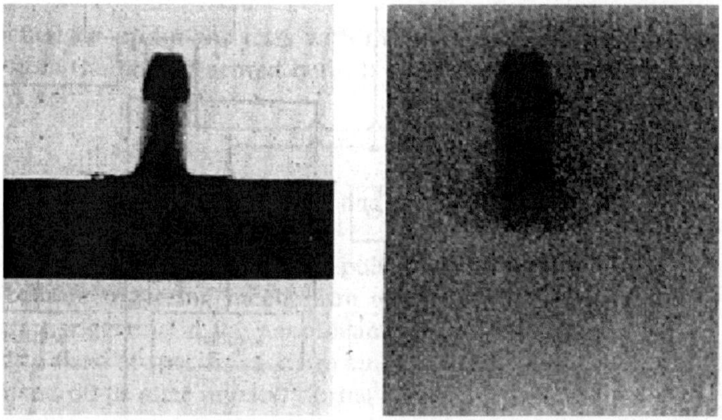

Figure 6.9. X-ray images, taken at the same time, of a 7.62 AP projectile 16 μs after initial contact with an alumina-aluminum panel. The image on the left was taken with the radiation beam in the plane of the panel. The image on the right was taken with the beam at an oblique angle (den Reijer 1991).

Den Reijer used this device to determine the positions of the projectile's tail and nose and the deflection of the metal plate for six different impacts with a cylindrical projectile similar to a 7.62 AP. Three different armor configurations were tested, using 8.1 mm thick 95% purity alumina tiles with 4.0, 6.0 and 8.0 mm thick 6061-T6 aluminum backing. Two series of shots were fired at each configuration: one above and one below the ballistic limit. Images were captured at different moments in each series (around 8 shots). These images were used to construct a photo history of the average positions of the projectile's tail and nose and the deflection of the plate. These results have been used by Den Reijer and other authors to validate simulation models (den Reijer 1991, Zaera and Sánchez-Gálvez 1998a, Fernández-Fdz and Zaera 2008).

7 Modeling the Impact Behavior of Fiber Armors and Ceramic-Faced Armors

The design of structures subjected to impact was traditionally carried out empirically, relying on real impact tests using the given projectile-target in each test. This method is very reliable but the results are valid only for the specimen tested, and they are not readily extrapolated; any variation of the impact velocity or of the characteristics of the projectile or of the target invalidates extrapolation of previous test results. It is also a costly procedure; testing in these conditions requires sophisticated installations and equipment. It becomes even more costly if the aim of the tests goes beyond the discovery of the resistance quality of a target, or the characteristics of the deformation of a projectile capable of perforating an armor or destroying a structure; to obtain information on the deformation of the target or on the position of the projectile during penetration, ultra-rapid photographs -either optical or X-ray- are required, and this again increases the cost of the tests.

With a view to extending the validity of real-fire tests to other projectile-target configurations, some so-called *semi-empirical* techniques were developed; these proposed algebraic equations with a certain theoretical base, calibrated from a large number of experimental data, allowed interpolation or extrapolation of the results. In general, these techniques contributed very little to the knowledge of the physical phenomena of the impact process. The need for design tools to simulate this process triggered the development in recent years of a large number of models of different types; all of them belong to two families: those of *analytical modeling* and those of *numerical modeling*.

In *analytical modeling,* a series of simplifying hypotheses are assumed in the physical mechanisms that define the behavior of the solids to produce, by means of simple equations, a model that enables each problem to be solved in a few minutes or even seconds. These methods are not as precise as the numerical methods, but they are so rapid that a great number of calculations can be made in a short time. Also known as *engineering models*, these are useful in the design of armors; not only are they quick and easy to use but they give detailed information about the process of penetration. The representation of the materials in these models is expressed in simple constitutive equations using parameters such as the elastic modulus, yield stress or failure strain. The equations used in continuum mechanics are greatly simplified by the introduction of hypotheses obtained from the analysis of the global behavior of the phenomenon. The approximation obtained with this type of model depends largely on its complexity. The introduction of a larger number of variables into the model gives a more precise representation of the physical phenomena involved in the impact although this complicates the solution of the system of differential equations and delays the computation time. Therefore, a compromise must be made between the capacity of the model to predict the experimental results and the need to avoid complexity in the calculations. The following hypotheses are usually considered for the simulation of high-speed impacts (Zukas 1982):

i) Localized influence: it is assumed that only a small area of the target, surrounding the impact point, is affected by the impact. The target beyond this area is not considered.

ii) Thermodynamic phenomena are not considered. This excludes friction, increases of temperature due to deformation, and variation of the mechanical properties of the material due to temperature changes.

iii) The materials are considered rigid until their yielding. This avoids the predicament of assuming phenomena of transmission of elastic waves, since most of the impact energy is dissipated in plastic deformation.

With these hypotheses, the predictions of the analytical models are usually precise enough, although each model is valid only for the specific problem for which it was developed. However, the last hypothesis can't be used for the analytical modeling of fiber composite armors since the energy absorption is mainly due to elastic deformation.

The use of *numerical simulation* leads to a solution of an impact problem through the integration of the differential equations of the thermomechanics of continuous media, much more reliable than that obtained with the analytical model. A number of commercial programs of finite elements or finite differences are available, and these are very successful in simulating impact problems. However, their capacity of close approximation to reality depends more on the constitutive equations used for the materials than on the errors inherent in the numerical method itself. The impact produces temperature increases due to irreversible processes -plastic flow, shock waves- and deformations at a very high rate that can be simulated correctly only by constitutive equations whose expression is often unknown. The main problem with these models is the uncertainty with regard to the input data required for the materials, and this means that the model has to be calibrated for each material to be used. Another drawback is the amount of time required to solve each impact problem, usually several hours or even days of CPU time.

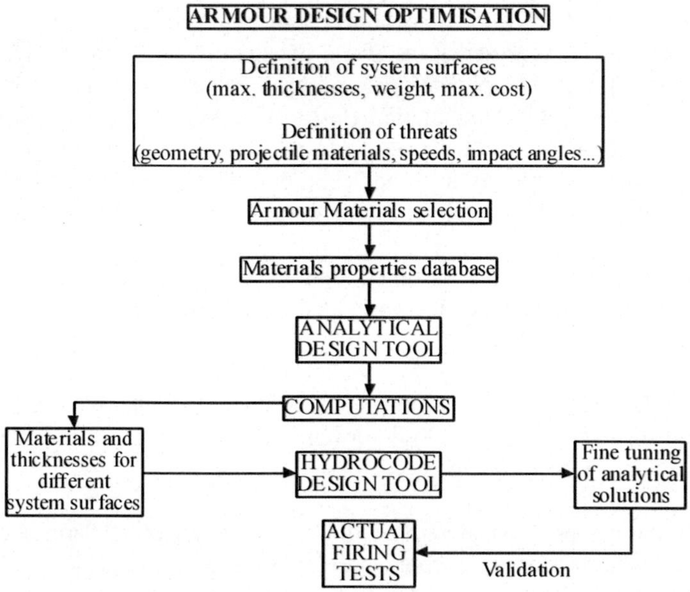

Figure 7.1. Sketch of the armor design process (Zaera and Sánchez-Gálvez, 1998b).

Analytical and numerical models must be considered as complementary in design work. According to the speed of their calculation, analytical models should be used at the initial stages when normally a large number of cases must be modeled to select the geometries and materials. When the conditions of design have been determined (type of projectile, speed and angle of impact, maximum weight, cost of the armor, etc.) the materials are selected and the thicknesses are specified for each zone of the system to be protected, using analytical models. After this first selection, numerical codes are used, and finally, actual firing tests are essential to confirm the results and, if necessary, to calibrate the models used in order to repeat the calculations (Figure 7.1).

7.1 Analytical Modeling of Fiber Armors and Ceramic-Faced Armors

7.1.1 Fiber armors

There are several analytical models of impact on fiber reinforced polymer composites. Navarro et al. (1994) proposed a simplified way to capture the wave propagation phenomena in a woven FRP armor, partially based on the analysis of the response of a single fiber due to Jameson (1962) and Smith (1965). The transverse impact on an elastic fiber gives rise to two waves at different velocities: one longitudinal tensile wave transmitted at velocity c_l (Eq. 5.1)

$$c_l = \sqrt{E_f / \rho_f} \tag{7.1}$$

and another at a lower velocity c_t (Eq. 5.4 assuming a small plastic strain)

$$c_t = \sqrt{\sigma_f / \rho_f} \tag{7.2}$$

that produces a displacement in the transversal direction (Figure 5.1). E_f is the Young modulus of the fiber, ρ_f its density, and σ_f the stress on the fiber. In laminates used in protection against impact, the behavior of individual yarns is much more complex since they interact on each other through the fabric and the matrix. The analytical model proposed by Navarro et al. (1994) simplifies this behavior by taking the laminate as a set of yarns that react individually to the impact, and this simplifies the problem. A two-dimensional field of displacements is proposed, extrapolated from that shown in Figure 5.1 top, and the impact resistance of the laminate is taken to be the sum of the contributions of each of the yarns, each with a different deformation. According to Figure 7.2, the following geometric estimations can be made

$$R'_t = \sqrt{\left(\frac{D_t}{2}\right)^2 - d^2} \qquad R'_{tc} = \sqrt{\left(\frac{D_{tc}}{2}\right)^2 - d^2} \tag{7.3}$$

$$w'_t = w_t\left(1 - \frac{2d}{D_t}\right) \qquad \varphi = arctg\left(\frac{R'_t}{w'_t}\right) \tag{7.4}$$

w_t being the transverse displacement of the composite panel at the impact point. Then, the yarn strain and yarn force can be computed as

$$\varepsilon_y = \frac{\sqrt{R_t'^2 + w_t'^2} - R'_t}{R'_{tc}} \tag{7.5}$$

$$f_y = 2E_f \varepsilon_y \Omega_y \cos\varphi_y \tag{7.6}$$

Ω_f being the fiber cross section. The whole resisting force is given by

$$F_t = N \sum_{yarns} f_y \tag{7.7}$$

N being the number of layers in the laminate. Perforation occurs when the deformation of the fibers of the laminate reaches a limit value.

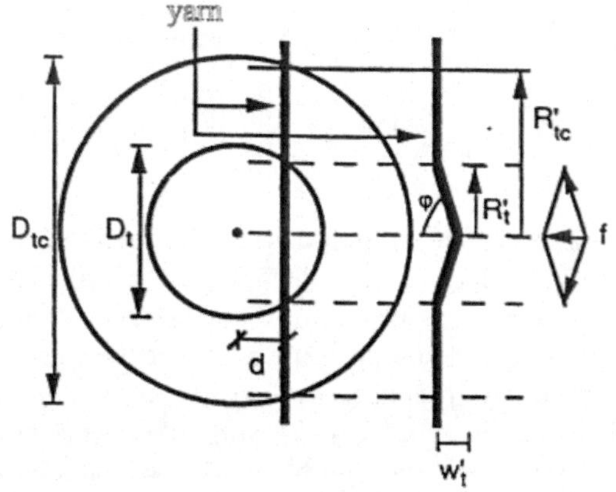

Figure 7.2. Simplified model of laminate deformation (Navarro et al., 1994).

This model gives ground for an easy-to-program tool that can be utilized to simulate the impact on a woven armor. More sophisticated models were proposed later by Chocrón-Benloulo and Sánchez-Gálvez (1998) who added a strain energy based damage variable to predict failure, Morye et al. (2000) who proposed a model based on an energy balance, Billon and Robinson (2001) who developed a model to predict the ballistic limit for fabric armors constructed as hybrids of two or more materials, Phoenix and Porwal (2003) who developed a 2D membrane model, Gu (2003) who developed a model for multi-layered planar plain-woven fabrics based on the energy conservation law (strain energy and kinetic energy of the fabric), or Naik et al. (2006) who incorporated different damage and energy absorption mechanisms (cone formation on the back face of the target, tension in primary yarns, deformation of secondary yarns, delamination, matrix cracking, shear plugging and friction during penetration) to model woven laminates. A review of analytical models previously developed for fiber reinforced armors can be found in Navarro (1998).

7.1.2 Ceramic-faced armors

The first analytical model for the design of ceramic-faced armors was that of Florence (1969) for ceramic/metal panels. It conceived the projectile as a rigid cylinder impacting on the armor, and admitted the progressive fragmentation of the ceramic under impact to form a cone, which distributes the load to the metal plate. The analysis is in terms of energy: if the kinetic energy of the projectile is greater than that absorbed by the plastic deformation of the metal plate, the projectile will perforate the plate; otherwise it will be arrested. To handle the problem analytically, Florence adopted the following simplifications:

i) The diameter of the circular zone of the metallic plate affected by the impact is equal to the base of the cone of fractured ceramic; the radius of the base of the cone is equal to that of the projectile plus double the thickness of the ceramic (Figure 7.3). This is a hypothesis coincident with the observation of the formation of a conoid mass of about 65° semiangular to the axis, in the course of an impact.

ii) To account for deformation, the plate is modeled as a circular membrane of radius a supported at the edge. The plate breaks when its maximum deformation reaches a limit value equal to the uniaxial failure strain deformation of the plate.

iii) No consideration is given to the energy consumed in the fragmentation of the ceramic tile.

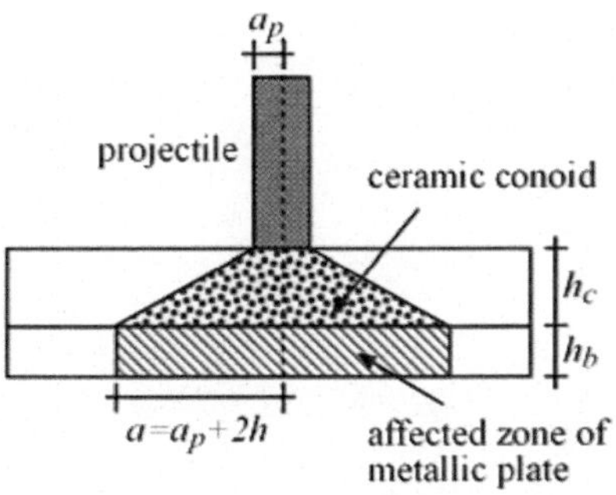

Figure 7.3. Fragmented ceramic conoid and extension of plate zone affected by the impact in the Florence model (1969).

With these hypotheses, and equalizing the kinetic energy of the projectile with that absorbed by the plate up to breakage, an equation is obtained that gives the ballistic limit of the projectile

$$V_{50} = \sqrt{\frac{\varepsilon_c \sigma_c h_2}{0.91 f\ M_p}} \tag{7.8}$$

M_p being the projectile mass, σ_c and ε_c the tensile strength and failure strain of the metallic plate, h_2 the metal plate thickness and f a function given by

$$f = \frac{M_p}{\left(M_p + (h_1\rho_1 + h_2\rho_2)\pi\left(a_p + 2h_1\right)^2\right)\pi\left(a_p + 2h_1\right)^2} \tag{7.9}$$

where a_p is the radius of the projectile, projectile, h_1 is the ceramic thickness, h_2 is the backing plate thickness, and ρ_1 and ρ_2 are the densities of the ceramic and backing plates respectively. Due to its simplicity, Florence's expression has been improved and used for optimization of ceramic/metal armors (Ben-Dor et al. 2005, Fawaz et al. 2006, Shi and Grow 2007, Ben-Dor et al. 2008). The model offers the advantages of simplicity and ease of calculation but it does not consider the type of ceramic, an essential factor of the efficacy of the armor, nor the kinetic energy of the projectile that is dissipated in its erosion. Thus the results tend to be too conservative, with ballistic limit velocities that are too low (Figure 7.4).

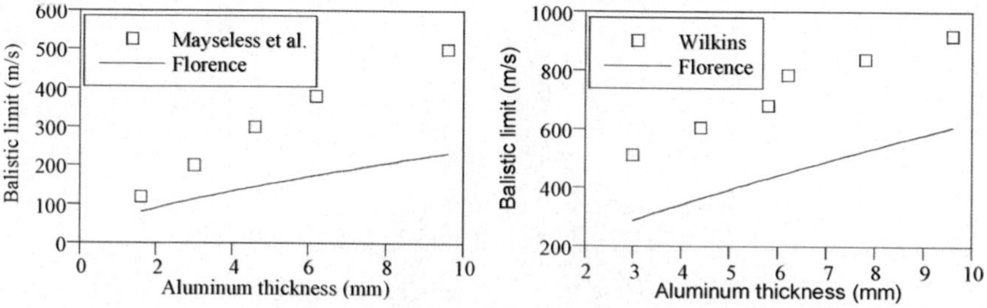

Figure 7.4. Left: Comparison of the predictions of Florence's model and the experimental results of Maysless et al. (1987), 12.70 AP against 6.35 mm alumina 85% purity backed by 6061-T6 aluminum plates of different thicknesses. Right: Comparison of the predictions of Florence's model and the experimental results of Wilkins (1978), 7.62 AP against 7.85 mm alumina backed by 6061-T6 aluminum plates of different thicknesses.

Other models were proposed later (Woodward 1990, den Reijer 1991, Zaera and Sánchez-Gálvez 1998a), with equations that incorporate the essential physical processes inherent in the process of penetration, and when integrated in time, indicate the evolution of the system from the instant of projectile/armor contact up to that of perforation or arrest. They are all *one-dimensional* models in the sense that they study only the displacement in the impact direction. Each of the solids of the problem (projectile, ceramic and backing plate) is studied separately (Figure 7.5). The equations for the projectile and the ceramic cone are of variation of the linear momentum in the direction of the impact, both systems being of variable mass since they both lose mass due to erosion with the advance of the projectile. Together with these equations of linear momentum variation, others are used which take into account the phenomenon of erosion in the projectile/ceramic interface and the penetration resistance of the ceramic.

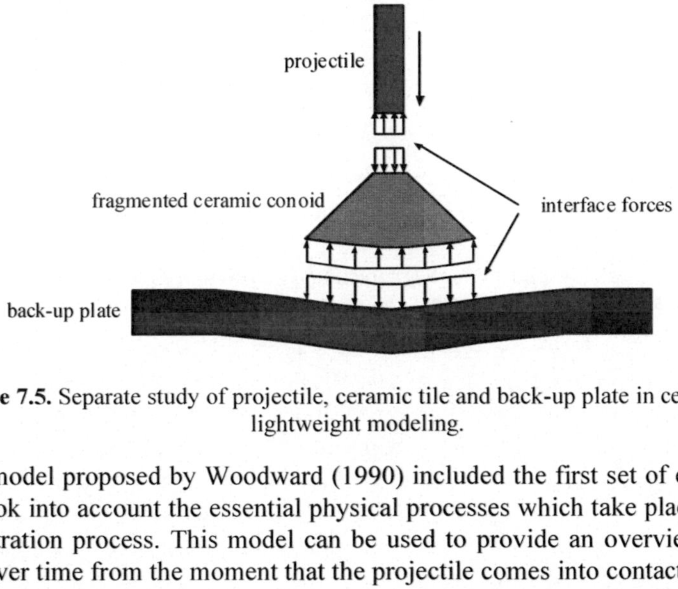

Figure 7.5. Separate study of projectile, ceramic tile and back-up plate in ceramic lightweight modeling.

The model proposed by Woodward (1990) included the first set of equations which took into account the essential physical processes which take place during the penetration process. This model can be used to provide an overview of the system over time from the moment that the projectile comes into contact with the ceramic/metal armor until it perforates the armor or stops. The model also examines different stages of the penetration process, from the formation of a cone of fragmented ceramic in the first few microseconds to the failure of the metal plate for whatever reason. It is also useful to examine the process of erosion of the projectile and the ceramic at the contact interface, which can be done using momentum equations. Figure 7.6 shows the results obtained using Woodward's model and the results of experiments carried out by Wilkins (1978) and Mayseless (1987).

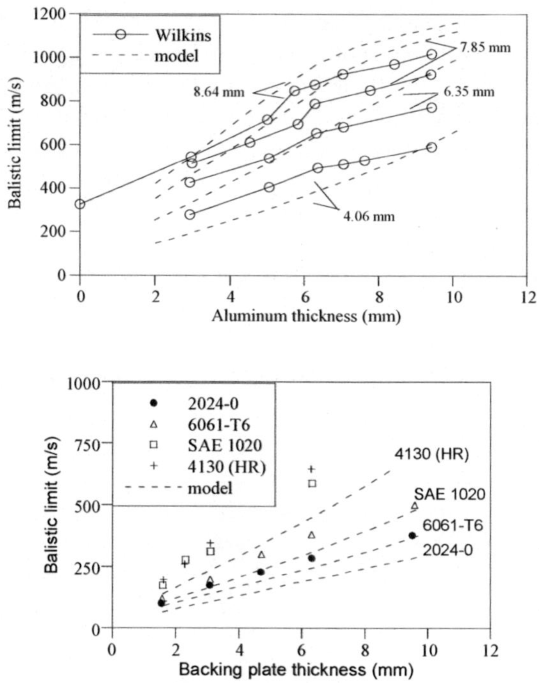

Figure 7.6. Top: Comparison of the predictions of Woodward's model (1990) and the experimental results of Wilkins (1978). Bottom: Comparison of the predictions of Woodward's model (1990) and the experimental results of Mayseless et al. (1987).

The model proposed by Den Reijer (1991), which goes by the name ALARM (Analytic Lightweight Armor Response Model), is another important step in the analytical simulation of ceramic-metal armors. The model allows the projectile to erode mass, to mushroom or to be rigid. Most emphasis is placed upon the interaction between the ceramic and the metal plate and the response of the plate itself, after the ceramic cone has been formed. As the projectile penetrates the ceramic facing, the impact load is distributed over a decreasing area of back-up plate. The response of the plate is simulated by different deformation models: initially the strength of the plate is achieved essentially by bending, but as deformation progress membrane forces are taken into account; moreover, when the projectile strikes the plate after penetration of the ceramic layer, the development of a plug is also considered. The plate is allowed to fail either by plug shearing or membrane deformation. This model includes a higher number of variables and can therefore be used to examine the projectile and armor deformation processes in more detail, analyzing both the projectile and ceramic erosion process and the deformation of the metal plate. It is also a one-dimensional model which can be used only in cases where the impact on the surface of the armor is normal. This model provides precise results for low caliber projectiles (Figure 7.7).

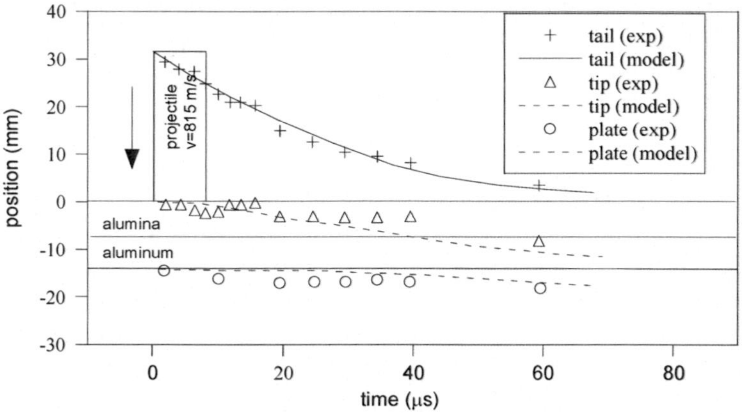

Figure 7.7. Comparison of the simulation predictions and the experimental results (den Reijer, 1991). Position time history for the projectile tail, projectile tip and back face of the metal plate. 7.62 AP fired at 815 m/s against an 8.1 mm alumina 95% purity tile backed by 6.0 mm aluminum 6061-T6.

The results achieved with the analytical models of Woodward (1990) and den Reijer (1991) are fairly accurate for the impact simulation of low caliber projectiles such as the 7.62 AP. However, when they are used for the simulation of the impact behavior of medium caliber projectiles (12.70 AP, 20 APDS, 25 APDS, 30 APDS) their results do not correspond to experimental ones. Zaera and Sánchez-Gálvez (1998a) proposed an analytical model in which the erosion process was simulated with the Tate (1967) and Alekseevskii (1966) equations of penetration. The model has been utilized both for small caliber projectiles and medium caliber projectiles. Analytical results have demonstrated good correlation to experimental data.

Fellows and Burton (1999) developed a model which predicts the penetration of projectiles into ceramic tiles backed by semi-infinite metal plates, including a detailed analysis of the projectile/ceramic, ceramic/metal and projectile/metal interfaces. The model results correlate well with experimental data for long-rod penetrator impacts onto ceramic backed by thick metal plates.

Each of the models described, whether applicable to fiber armor or to ceramic-faced armour, shows its predictive capacity for a certain range of variation in the geometry of the projectile, the geometry of the armour plating, or the properties of the materials. Thus, they can provide the trends associated with specific impact events, and it is necessary to prevent the variations in these parameters from mis-defining projectile/armor configurations in such a way that the mechanisms of deformation and damange assumed in the development of the model become invalid.

The process of projectile penetration into the armor is characterized by the occurrence of diverse phases in which the behavior of both solids and its interaction change markedly. As a general rule the predictive capacity of an analytical model and the range of its field of application are further reduced when they represent the impact over a single stage. On the other hand, the most sophisticated models are fundamentally characterized by considering various modes of deformation and damage, energy absorption mechanisms and transmission of load between solids, which acquire distinct importance as the process of simulation evolves. In these cases, the model acquires greater accuracy and permits the prediction of the behavior against impact for armors with more diverse characteristics.

7.2 Numerical Modeling of Fiber Armors and Ceramic-Faced Armors

Numerical codes for the simulation of high speed impact have made enormous progress with the increased capacity of computers which help to solve the set of differential equations of the continuum mechanics by numerical methods, normally *finite elements* or *finite differences*. The ballistic impact codes were first known as *hydrocodes* -an abbreviation of "hydrodynamic computer codes"- since they were used to simulate impact at hypervelocity and the materials were treated as fluids (Anderson 1987). Codes were later developed to include the strength of materials. There are two possible methods of formulating kinematics of continuum mechanics, with a direct influence on the numerical method used to solve the equations: the material or *Lagrangian* and the spatial or *Eulerian*. Apart from the mathematical differences, it is easy to understand the implications of these two approaches. In the Lagrangian, each node of the mesh is associated with a material particle of the solid, so the mesh moves with the solid (Figure 7.8), but in the Eulerian system the mesh is fixed in space and the material flows over it (Figure 7.9). In the Lagrangian mesh, therefore, the mass in each cell is constant while its volume varies; in the Eulerian, it is the mass that varies in each cell, not the volume.

Both types of formulation have advantages and drawbacks. In the Lagrangian mesh, the boundary of each solid is clearly defined, while in the Eulerian the contour can be spread over several cells. So the lagrangian formulation is preferable if boundary conditions have to be set, if there are contacts between the solids, or if a simulation is to be made of the solids formed of several layers of material. On the other hand, if the problems may involve a mixture of gaseous materials, as would be the case of explosions, then Eulerian meshes would be used so that different materials could be accommodated in one cell.

Another advantage of the Lagrangian mesh is that it allows the history of the material to be followed since each node represents a particle. This is particularly useful if damage models are required to describe the behavior of the material. Hydrocodes normally allow the use of models of damage to materials only where the description is by a Langrangian mesh due to the difficulty of "follow the track" of the material on the Eulerian mesh.

The main drawback when using the Lagrangian formulation appears when the deformations of the solids are very pronounced and the mesh is distorted. For the stability of an explicit numerical integration method, a time increment must be adopted which fulfils the Courant condition

$$\Delta t \leq \frac{kl}{c} \tag{7.10}$$

l being the size of the smallest cell, c the elastic wave propagation velocity in the material, and k a factor of stability below unity. If the mesh is very distorted, the parameter l becomes very small, as does the time integration interval, which slows the process of calculation. To overcome this problem, the calculation may be interrupted and the mesh remeshed at the point of greatest distortion before continuing the integration. Another possibility that avoids this interruption is to introduce an *erosion criterion* into the material. This automatically eliminates the badly distorted cells, but also eliminates mass, momentum and kinetic energy from the system. With the Eulerian formulation, the size of the mesh is unchanged and the integration time varies only slightly on account of the variation of c with the density in the material.

In the last decade, a new class of meshless computer codes gained popularity in the simulation of high-speed impact problems with extremely large deformations, moving boundaries and fragmentation processes. Modeling with these methods only requires a set of unstructured points that cover the domains of interest, and hence no connectivity concept is involved. There is no need to provide a priory any information about the relationship of the nodes and this provides flexibility in adding or deleting points whenever needed. Among the different meshless methods used in engineering, the Smoothed Particle Hydrodynamics (Gingold and Monaghan, 1977) and the Element Free Galerkin (Belytschko et al., 1994) are probably the most used in the field of ballistics.

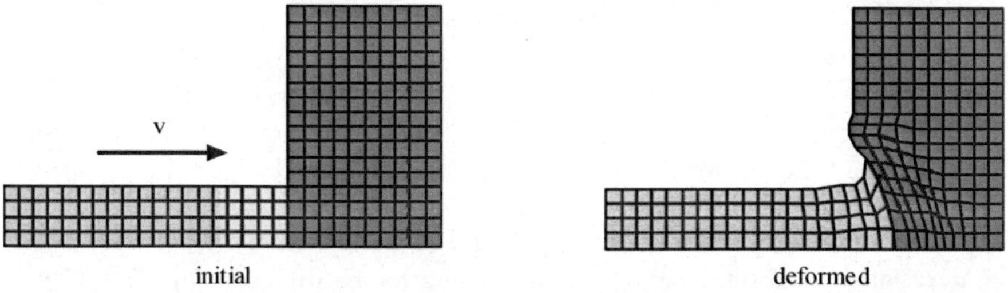

Figure 7.8. Normal impact of a projectile on a panel. Behavior of the lagrangian mesh.

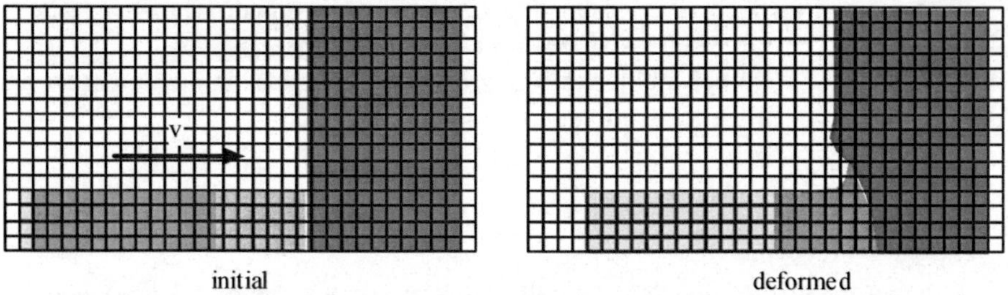

Figure 7.9. Normal impact of a projectile on a panel. Behavior of the eulerian mesh.

7.2.1 Fiber armors

There are different approaches to simulate the impact of a projectile onto a fiber reinforced composite plate. The most common is to use 3D finite element or meshless codes, and constitutive equations that capture the mechanical behavior of fiber reinforced laminates under impact loading. These models usually assume that the material is represented as a homogenized continuum. The non-damaged behavior of the composite is usually considered by a linearly elastic orthotropic behavior so that the total stress can be related to the total strain through the orthotropic stiffness matrix. The nonlinear behavior of the composite ply is caused by different failure modes (delamination, fiber failure…). A well-known model in the field of impact is that due to Chang and Chang (1987) that considers three failure criteria, namely, matrix cracking, compression failure and fiber breakage. The matrix cracking failure criterion is determined by

$$\left(\frac{\sigma_2}{S_2}\right)^2 + \left(\frac{\tau_{12}}{S_{12}}\right)^2 = 1 \tag{7.11}$$

where S_2 is the transverse tensile strength and S_{12} is the shear strength. When this condition is satisfied, the material constants E_2, G_{12}, ν_{12} and ν_{21} are set to zero. Compressive failure is predicted by the Hashin criterion

$$\left(\frac{\sigma_2}{2S_{12}}\right)^2 + \left[\left(\frac{C_2}{2S_{12}}\right)^2 - 1\right]\frac{\sigma_2}{C_2} + \left(\frac{\tau_{12}}{S_{12}}\right)^2 = 1 \tag{7.12}$$

where C_2 is the transverse compressive strength. When this condition is satisfied, the material constants E_2, ν_{12} and ν_{21} are set to zero. Fiber failure criterion is determined by

$$\left(\frac{\sigma_1}{S_1}\right)^2 + \left(\frac{\tau_{12}}{S_{12}}\right)^2 = 1 \tag{7.13}$$

where S_1 is the tensile strength in the fiber direction. When this condition is satisfied, the material constants E_1, E_2, G_{12} ν_{12} and ν_{21} are set to zero.

Other damage models used for ballistic applications are those developed by Matzenmiller et al. (1995) and by Hiermaier et al. (1999). The first one describes the elastic-brittle behavior of a fiber-reinforced lamina by considering four damage modes: fiber failure, fiber buckling, matrix cracking and matrix crushing. Delamination of individual layers, typically encountered in composite structures, is not considered in this model. The elastic region is bounded by a series of surfaces associated with the different lamina failure modes. The model due to Hiermaier et al. (1999), based on an approach by Anderson and coworkers (1994), considers anisotropic strength degradation by modifying material stiffness and strength properties depending on the failure modes. It also considers decrimping effects by using coefficients of the stiffness matrix depending on the strain. The Mie-Gruneisen equation of state is chosen to reproduce shock response.

Practical applications of the Chang-Chang model (1987) can be found in the work of Gu and Xu (2004) who simulated 3D braided twaron™/epoxy composites subjected to impact by a conically cylindrical steel projectile, and in Aare and Kleiven (2007) who analyzed the performance of a woven-aramid helmet (Figure 7.10). On the other side, Nandlall et al. (1998) simulated the impact of an FSP onto an S2-glass fiber-reinforced plastic laminate using a modified version of the Matzenmiller et al. (1995) model, and Gower et al. (2008) reproduced the impact response of a Kevlar™ woven laminated panel using the Matzenmiller et al. (1995) model. It is worth noting, in this last work, the method used by the authors to simulate delamination by modeling each ply as a separate entity with a tie-break interface (Figure 7.11). The Hiermaier et al. (1999) model was used by Silva et al. (2005) and by Tham et al. (2008) to simulate the ballistic response of Kevlar™ laminates.

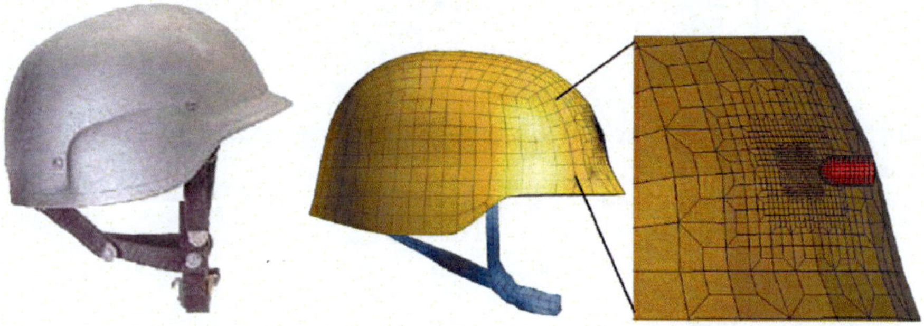

Figure 7.10. From left to right: real helmet, FE model of the helmet and detail of the mesh at the impact point (Aare and Kleiven, 2007).

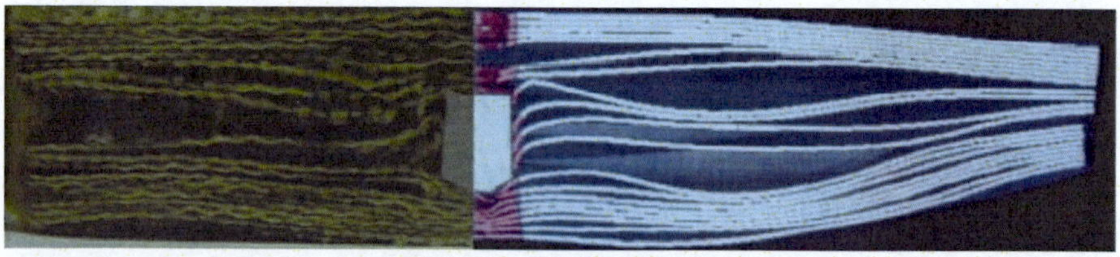

Figure 7.11. Cross-section comparing numerical (right) with experimental (left) results (Gower et al. (2008).

A more detailed description of the behavior of fiber reinforced fabrics could be done by modeling each yarn. This would allow for a better consideration of friction between yarns, decrimping and other geometric-based non-linearities in the material behavior, but it is only practical to model small portions of the panel due to the high computational cost. Duan et al. (2005) simulated the impact of a rigid sphere into a square patch of single-layer plain-wave fabric (Figure 7.12) in order to examine the role of friction during ballistic impact of high-strength fabric structures. Friction was found to delay fabric failure and to increase impact strength, allowing the fabric to absorb more energy.

Modeling the structure of woven fabrics at meso-scale is also used to capture the essential dynamic characteristics of these materials, analyze their response and propose constitutive models. Shahkarami and Vaziri (2007) used a unit-cell (smallest repeating unit in the fabric) model to obtain a physics-based relationship to capture the biaxial coupled response of the fabric's warp and weft yarns under general applied displacements in these directions. The results of this analysis allowed development of a special membrane finite element which provides a continuum representation of the mechanical attributes of the unit-cell. Layers of fabric constructed from these specialized elements are stacked together to create fabric targets that are then analyzed under projectile impact.

Figure 7.12. Finite element model for a plain-weave fabric structure (Duan et al. 2005).

The simulation of fabric armors using membrane elements was also proposed by other researchers to replace the more complicated and numerically costly 3D solid models. Lim et al. (2003) used a membrane finite element which incorporates a material model with viscoelasticity effects and a strain-rate-sensitive failure criterion. The viscoelastic nature is obtained through mathematical manipulation of the three-element spring-dashpot Zener model. The model predicts the ballistic limit, residual velocity, energy absorption and transverse deflection profiles of fabrics submitted to ballistic impact.

With the same approach but using 1D bar finite elements to model woven fabric, Tan and Ching (2006) proposed a one-dimensional material model that reflects the viscoelastic nature of the twaron™ yarns by a 3-element Zener model. The fabric is modeled as a network of nodal masses connected by one-dimensional viscoelastic elements, and considers both sliding contact between yarns and yarn breakage. Since the model reproduces the structure of the fabric, it allows decrimping to be captured, as well as the "wedge-through" effect by which, once a small opening is formed by breaking several yarns, the projectiles wedge their way through the opening by pushing aside and slipping past the remaining yarns.

Another way to numerically simulate the behavior of a fabric armor could be done by considering each ply as nodal masses interconnected by extensible fiber elements. These nodal masses represent the areal density (inertia) while the strings represent the stiffness properties of the fabric. The nodal positions and velocities are updated through an explicit time integration scheme. Tan et al. (2005) developed a model with this methodology, considering the viscoelastic behavior of the yarns through a three-element Zener viscoelastic constitutive model. Decrimping was incorporated into the model by embedding the effects of crimp into the viscoelastic constitutive equation; the following relationship between yarn strain ε_{yarn} and fabric strain ε_{fabric} was considered

$$\varepsilon_{yarn} = \varepsilon_{fabric} - \varepsilon_{crimp}\left(1 - e^{-\varepsilon_{fabric}/\varepsilon_{crimp}}\right) \tag{7.14}$$

ε_{crimp} being a constant accounting for the crimping effect. This equation means that fabric strain (i.e. strain in the plane of the fabric) is greater than the strain of the yarns until the yarns have straightened out (Figure 7.13). Novotny et al. (2007) modeled the fabric panel as a two-dimensional assembly of pin-jointed yarns that results in a net-like structure. Discrete nodal masses are located at the crossings of the yarns, and are free to translate in all three global cartesian directions. The masses are inter-connected by cable elements in orthogonal in-plane directions representing the warp and weft yarns of the fabric. Multi-layer fabrics are modeled by stacking layers of masses and strings with a prescribed interlayer gap. The nodal impulse-momentum equations are solved for the current nodal velocities. Satisfaction of the momentum balance in all directions results in a complete velocity field response for the fabric, to be used to update the nodal coordinates.

7.2.2 Ceramic-faced armors

Modeling ceramic materials using numerical codes is carried out using damage models which consider a damage variable to gather data about the degradation undergone by the material as a result of its fragmentation. Examples of these models include the micromechanical fragmentation and granular flow model designed by Curran et al. (1993) and the multi-scale model of dynamic fragmentation defined by Denoual and Hild (2002). These examples are particularly noteworthy due to the solid physical foundations on which they are based and the detailed analysis of the cracking process carried out by the authors who developed them. From a practical point of view, however, these models are difficult to implement and, as a result, more simple phenomenological models tend to be used, such as the model designed by Cortés et al. (1992) and the model by Johnson and Holmquist (1994). Both of these models were designed specifically to model ceramic materials which are subjected to high-speed impacts, and both make use of a damage parameter to consider degradation in the material. This damage parameter is assumed to not change the isotropic elastic behavior, but to decrease the yield stress of the ceramic. The difference between both models lays in the damage growth law and in the yield function.

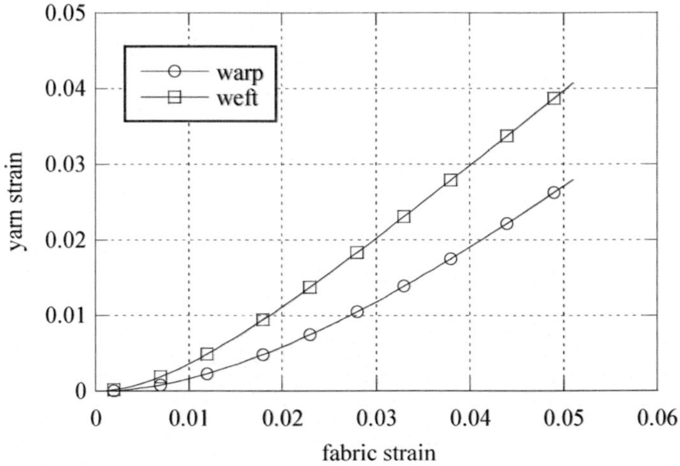

Figure 7.13. Relation between fabric strain and yarn strain in warp and weft directions (Tan et al. 2005).

The model designed by Cortés et al. (1992) was developed specifically to be used in the application that is being studied here. The model is based on a monolithic material that fragments, gradually behaving more and more like a granular material. The model represents ceramic degradation using a scalar damage variable D which varies in range [0,1] (intact-pulverized). This damage variable should be seen as a measurement of the level of fragmentation of the material, just as damage in ductile metallic materials is commonly associated with void volume fraction within the material. The model includes cohesion and assumes that there is a linear relationship between strength and pressure, like the Drucker-Prager model, for the intact material, and a frictional linear relationship, like the Coulomb frictional model, for the pulverized material, passing through intermediary states with a different level of fragmentation in each situation (Figure 7.14 left). The yield function consists of two terms which include both damage-weighted states using a direct rule of mixtures

$$f = \bar{\sigma} - (1-D)\left(3\alpha p + \sigma_{0y}\right) - \mu p D \tag{7.15}$$

where $\bar{\sigma}$ is the equivalent stress, p is hydrostatic pressure and α , σ_{0y}, μ are the material parameters. Damage evolution is linearly dependent on hydrostatic tension $\sigma_H = -p$

$$\frac{dD}{dt} = D_0 \left\langle \sigma_H - \sigma_{H0} \right\rangle \tag{7.16}$$

where $\langle \circ \rangle$ are the MacCauley brackets (positive part of the argument), σ_{H0} is the damage threshold tension and D_0 is a damage growth parameter.

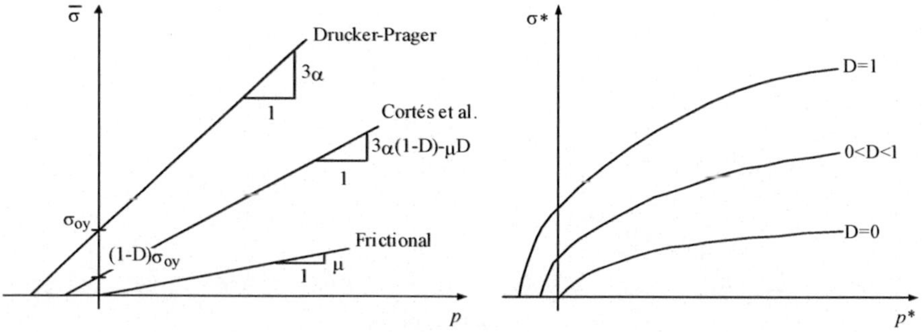

Figure 7.14. Variation of equivalent stress with pressure. Left: Cortés et al. (1992) model. Right: Johnson-Holmquist (1994) model.

Fernández-Fdz and Zaera (2008) modified this model and replaced the hydrostatic stress in the damage evolution law by the maximum principal stress. As hydrostatic stress is more appropriate as the governing variable for void nucleation and growth processes in ductile materials, this modification was considered consistent with crack apparition and propagation phenomena in ceramic materials.

The model designed by Johnson and Holmquist (1994) examines a similar fragmentation process expressed as a variable of damage D. Just as in the model designed by Cortés et al. (1992), D is used as a mixing parameter to obtain the yield function (Figure 7.14 right), as follows

$$f = \sigma^* - (1-D)\sigma_i^* + D\sigma_f^* \qquad (7.17)$$

where σ_i^* and σ_f^* are the non-dimensionalized equivalent tensions for the intact materials and the completely pulverized material, respectively, determined by

$$\sigma_i^* = A\left(p^* + T^*\right)^N \left(1 + C\ln\dot{\overline{\varepsilon}}\right) \qquad (7.18)$$

$$\sigma_f^* = B\left(p^*\right)^M \left(1 + C\ln\dot{\overline{\varepsilon}}\right) \qquad (7.19)$$

In the equations above, the variables $(\circ)^*$ have been non-dimensionalized with the values corresponding to the Hugoniot Elastic Limit (HEL) where $\sigma^* = \overline{\sigma}/\overline{\sigma}_{HEL}$ for σ^*, σ_i^*, σ_f^*, $T^* = T/p_{HEL}$, T represents the maximum hydrostatic tensions that the material can withstand and $p^* = p/p_{HEL}$. A, B, C, N and M are parameters of the material.

Furthermore, the model uses a cumulative damage law which is dependent upon the equivalent plastic strain expressed as

$$D = \frac{\sum \Delta\overline{\varepsilon}^p}{\overline{\varepsilon}_f} \qquad (7.20)$$

where $\overline{\varepsilon}_f$ is the plastic deformation fracture threshold which depends on pressure.

7.3 Prediction Using Artificial Neural Networks

In the current multidisciplinary framework of engineering, with the vast increase in artificial computation techniques, Neural Networks are providing low-cost solutions to mechanical problems that are characterized by a high nonlinearity, a dependence on a high number of parameters and a wide range of variation of these parameters. These characteristics certainly apply to high-velocity impact on lightweight armors. Multilayer Perceptron (MLP) is the most commonly applied neural network in the field of Mechanics (Adeli and Yeh al. 1989, Waszczyszyn and Ziemianski 2001, Liu et al. 2002, Ince 2004, El Kadi 2006). MLPs have also been used to analyze the response of ceramic-metal armors against high-velocity impact of projectiles, showing a remarkable predictive ability (Fernández-Fdez and Zaera 2008).

In an MLP, a processing element or artificial neuron (Figure 7.15 left) receives outputs x_j from neurons of the previous layer which are weighted by synapse weights φ_{ji}. Neuron activation occurs when the sum of these weighted signals exceeds a pre-set value known as threshold activation b_i. If the neuron is fired, it generates an output v_i determined by the expression

$$v_i = f_i\left(\sum_{j=1}^{K} \varphi_{ji} x_j - b_i\right) \tag{7.21}$$

f_i being the activation function. Some examples of functions f_i are shown in Figure 7.15 right. The topology of an MLP (Figure 7.16 left) is characterized by grouping neurons in the input, hidden, and output layers. In the input layer the identity function is used (linear with slope equal to one). In the hidden and the output layer, although logistic or threshold functions can be employed, the hyperbolic tangent is most common.

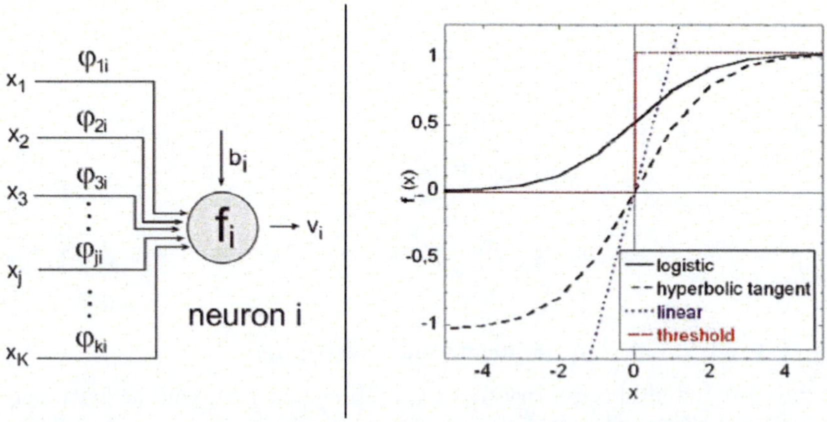

Figure 7.15. Left: Computational model of the artificial neuron. Right: Some common activation functions f_i, logistic, hyperbolic tangent, linear and threshold.

This network presents total connectivity, which means that each neuron is connected to all the neurons of contiguous layers but not with those located in the same layer. Connections are made from the input to the output layer, mapping an n-dimensional vector x containing the input variables to a m-dimensional vector y containing the output variables, so that an MLP defines a nonlinear continuous function F from R^N to R^M

$$y = F(x, \Phi, \beta) \tag{7.22}$$

Φ and β being the set of weights and thresholds. These values are adjusted during the training stage to minimize the error when the MLP produces an output y corresponding to an input x.

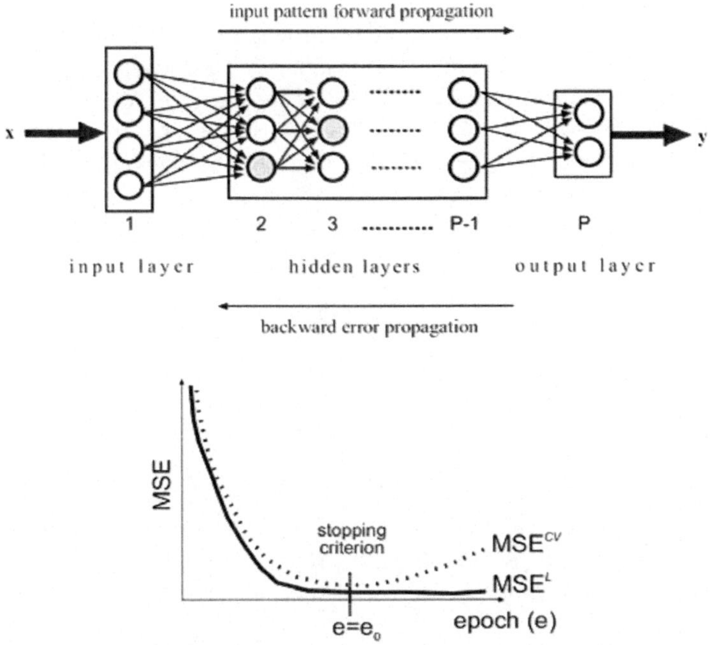

Figure 7.16. Top: Multilayer Perceptron Topology: feedforward network with backward propagation error and p layers (1 input layer, 1 output layer and p–2 hidden layers). Bottom: MSE vs. iteration number (epoch) for learning and cross-validation.

The patterns $(x, y)_s$ form the variability space $\mathcal{R}$ of the excitation and response of the system studied. During the training process the MLP extracts the characteristics of the system from a reduced set of training patterns $\mathcal{K}$ in which both inputs and outputs are known

$$\mathcal{K} = \left\{ (x,y)_s^{\mathcal{K}} \,\middle|\, s = 1,\ldots,K \right\} \subset \mathcal{R} \tag{7.23}$$

which is randomly divided into the subset of learning $\mathcal{L}$ and the subset of cross-validation $\mathcal{V}$

$$\mathcal{L} = \left\{ (x,y)_s^{\mathcal{L}} \,\middle|\, s = 1,\ldots,L \right\} \subset \mathcal{K} \tag{7.24}$$

$$\mathcal{V} = \left\{ (x,y)_s^{\mathcal{V}} \,\middle|\, s = 1,\ldots,V \right\} \subset \mathcal{K} \tag{7.25}$$

The first set $\mathcal{L}$ is used to determine the values of Φ and β through an algorithm called Backpropagation, abbreviation of "Backwards propagation of the errors" (Rumelhart and McCleland 1988). The algorithm iteratively modifies the initial values of weights and thresholds to arrive at the minimum value of a function which measures the predictive error of the network, following the direction of the gradient descent of this function. Although there are many definitions of the error, the most common is the mean squared one

$$MSE = \frac{1}{P}\sum_{s=1}^{P} e(s) \tag{7.26}$$

$e(s)$ being the squared error for a pattern s

$$e(s) = \frac{1}{2}\sum_{i=1}^{M}\left(y_i^*(s) - y_i(s)\right)^2 \tag{7.27}$$

where $y_i^*(s)$ and $y_i(s)$ are, respectively, the desired and predicted outputs for parameter i in the output pattern s (of dimension M).

The training process can be summarized as follows: weights and thresholds are randomly initialized with values close to zero; the input $x^{\mathcal{L}}(1)$ (from first learning pattern) is propagated, an output $y(1)$ is determined, and error $e(1)$ is computed; the Backpropagation algorithm is applied and weights and thresholds are incremented in the negative direction of the error gradient; the two previous steps are repeated for the rest of the learning patterns $(x,y)_s^{\mathcal{L}}$, updating Φ and β each time; the global learning error $\mathrm{MSE}^{\mathcal{L}}$ is computed and an epoch (learning cycle) is completed; consecutive epochs are repeated until a stable value of $\mathrm{MSE}^{\mathcal{L}}$ is reached (Figure 7.16 right).

The cross-validation subset $\mathcal{V}$ of training patterns is used during the application of the Backpropagation algorithm to prevent overlearning of the MLP. This spurious effect consists of an accurate prediction of the output belonging to the patterns of the learning subset $\mathcal{L}$ but not for independent inputs. Thus, the cross-validation global error $\text{MSE}^{\mathcal{V}}$ is calculated at the end of each epoch and the learning algorithm is stopped when this error starts to increase (Figure 7.16 right). Once the training algorithm has finished, a testing set $\mathcal{T}$

$$\mathcal{T}=\left\{(x,y)_s^{\mathcal{T}}\,\middle|\, s=1,\ldots,T\right\}\subset\mathcal{R} \tag{7.28}$$

is used to check the predictive ability of the MLP with patterns independent from those employed during training. If the test results are precise enough, the network is ready to generate reliable output data.

Fernández-Fdez and Zaera (2008) developed a tool based on Artificial Neural Networks for the design of lightweight ceramic-metal armors. The tool developed predicts, in real-time, the response of the armor: impacting projectile arrest or target perforation are determined and, in the latter case, the residual mass and velocity of the impacting body are calculated. The problem of high velocity normal impact of cylindrical projectiles on ceramic/metal armors was defined by the following variables: projectile length L_p, projectile diameter D_p, impact velocity v_i, ceramic tile thickness H_c, metal plate thickness H_m, projectile material, ceramic tile material and metal plate material. Two different materials have been used for the projectile (tungsten and high-strength steel), for the ceramic tile (alumina and aluminum nitride) and for the metal plate (aluminum and mild steel). A wide variation range for the input parameters has been established to cover most cases of the high-velocity impact with low and medium caliber projectiles:

$$\begin{aligned}
&4\text{ mm} \le D_p \le 12\text{ mm}\\
&3D_p \le L_p \le 6D_p\\
&0.3L_p \le (H_c+H_m) \le 0.6L_p\\
&H_m \le H_c \le 3H_m\\
&500\text{ m/s} \le v_i \le 1200\text{ m/s}
\end{aligned} \tag{7.29}$$

Within this range of input variables, a set of 200 impact cases has been generated (185 for training and 15 for testing), by Finite Element simulation, in order to train and test two Multilayer Perceptrons. The first one is used to predict the occurrence of perforation, and acts as a classifier deciding whether the pattern belongs to the class of armor perforation cases or to the class of projectile arrest cases. The classification is done by means of two outputs given by the network, varying in a continuous range [0, 1], which represents the probability of belonging to both classes, these variables being complementary. The determination, in the case of perforation, of the residual mass and the residual velocity of the projectile is considered a regression problem in which the network predicts these variables as a function of the input pattern. The second network was developed to perform this task.

Once trained, the networks were tested with independent data. The first network has correctly classified 14 of 15 test impact cases, the one incorrectly classified being defined by input variables close to the limits imposed in the definition of the problem. In the calculations of the residual variables, the second network provided, for the residual velocity, an average relative error of 7.5% and a maximum error of 13.9%. Concerning the residual mass the network provided an average relative error of 2.9% and a maximum error of 6.8%.

Once the networks are tested, they can be used to simulate the behavior of ceramic-metal armors against high-velocity impact. The curves of probability of perforation, in order to determine the ballistic limit v_{50}, are one of the classical ways to study this problem. The first network can be used to reproduce these curves taking into account that the output "PERFORATION YES" varying in the range [0,1] can be understood as the probability of perforation for a system projectile-armor. Figure 7.17 shows these curves for three different armors reinforced with a tile of 8.1 mm alumina backed with a plate of aluminum 6061-T6 of different thicknesses: 4, 6 and 8 mm. The armors were impacted with a cylindrical hard-steel projectile of 31.5 mm in length and 6 mm in diameter. Table 7.1 shows the experimental results of den Reijer (1991) and the values of V_{50} determined with the neural network. The predictions for the amors of 8.1 mm alumina/4 mm aluminum and 8.1 mm alumina/6 mm aluminum proved to be precise. For the thickness 8.1 mm alumina/8 mm aluminum the error increases, again because the ratio of ceramic thickness to metal thickness corresponds to an extreme value in the definition of the impact cases used to train and test the network.

Multilayer perceptrons provide rapid and reliable predictive results if the analyzed cases are within the range used during the training phase, and may constitute a complementary tool to the conventional design methodologies.

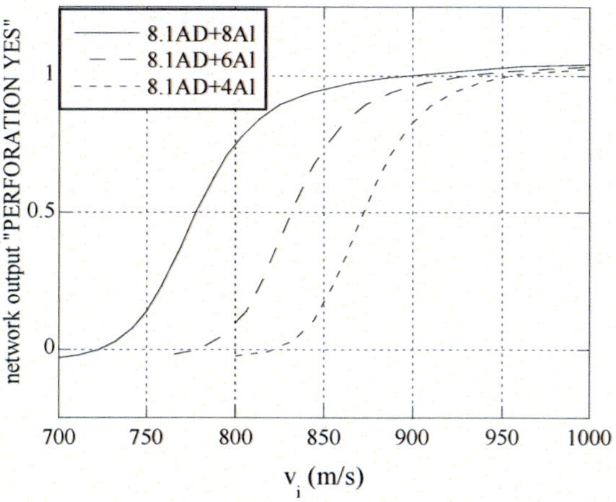

Figure 7.17. Curves of probability of perforation for three bi-layered armors alumina/aluminum of different thicknesses (Fernández-Fdez and Zaera 2008).

Table 7.1. Values of the ballistic limit obtained experimentally [den Reijer 1991], and given by the multilayer perceptron (Fernández-Fdez and Zaera 2008).

Armor	Experimental V_{50} (m/s)	Neural network V_{50} (m/s)
8.1 mm AD + 4 mm Al	786-829	778
8.1 mm AD + 6 mm Al	815-916	832
8.1 mm AD + 8 mm Al	995-1091	872

8 Oblique Impact and Its Simulation

8.1 Introduction

Most impacts are oblique, and actually the impact at normal incidence is an infrequent case. Moreover, the armors are frequently sloped with respect to the plane normal to the firing direction. This increases the thickness of the armor in the plane of attack and the projectile must travel through a longer distance to penetrate, improving the protection. But this may also increase the weight of the armor in terms of areal density relative to the attack plane, defined as (Figure 8.1)

$$AD_{(\theta)} = \frac{h_{(\theta)}\rho}{\cos\theta} \tag{8.1}$$

So that two plates, one sloped and one normal to the firing direction would have the same areal density relative to the attack plane if

$$\frac{h_{(\theta)}}{h_{(n)}} = \cos\theta \tag{8.2}$$

The perforation resistance of armors impacted at a certain obliquity is commonly described by the Equivalent Protection Factor (EPF), defined as the ratio of the areal density of the armor providing protection against oblique impact $AD_{(\theta)}$ to the areal density of that providing protection against normal impact $AD_{(n)}$. The EPF is commonly plotted against the angle of obliquity to indicate the reduction in armor thickness required with increased obliquity of attack. If this curve is below the cosine curve (Eq. 8.2), the sloped armor is lighter in terms of areal density relative to the attack plane; if the EPF curve is above the cosine curve a weight penalty is incurred.

Sadanandan and Hetherington (1997) found that for ceramic faced armors, a good correlation with experimental data of the equivalent protection factor when using an empirical relationship based on the root cosine expression is

$$\frac{AD_{(\theta)}}{AD_{(n)}} = \sqrt{\cos\theta} \tag{8.3}$$

This has been validated with alumina/steel and alumina/aluminum armors impacted by AP or ball (soft core) 7.62 mm projectiles and may prove useful in armor design. Since the root cosine is above the cosine curve, inclining ceramic/metal armor normally incurs a weight penalty.

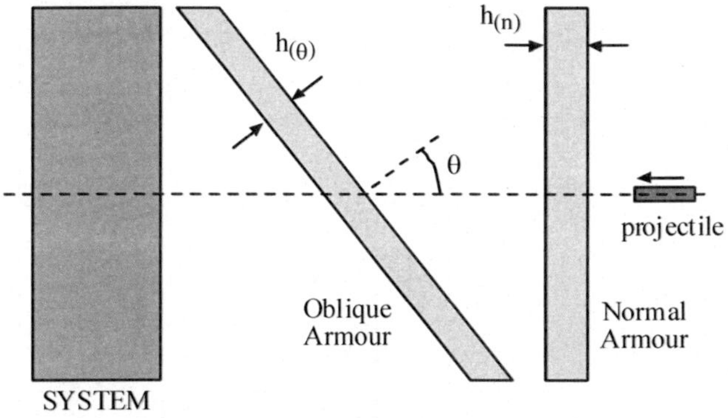

Figure 8.1. Configuration of a normal armor and an oblique armor.

8.2 Modeling Oblique Impact

Oblique penetration is essentially a three-dimensional problem, much more difficult to model than a normal impact. Numerical modeling of an oblique impact is only possible if the code can generate three-dimensional meshes. Fawaz et al. (2004) performed 3D numerical simulations of oblique impacts on alumina/CFRP panel, and Shokrieh and Javadpour (2008) did the same for boron carbide/Kevlar™ 49 panels to analyze the effect of obliquity on the performance of ceramic faced amors. Difficulties also arise in the formulation of analytical models. Therefore, it is usual to transform the oblique problem into an equivalent normal one, applying a Ballistic Equivalent Rule (BER) whose reasoning is outlined in Figure 8.2. The oblique problem is transformed into a normal equivalent by a series of conversions in defining parameters of the impact problem. Once the characteristics of the new problem are determined, this is solved with a model for normal impact. The last step is to transform this solution again by applying a second group of the BER, which gives the result for the original problem.

The transformation is normally carried out only on the geometric parameters defining the problem. A very simple rule that is often used in the design of metal targets is the so called Cosine Rule. This consists of transforming the oblique impact of the projectile on a plate of thickness h_b into a normal impact of the same projectile on a target of the same material but of a greater thickness h_b' as expressed in the equation

$$h_b' = \frac{h_b}{\cos\theta} \tag{8.4}$$

θ being the angle between the direction of fire and that perpendicular to the target. The results of the equivalent normal impact are then taken directly as the solution for the oblique impact.

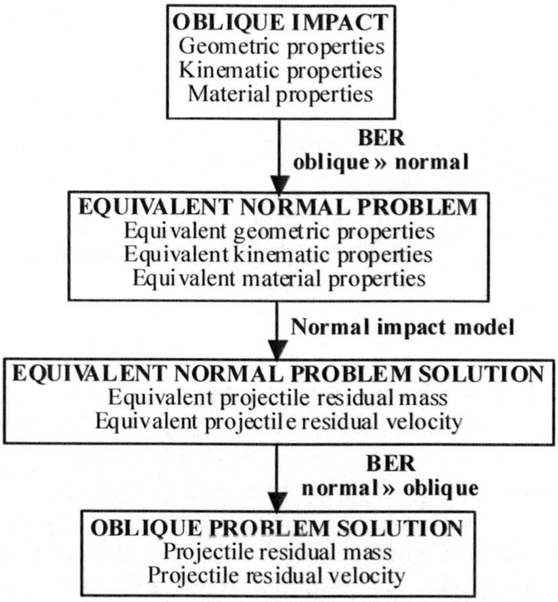

Figure 8.2. Diagram of a Ballistic Equivalence Rule (BER).

Hetherington and Lemieux (1994) apply a geometric BER to the ceramic-metal model developed by Florence (1969). This consists of assuming that there is an elliptical contact surface between the projectile and the ceramic, whose semi-minor axis is the same as the projectile radius a_p and the semi-major axis A is the equal to

$$A = \frac{a_p}{\cos\theta} \tag{8.5}$$

where θ is the angle of obliquity. The fractured ceramic truncated cone is surrounded by a surface which forms an angle α with the normal armor surface (Figure 8.3) and distributes the load over the metallic plate via an elliptical surface of semi-axes B and C with the values

$$B = a_p + h_1 \, tg\alpha \quad (8.6)$$

$$C = \frac{a_p}{\cos\theta} + h_1 \, tg\alpha \quad (8.7)$$

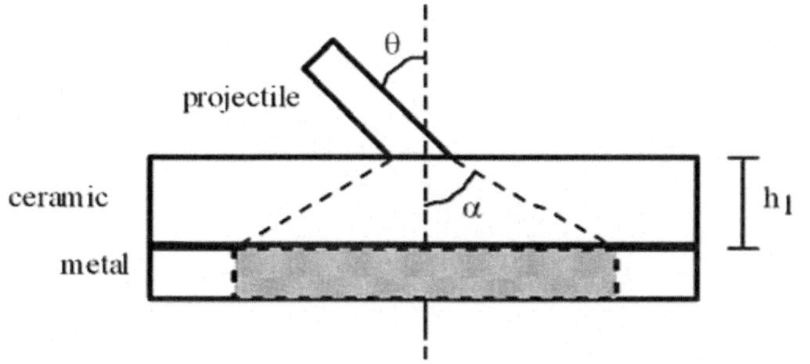

Figure 8.3. Applying the Florence model (1969) to an oblique impact (Hetherington and Lemieux, 1994).

The energy that can be absorbed by the elliptical area of the plate, through plastic deformation until it breaks, is the same as the energy of a circular membrane of the same area, so the Florence expressions are modified by one geometric factor only. The results obtained through this BER are compared with the experimental data in Figure 8.4. This case shows that using a simple BER can provide suitably accurate results.

Application of the simple BER described by Equation (8.4) for numerical or analytical models of ceramic/faced armors would provide an equivalent protection with a strength higher than that of the actual protection. The explanation of the discrepancy is the duration of the fragmentation stage, which is shorter in actual cases than that assumed by the equivalent normal impact problem, since the actual ceramic tile is thinner than the equivalent one. Zaera and Sánchez-Gálvez (1998a) used a slightly different approach for their analytical model: the cosine rule was not applied to the fragmentation stage, for which the actual projectile trajectory is simulated. Although oblique impact is tridimensional, the stress field produced during the first microseconds keeps the axial symmetry observed in normal impact, since stress wave propagation is radial from the impact point. The ceramic fragmentation process is thus very similar to that occurring in normal impact cases, with conical cracks propagating from the rear face of the tile (Figure 4.1). The fragmentation stage is then finished when the projectile tip meets the radial cracking front propagating from the rear face of the tile.

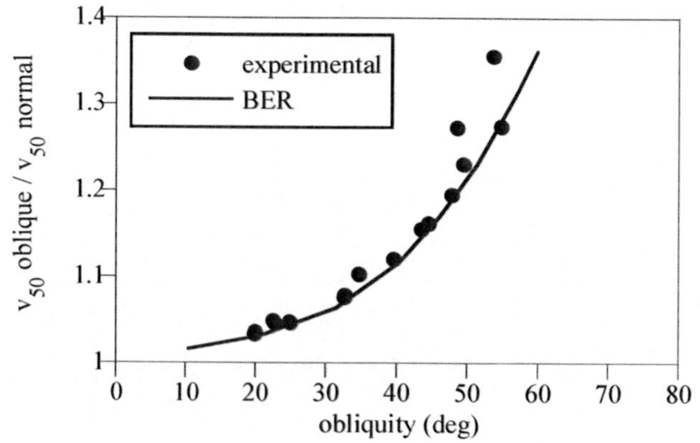

Figure 8.4. Variation of V_{50} with impact obliquity. Projectile 7.62 mm with lead core (Hetherington and Lemieux, 1994).

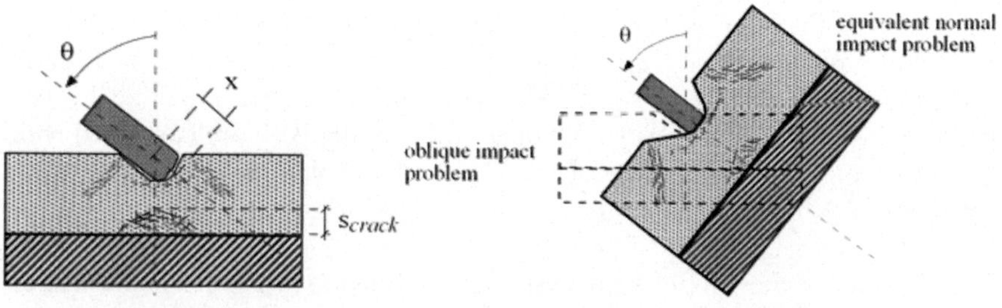

Figure 8.5. Left: Oblique penetration of projectile during ceramic fragmentation stage. Right: Conversion to normal impact by the Ballistic Equivalence Rule. (Zaera and Sánchez-Gálvez, 1998a).

The simple BER is applied thereafter utilizing the analytical model with equivalent thicknesses (Figure 8.5). Figure 8.6 illustrates residual velocities of 20 APDS projectiles after perforation of ceramic/metal targets at 50° obliquity, showing a good agreement between experimental and analytical results (Zaera and Sánchez-Gálvez, 1998a).

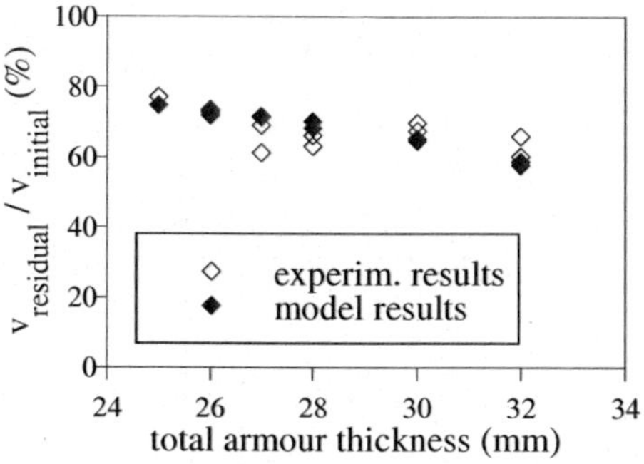

Figure 8.6. Analytical and experimental results of projectile residual velocities and residual masses of 20 APDS projectiles onto alumina 99.5 % / aluminum alloy 6082; Impact angle: 50° (Zaera and Sánchez-Gálvez, 1998a).

8.3 Ricochet Induced by Brittle and Lightweight Materials

The ricochet is the rebound of the projectile after impacting the target surface. This effect is also induced by placing a sloped armor, so that the projectile is deviated from its original trajectory. In this case, the most interesting armor is that with a lower critical ricochet angle (the angle at which ricochet is expected to onset), so that the protection easily deflects the thread. Rosenberg et al. (2005) analyzed the oblique penetration of the projectile into the target in terms of asymmetric forces taking place during the penetration, and demonstrated how a high dynamic compressive strength, low density and brittle material could exert strong enough forces on the projectile to deflect its course of penetration. The high compressive strength is needed in order to exert an asymmetric force on the projectile tip (Figure 8.7). The high brittleness is needed to produce failure of the material near the entrance hole.

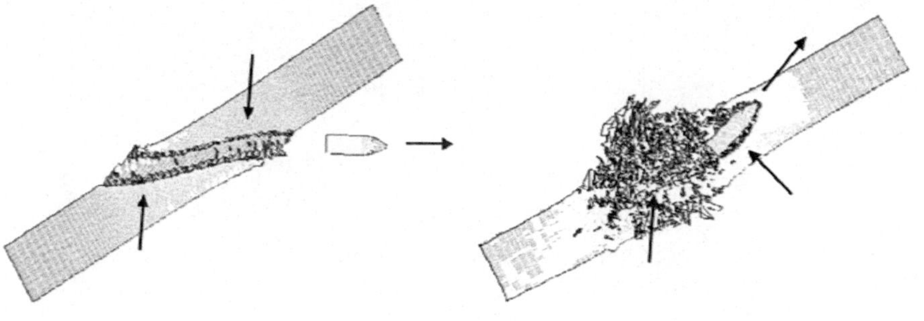

Figure 8.7. Left: Trajectory of a projectile in a ductile target. Right: Trajectory of a projectile in a brittle Plexiglas plate (Rosenberg et al. 2005).

9 New Developments in Ballistic Protection Armors

The efficiency of lightweight armors has increased considerably since this type of armor first emerged, not just through the introduction of new compositions and manufacturing methods, but also through the use of new materials which improve multi-hit capability, and through the incorporation of new layers and optimization of existing ones.

One of the greatest drawbacks of using ceramic materials for impact protection is their low multi-hit resistance. Because they are so brittle, the ceramic tiles which receive the impact, and often the adjacent tiles as well, are completely fragmented (Figure 9.1 left) and this means that their capacity to protect against subsequent impacts is significantly reduced. The traditional method used to improve the multi-hit capability of the armor is to use multi-tile configurations with hexagons or squares of different sizes. Small tiles also help to stop the damage from spreading, although this can result in a decreased capacity to absorb energy (Hazell et al., 2008).

Recently, a number of different authors have proposed the use of composite materials which combine ceramics with more ductile materials, as these can considerably increase the multi-hit capability of the armor. Furthermore, the ballistic efficiencies achieved with these new materials are competitive enough for them to be used to replace conventional monolithic tiles.

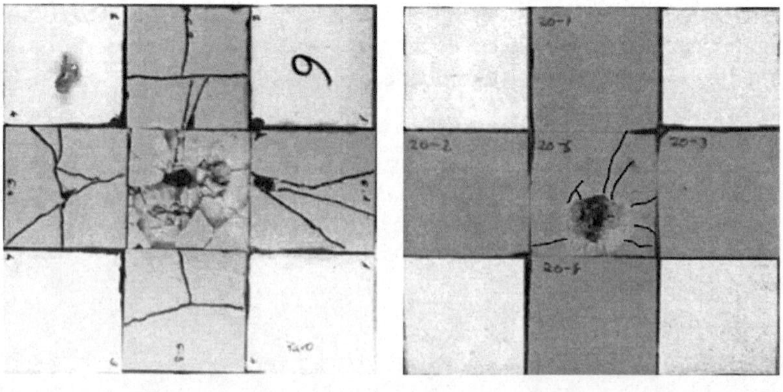

Figure 9.1. Alumina/aluminum panels after suffering the impact in the central tile. Left: Backed by polyurethane. Right: Backed by rubber-toughened epoxy (Zaera et al., 2001).

The first of these developments was due to Song et al. (1997), who developed a Gradient Designed Composite embedding alumina spheres in an epoxy matrix. A proper distribution of spheres of different sizes allows maximization of the volumetric content of the harder phase in the composite material. The ceramic spheres have the capability of defeating the projectile, similar to ceramic monolithic tiles, with a reduced fragmentation after impact.

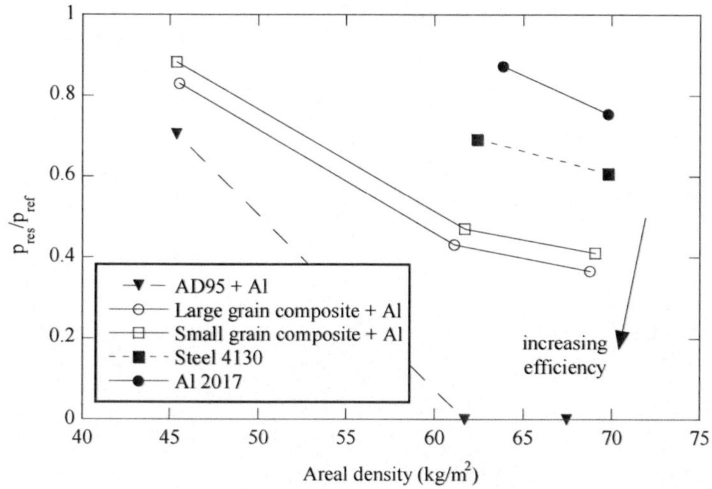

Figure 9.2. Non-dimensional residual penetration in a reference aluminum block (Depth of Penetration Test) versus areal density for different materials (Arias et al., 2003).

Following the work of Song et al. (1997), Arias et al. (2003) developed a low-cost process to manufacture ceramic/polymer tiles. Angular alumina particles, cheap and readily obtainable as an abrasive product, are classified by size, and blended and mixed with a polymeric matrix. During polymerization, the mixture is pressed by uniaxial compaction towards reaching a high content of ceramic. After unmolding, the polymerized composite is cured. The manufacturing process demands few operations, and allows good dimensional precision which facilitates molding and produces a material with an 80% ceramic weight content and low porosity. The efficiency against impact afforded by tiles of the composite with a backing of aluminum is halfway between that of the metals commonly used in these applications and the armors of monolithic ceramic/aluminium (Figure 9.2). These qualities justify the use of the composite for protections in which weight is not the primary concern and a cost savings is desirable. As in the case of the ballistic material developed by Song et al. (1997) the ductile polymeric matrix reduces the fragmentation in the tile to an area close to the impacted point, thus improving its performance for successive hits.

Louvigné and Schultz (2001) proposed a ceramic/metal composite material to improve the multi-hit capability. Tiles of sintered SiC with a high open porosity were encased in a steel canister perforated by numerous holes. The canister with the ceramic was preheated in a furnace, placed in a die of squeeze-casting equipment, and finally infiltrated with liquid aluminum or an aluminum–silicon alloy by applying high pressure. The composite material obtained with this process is weaker than monolithic SiC ceramics, having a 25% lower Young's modulus and a 40% lower tensile average stress (Forquin et al., 2003), but the aluminium skeleton provides sufficient cohesion and shows a significant reduction in cracking density (Figure 9.3).

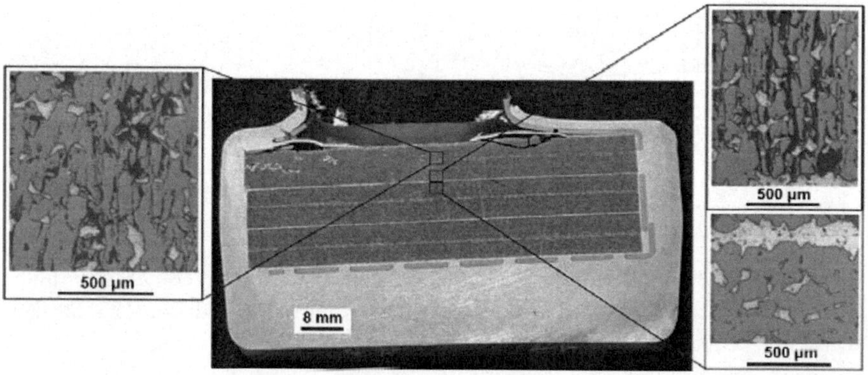

Figure 9.3. Porous SiC infiltrated with aluminum after impact of a 7.62 AP projectile (Forquin et al., 2003).

The influence that the layer joining the ceramic and the backing material can have on the ballistic efficiency of the armor has also been discussed by a number of different authors. Full-scale fire tests (Marshall 1994, Zaera et al. 2000) have shown that the armor behavior may be influenced significantly by the type of adhesive used. One of the most important stages of the impact process in terms of efficiency of the ceramic tile is the initial phase immediately after contact with the projectile. A premature fragmentation of the ceramic tile reduces its erosive capacity and consequently the ballistic efficiency of the armor. One of the most important phenomena during this fragmentation is the appearance of tensile stresses in the rear face of the tile, and thus the subsequent cracking of the ceramic material. As stated by Zaera et al. (2000) the speed at which these stresses arise, and their intensity, depend basically on the speed at which the compressive waves reach the rear face of the ceramic tile as well as on the amplitude of the traction waves reflected onto the ceramic-adhesive interface. An adhesive with a low mechanical impedance (defined as the product of its density times the elastic wave velocity) results in a low transmission of energy from the ceramic tile to the adhesive layer. This means that nearly all the incident energy is reflected back to the ceramic, increasing the tensile stresses in the tile and its subsequent cracking. It is therefore best to use adhesives with a mechanical impedance similar to that of the ceramic material (Figure 9.1 right).

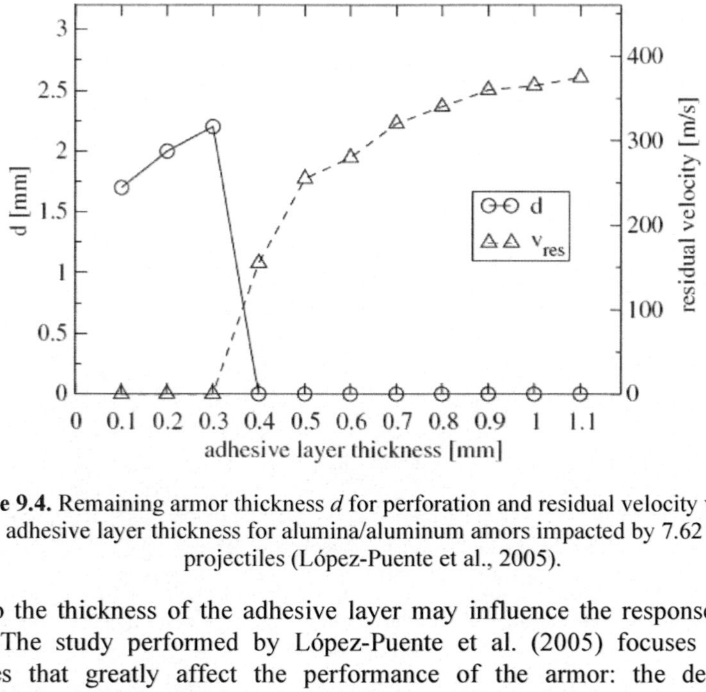

Figure 9.4. Remaining armor thickness *d* for perforation and residual velocity versus adhesive layer thickness for alumina/aluminum amors impacted by 7.62 AP projectiles (López-Puente et al., 2005).

Also the thickness of the adhesive layer may influence the response of the armor. The study performed by López-Puente et al. (2005) focuses on two variables that greatly affect the performance of the armor: the degree of fragmentation of the ceramic material, and the transmission of the impulsive energy to the backing plate. The first of these has a negative effect on the erosive capacity of the ceramic, and the second has the positive effect of helping to dissipate the energy in plastic deformation of the metal. The study showed that the thicker layer of adhesive leads to a wider area of plastic deformation of the metallic backing plate, which helps to absorb the kinetic energy of the projectile. On the other hand, the ceramic tile is shattered earlier when the adhesive layer is thicker. These have contrary effects on the resistance capacity of the armor plating but the fragmentation of the ceramic material is expected to be of greater significance. The thickness of the layer of adhesive should be reduced as far as possible, particularly with thicker tiles, so as to delay their fragmentation and augment the erosion of the projectile. A value of 0.3 mm was found to be optimal for the alumina/aluminium configuration considered (Figure 9.4). On the other hand, experimental results showed that a thicker layer of adhesive cushions the impact on the adjacent tiles and reduces the risk of their fragmentation.

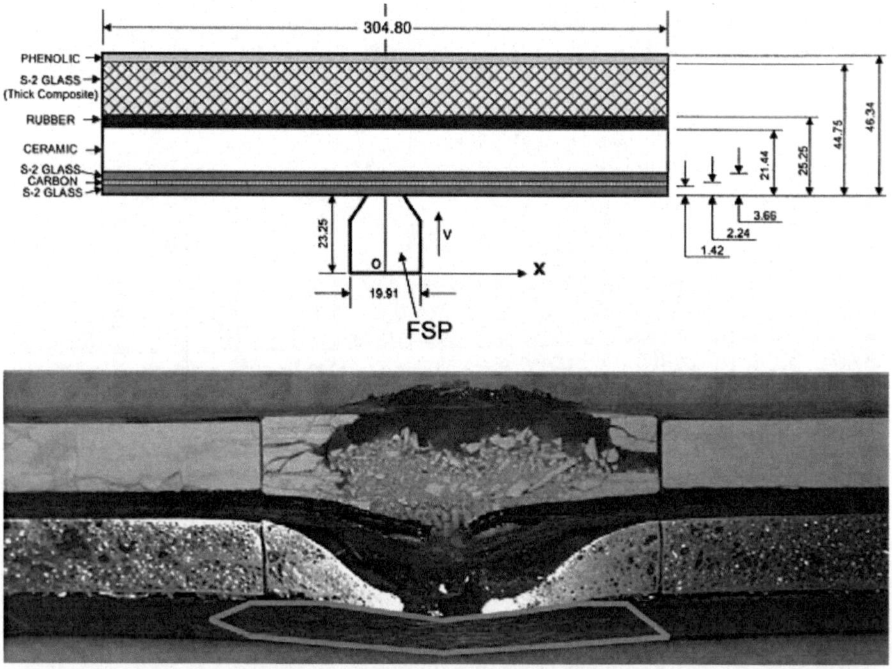

Figure 9.5. Top: Components of an integral armor (Mahfuz et al., 2000). Bottom: Multi-layer armor: cover/alumina/rubber/aluminum foam/fiberglass/vinylester after impact (Gama et al., 2001).

The development of two-component armors (ceramic/metal or ceramic/fiber reinforced composite) first began a few decades ago. Since then, significant improvements have been made and these armors have been the subject of a number of different studies. Over recent years, the complexity of the armor has increased as additional layers have been added to improve the protection efficiency of the armor and to add new functionalities. Mahfuz et al. (2000) considered the response of an integral armor, composed of S2-glass fiber/Vinylester, carbon fiber/epoxy, AD-90 ceramic, rubber and a phenolic layer (Figure 9.5 top). The rubber layer behind the ceramic tile, with a high compressive stiffness, transmits the impact energy and increases multi-hit capability. The thin cover composite layers are placed for radar signature and corrosion purposes, the thick polymer composite backing contains spalled ceramic fragments, and the phenolic layer is incorporated into the armor as a fire-toxicity retardant. Mahfuz et al. (2000) analyzed the stress distributions during impact along the armor thickness to detect the tensile stresses due to the impedance mismatch between layers. Gama et al. (2001) included a closed-cell aluminum foam behind the rubber layer (Figure 9.5 bottom) to attenuate the propagating stress waves. The foam layer serves to provide less ceramic fragmentation and less delamination and deflection on the composite backing plate.

Mines (2004) applied classical one-dimensional stress wave theory to the analysis of the stress-wave behavior of an integral armor. This simplified analysis, which showed good agreement with the numerical simulation of Gama et al. (2001), provided insight into the effects of layer materials and thicknesses.

Concerning the effect of covering the armor, Sarva et al. (2007) placed a thin membrane of suitable tensile strength (fiber/epoxy or titanium) on the front face. This improved the ballistic efficiency of the armor by 25% for a 2.5% increase in weight. When the movement of the ceramic fragments was restrained, the confinement increased, making penetration more difficult. Inclined X-ray shadowgraphs provided a view of the mushrooming process of the projectile (Figure 9.6): the projectile in the restrained armor exhibits a larger mushroom, confirming the increased penetration resistance.

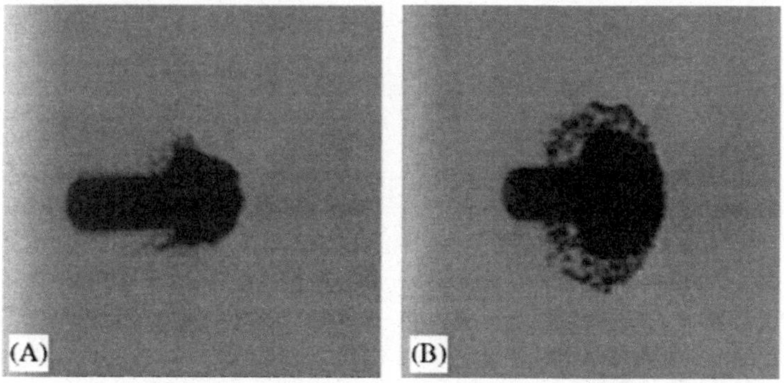

Figure 9.6. Left: Comparison of projectile mushrooming. Left: Bare ceramic tile. Right: Restrained ceramic tile. Sarva et al. (2007).

Another promising armor possibility is the use of functionally gradient materials (FGM). An FGM is a multi-phase material where the constituents vary gradually in a pre-determined way, resulting in a material with continuously varying mechanical properties. Pettersson et al. (2005) developed such a material for ballistic applications by combining ceramic (TiB_2) and metal (Ti) in such a way that an optimum combination of hardness and ductility could be achieved. The ballistic results indicate a suitable composition, which does not have to be a pure ceramic, for the outer high hardness armor layer.

10 Concluding Remarks

The use of composites for load-bearing applications advances slowly in the civil engineering and automobile sectors, and somewhat more rapidly in the aeronautics sector, which traditionally has served as the field of development and experimentation for these materials. However, the structures of primary responsibility continue to be made of metal alloys, due to greater confidence in their mechanical properties and the methods used for the design of manufactured elements with them. In the case of ballistic protection, confidence in composite armors has reached a notable level; its use has extended among security and defense forces, both military and civilian, and the sector dedicated to its production and the market are mature.

In any case, the development of composite materials for protection against ballistic impact still has great development potential. Security and defense forces are experiencing a rapid transformation towards the incorporation of units that permit its easy transportability and deployment with the aim of gaining mobility but losing neither efficacy nor survivability. This transformation implies, in great measure, the development of new materials with better specific mechanical properties (stiffness to weight and strength to weight), the design of configurations that permit obtaining maximum performance of the materials, and the improvement of tools for modeling its response to impact.

The greatest advances in the field of ballistic protection during the last 50 years have had their origins primarily in the appearance of new materials whose mechanical properties are notably superior to those of their predecessors. Polymer fibers constitute, without a doubt, the best example, since they have allowed the achievement of levels of efficiency and low weight that were unthinkable with the metal alloys used before. The recent development of the M5 fiber in 1999 has opened a wide path for the improvement of armors, and the results will be seen within a few years. Although the synthesis of M5 is relatively recent, it is not the latest of the advances in the development of new materials for ballistic protection. In the year 2002 the University of Delaware's Center for Composite Materials, in partnership with the Weapons and Materials Research Directorate of the U.S. Army Research Laboratory, presented a special fluid called Shear Thickening Fluid with colloidal silica nano-particles that form clusters when deformed at high strain rates. Thus, under impact loadings its viscosity strongly increases, acting as a rigid phase. This special material can be applied to conventional ballistic fabrics used in personal protection, lubricating them to remain flexible under normal wear, but simultaneously becoming rigid and increasing energy absorption when impacted by a knife or high velocity projectile or fragment. The technology based on nano-science allows us to hope for the appearance of new materials capable of increasing the level and quality of ballistic protection without compromising its weight. It is also necessary to remember the requirements of wearability and temperature isolation (for personal protection) and low flammability (for fighting vehicles) for ballistic protection. Compliance with all of these requirements will be difficult to achieve with a single material and the multifunctional integration of metals, ceramics, and polymers will be necessary.

Another challenge in the field of ballistic protection consists of the use of armors as structural material for fighting vehicles. The composite materials used in appliqué armour serve only to increase impact protection. The low resin content prepares them to absorb impact energy but not to support load. Multifunctional materials that are currently being considered for the design of future combat systems must carry typical vehicle kinematic loads, yet absorb energy like an armor-grade composite under ballistic impact.

In relation to the design methodologies for ballistic protection, an experimental approach based on the trial and error philosophy has been progressively replaced by techniques of virtual prototyping. The notable improvement in the algorithms reached in recent decades for modeling solid mechanics problems has allowed simulation of highly non-linear impact problems. However, the treatment of the contacts and of the distortions that appear in the mesh due to the large deformations that characterize this type of problem present challenges for which the recent lagrangian particle techniques present certain advantages. Studies conducted in this field more and more often deal with it using this type of technique, in which the discretization of the continuum permits very efficient treatment of large deformations. Although its computational cost is still quite high, compared with the techniques of finite elements or finite differences, they will continue to be developed over the next few years and will extend to the modeling of ballistic problems, in the same way they are being used for other applications in solid mechanics.

In any case, the level of sophistication and reliability achieved by these numeric methods is highly dependent on the correct modeling of the materials used in ballistic protection. A good portion of the constitutive equations currently available have a phenomenological character and advances are required in the development of constitutive equations formulated on the basis of first principles that permit consideration with greater rigor the complex damage phenomena, in metals, fiber composites and ceramics.

The rapid development of new products, a growing trend in all industrial sectors and also in those related with ballistic protection, requires design and evaluation in shorter and shorter timeframes. To that end, virtual testing is essential in this process and the aspects of reliability in numeric modeling have great importance for design engineers. Without reasonable confidence in these methods, their utility will be seriously questioned. Therefore, it is predicted that this concern for reliability will foster the development of modeling tools that are ever more rapid and precise.

Annex: Basic Ballistic Terminology

Anti-material projectile: ballistic threat designed to defeat armors and materials of construction.

Anti-personnel projectile: ballistic threat designed to injure or kill animals or people.

Areal density: area density of a two-dimensional object, calculated as the mass per unit area.

Armor penetration: event during which a projectile creates a discontinuity in the original surface of the armor.

Armor Perforation: event during which a projectile emerges totally or partially from the back face of the armor.

Armor Piercing (AP) projectile: ballistic threat with a core made of high-strength and high-density material, designed to penetrate hard targets.

Ballistic limit: minimum velocity at which a particular projectile is expected to completely penetrate an armor of given thickness and physical properties at a specified angle of obliquity

Barrel: tube through which a controlled explosion or rapid expansion of gases is released in order to propel a projectile out of the end at great speed.

Behind Armour Blunt Trauma: non-penetrating injury resulting from the rapid deformation of personal body armors covering the body.

Caliber: inside diameter of the barrel of a firearm.

Cartridge: unit of ammunition, composed of a metallic case, a bullet, gunpowder and primer, precisely made to fit the firing chamber of a firearm.

Discarding Sabot (DS) projectile: ballistic threat with a diameter smaller than the caliber of the barrel, guided during travel in the bore by a sabot to increase muzzle velocity.

Dual hardness armor: ballistic protection made of two layers of different materials with different hardnesses, the hardest being the first impacted by the projectile.

Fragment Simulating Projectile (FSP): shrapnel made with standardized shape and material, used to perform impact tests reproducing the effect of fragmenting explosive weapons under unvarying conditions.

Grenade: small anti-personnel bomb designed to be thrown by hand, to explode a short time after release and to break into small fast-moving fragments.

Handgun: firearm designed to be held in one hand when used.

Heavy gun: larger-caliber, high-power firearm fired from heavy mounts or static positions.

Mortar: projectile containing an explosive, usually launched by muzzle-loading indirect fire weapons, designed to break into small fast-moving fragments after impact.

Personal armor: helmet or vest designed to protect personnel against ballistic impact.

Residual velocity: velocity of the projectile after perforating the target.

Rifle: firearm designed to be held in one hand to support the barrel and in the other arm and shoulder to support the stock.

Rolled Homogeneous Armor (RHA): ballistic protection consisting of a rolled steel plate of a hardness between 35 and 40 Rockwell C.

Sabot: lightweight material surrounding a sub-calibrated projectile and divided into several pieces that are pulled away by the air once it reaches the muzzle of the barrel.

Shrapnel: ballistic threat consisting in a small metallic fragment, with a mass of between 0.1 and 3 g, launched at high velocity and generated after explosion of mortar ammunition or grenades.

V_{50}: critical impact velocity at which the probability of perforation is 50 %.

References

Aare M., Kleiven S. (2007). Evaluation of head response to ballistic helmet impacts using the finite element method. *Int J Impact Engng* 34:596-608.

Adeli H., Yeh C. (1989). Perceptron learning in engineering design. *Microcomp Civil Engng* 4:247–256.

Alekseevskii V.P. (1966). Penetration of a rod into a target at high velocity. In *Combustion, Explosion and Shock Waves 2*. Faraday Press, New York, USA.

Anderson C.E. Jr., Morris B.L. (1992). The ballistic performance of confined Al2O3 ceramic tiles. *Int J Impact Engng* 12:167-187.

Anderson C.E. (1987). An overview of the theory of hydrocodes. *Int J Impact Engng* 5:33-59.

Anderson C.E., Cox P.A., Johnson G.R., Maudlin P.J. (1994). A constitutive formulation for anisotropic materials suitable for wave propagation computer programs—II. *Comput Mech* 15:201–23.

Arias A. (2002). Desarrollo y modelización del comportamiento frente a impacto de materiales compuestos de matriz polimérica y carga cerámica. PhD Thesis, Universidad Carlos III de Madrid.

Arias A., Rodríguez-Martínez J.A., Rusinek A. (2008). Numerical simulations of impact behaviour of thin steel plates subjected to cylindrical, conical and hemispherical non-deformable projectiles. *Engineering Fracture Mechanics* 75:1635–1656

Arias A., Zaera R., López-Puente J., Navarro C. (2003). Numerical modeling of the impact behavior of new particulate-loaded composite materials. *Composite Structures* 61:151-159.

AS/NZS 2343 1997 (1997). *Bullet Resistant Panels and Elements*. Australian/New Zealand Joint Technical Committee MS/43.

Belytschko T., Lu Y.Y., Gu L. (1994). Element free galerkin methods. *International Journal for Numerical Methods in Engineering* 37:229–256.

Ben-Dor G., Dubinsky A., Elperin T. (2008). Improved Florence model and optimization of two-component armor against single impact or two impacts. *Composite Structures* 88:158-165.

Ben-Dor G., Dubinsky A., Elperin T. (2005). Optimization of two-component composite armor against ballistic impact. *Composite Structures* 69:89-94.

Bhatnagar A., Wagner L., Shahkarami A., Cepus E., Vaziri R., Poursartip A., Hamouda A.M.S., Risby M.S., Dunn D.R., Tam T., Song J.S., Lee B.L., Thomas H.L., Arvidson B., Pataki W., Hannibal A., Weir B., Scott B.R., Lang D., Salamé J.M., Quefelec B. (2006). *Lightweight ballistic composites, Military and law-enforcement applications*. Woodhead Publishing, CRC Press, Cambridge.

Billon H.H., Robinson D.J. (2001). Models for the ballistic impact of fabric armour. *Int J Impact Engng* 25:411-422.

Bless S.J., Rosenberg Z., Yoon B. (1987). Hypervelocity penetration of ceramics. *Int J Impact Engng* 5:165-171.

Bless S.J., Subramanian R., Partom Y., Lynch N. (1995). Effects of radial confinement on the penetration resistance of thick ceramic tiles. In *Proceedings on the 15th International Symposium on Ballistics*, Jerusalem.

Burkett M.W., Cort G.E., Parker R.B., Rollet A.D., Skaggs S.R. (1992). Ballistic Performance of 90 wt% Al2O3 Against Subscale Kinetic-Energy Long-Rod Penetrator. In *Proceedings on the 13th International Symposium on Ballistics*, Stockholm, Sweden.

CEN EN 1063 (1999). Glass in Building -Security Glazing- Testing and Classification of Resistance Against Bullet Attack. European Committee for Standardization.

CEN PrEN ISO 14876 (1998). *Test method for stab resistance of body armour.* International Organization for Standardization/European Committee for Standardization.

Chang F.K., Chang K.Y. (1987). A progressive damage model for laminated composites containing stress concentrations. *J Compos Mater* 21:834–55.

Chocron S., Pintor A., Gálvez F., Roselló C., Cendón D., Sánchez-Gálvez V. (2008). Lightweight polyethylene non-woven felts for ballistic impact applications: Material characterization. *Composites Part B: Engineering* 39:1240-1246.

Chocron-Benloulo I.S., Rodríguez J., Martínez M.A., Sánchez-Gálvez V. (1997). Dynamic tensile testing of aramid and polyethylene fiber composites. *International Journal of Impact Engineering* 19:135-146.

Chocron-Benloulo I.S., Sánchez-Gálvez V. (1998). A New Analytical Model to Simulate Impact onto Ceramic/Composite Armors. *Int J Impact Engng* 21:461-471.

Cork C.R., Foster P.W. (2007). The ballistic performance of narrow fabrics. *Int J Impact Engng* 34:495–508

Cortés R., Navarro C., Martínez M.A., Rodríguez J., Sánchez-Gálvez V. (1992). Numerical modelling of normal impact on ceramic composite armours. *Int J Impact Engng* 12:639-651.

Cunniff P.M., Auerbach M.A. (2005). High performance M5 fiber for ballistic/structural composites. *Course Mechanical Behavior of Polymers.* MIT, D. Roylance.

Cunniff P.M. (1999). Dimensionless parameters for optimization of textile-based armor systems. In *Proceedings 18th International Symposium on Ballistics*, San Antonio, Texas.

Curran D., Seaman L., Cooper T., Shockey D. (1993). Micromechanical model for comminution and granular flow of brittle material under high strain rate application to penetration of ceramic targets. *Int J Impact Engng* 13:53-83.

den Reijer P.C. (1991). *Impact on ceramic faced armours*. PhD thesis, Delft University of Technology.

Denoual C., Hild F. (2002). Dynamic fragmentation of brittle solids: a multi-scale model. *European J Mech A/Solids* 21:105-120.

Duan Y., Keefe M., Bogetti T.A., Cheeseman B.A. (2005). Modeling the role of friction during ballistic impact of a high-strength plain-weave fabric. *Composite Structures* 68:331-337.

El Kadi H. (2006). Modeling the mechanical behavior of fiber-reinforced polymeric composite materials using artificial neural networks a review. *Composite Structures* 73:1–23.

Ernst H.J., Hoog K. (1992). Protective power of several ceramics, correlated to some static material properties. In *Proceedings on the 13th International Symposium on Ballistics*, Stockholm, Sweden.

Ernst H.J., Hoog K., Wiesner V. (1994). Ballistic impact behaviour of some ceramics in different environments. *Journal de Physique* 3:677-682.

Ernst H.J., Hoog K., Wiesner V. (1995). Continuous measurements of the penetration depth in completely confined ceramic targets. In *Proceedings on the 15th International Symposium on Ballistics*, Jerusalem, Israel.

Fawaz Z., Behdinan K., Xu Y. (2006). Optimum design of two-component composite amours against high-speed impact. *Composite Structures* 73:253-262.

Fawaz Z., Zheng W., Behdinan K. (2004). Numerical simulation of normal and oblique ballistic impact on ceramic composite amours. *Composite Structures* 63:387-395.

Fellows N.A., Barton P.C. (1999). Development of impact model for ceramic-faced semi-infinite armour. *Int J Impact Engng* 22:793-811.

Fernández-Fdz D., Zaera R. (2008). A new tool based on artificial neural networks for the design of lightweight ceramic-metal armours against high-velocity impact of solids. *Int J Solids Structures* 45:6369-6383.

Florence A.L. (1969). *Interaction of projectiles and composite armor, Part II.* Stanford Research Institute, Menlo Park, CA, USA, AMRA CR 69-15.

Forquin P., Trana L., Louvigné P-F., Rota L., Hild F. (2003). Effect of aluminum reinforcement on the dynamic fragmentation of SiC ceramics. *Int J Impact Engng* 28:1061-1076.

Gama B.A., Bogetti T.A., Fink B.K., Yu C., Claar T.D., Eifert H.H., Gillespie Jr. J.W. (2001). Aluminum foam armor: a new dimension in armor design. *Int J Impact Engng* 52:381-395.

Gingold R.A., Monaghan J.J. (1977). Smoothed particle hydrodynamics: theory and application to non-spherical stars. *Monthly Notices of the Royal Astronomical Society* 181:375–389.

GOST R 50744-95 (1995). Armor clothes. Classification and general technical requirements. Standards Publishing House, Moscow.

Gower H.L., Cronin D.S., Plumtree A. (2008). Ballistic impact response of laminated composite panels. *Int J Impact Engng* 35:1000-1008.

Gu B. (2003). Analytical modeling for the ballistic perforation of planar plain-woven fabric target by projectile. *Composites Part B: Engineering* 34:361-371.

Gu B., Xu J. (2004). Finite element calculation of 4-step 3-dimensional braided composite under ballistic perforation. *Composites Part B: Engineering* 35:291-297.

Hauver G.E., Netherwood P.H., Benck R.F., Gooch W.A., Perciballi W.J., Burkins M.S. (1992). Variation of target resistance during long rod penetration into ceramics. In *Proceedings on the 13th International Symposium on Ballistics*, Stockholm.

Hazell P.J.,. Roberson C.J., Moutinho M. (2008). The design of mosaic armour: The influence of tile size on ballistic performance. *Materials & Design* 29:1497-1503.

Hetherington J.G., Lemieux P.F. (1994). The effect of obliquity on the ballistic performance of two component composite armours. *Int J Impact Engng* 15:131-137.

Hiermaier S., Riedel W., Clegg R., Hayhurst C. (1999). *Advanced material models for hypervelocity impact simulations.* Technical Report, ESA/ESTEC Contract No. 12400/97/NL/PA(SC).

Hohler V., Stilp A. J., Weber K. (1995-a). Ranking Methods of Ceramics and Experimental Optimisation of a Laminated Target with Ceramics. In *Proceedings Lightweight Armour System Symposium,* Royal Military College of Science, Cranfield, England.

Hohler V., Stilp A. J., Weber K. (1995-b). Hypervelocity penetration of tungsten sinter-alloy rods into alumina. *Int J Impact Engng* 17:409-418.

Ince R. (2004). Prediction of fracture parameters of concrete by artificial neural networks. *Engng Fract Mech* 71:2143–2159.

Jacobs M.J.N., Van Dingenen, J.L.J. (2001). Ballistic protection mechanisms in personal armour. *Journal of Materials Science* 36:3137-3142.

James B. (1995). The influence of the material properties of alumina on ballistic performance. In *Proceedings on the 15th International Symposium on Ballistics,* Jerusalem.

Jameson J.W., Stewart G.M., Petterson D.R., Odell F.A. (1962). Dynamic Distribution of Strain in Textile Materials under High-Speed Impact: Part III: Strain-Time-Position History in Yarns. *Textile Research Journal* 32:858-860.

Jearanaisilawong P. (2004). Investigation of deformation and failure mechanisms in woven and non-woven fabrics under quasi-static loading conditions. Master of Science Thesis, Massachusetts Institute of Technology.

Johnson G., Holmquist T. (1994). *An improved computational constitutive brittle material.* Technical Report, American Institute of Physics.

Karahan M., Kus A., Erenc R. (2008). An investigation into ballistic performance and energy absorption capabilities of woven aramid fabrics. *Int J Impact Engng* 35:499-510.

Lee L., Rudov-Clark S., Mouritz A.P., Bannister M.K., Herszberg I. (2002). Effect of weaving damage on the tensile properties of three-dimensional woven composites. *Composite Structures* 57:405-413.

Liable R.C., Henry M.C. (1969). *A review of the development of ballistic needle-punched felts.* Technical Report 70-32-CE, Clothing and Personal Life Support Equipment Laboratory, U.S. Army Natick Laboratories, Massachusetts, October.

Lim C.T., Shim V.P.W., Ng Y.H. (2003). Finite-element modeling of the ballistic impact of fabric armor. *Int J Impact Engng* 28:13-31.

Liu S., Huang J., Sung J., Lee C. (2002). Detection of cracks using neural networks and computational mechanics. *Comput Meth Appl Mech Engng* 191:2831–2845.

López Puente J., Arias A., Zaera R., Navarro C. (2005). The effect of the thickness of the adhesive layer on the ballistic limit of ceramic/metal armours. An experimental and numerical study. *Int J Impact Engng* 32:321-336.

Louvigné P-F, Schulz P. (2001). MMC for armour application: new concept. In *Proceedings of ICCE8 Conference*, Tenerife, Canary Islands, Spain.

Mahfuz H., Zhu Y., Haque A., Abutalib A., Vaidya U., Jeelani S., Gama B., Gillespie J., Fink B. (2000). Investigation of high-velocity impact on integral armor using finite element methods. *Int J Impact Engng* 24:203-217

Marshall J. (1994). Composite ballistic armour. In Proceedings of the 1st International Conference on Composites Engineering ICCE/1, 1057–8.

Matzenmiller A., Lubliner J., Taylor R.L. (1995). A constitutive model for anisotropic damage in fiber-composites. *Mech Mater* 20:125–52.

Mayseless M., Goldsmith W., Virostek SP., Finnegan SA. (1987). Impact on ceramic targets. *J Appl Mech* 54:373–378.

Mellgard I., Holmberg L., Gunnar L. (1989). An experimental method to compare the ballistic efficiencies of different ceramics against long rod projectiles. In *Proceedings on the 11th International Symposium on Ballistics*, Jerusalem.

MIL-STD-662F (1997). *V_{50} Ballistic Test for Armor*, US Department of Defense.

Mines R.A.W. (2004). A one-dimensional stress wave analysis of a lightweight composite armour. *Int J Impact Engng* 64:55-62.

Morye S.S., Hine P.J., Duckett R.A., Carr D.J., Ward I.M. (2000). Modelling of the energy absorption by polymer composites upon ballistic impact. *Composites Science and Technology* 60:2631-2642.

Moss G.M., Leeming D.W., Farrar C.L. (1995). *Military Ballistics. A basic manual.* Brassey's, London.

Naik N.K., Shrirao P., Reddy B.C.K. (2006). Ballistic impact behaviour of woven fabric composites: Formulation. *Int J Impact Engng* 32:1521-1552.

Nandlall D., Williams K., Vaziri R. (1998). Numerical simulation of the ballistic response of GRP plates. *Composites Science and Technology* 58:1463-1469.

NATO STANAG 4569 (2004). Protection Levels for Occupants of Logistic and Light Armoured Vehicles. North Atlantic Treaty Organization.

Navarro C. (1998). Simplified modeling of the ballistic behavior of fabrics and fiber-reinforced polymeric matrix composites. *Key Engng Mater* 141–143:383–400.

Navarro C., Zaera R., Cortés R., Martínez-Casanova M.A. (1994). The response of ceramic faced lightweight armours under projectile impact. In *Proceedings on Structures under Shock and Impact III Conference*, Madrid.

NIJ Standard 0101.04 (2001). *Ballistic Resistance of Personal Body Armor.* US National Institute of Justice.

NIJ Standard 0106.01 (1981). *Standard for Ballistic Helmets.* US National Institute of Justice.

NIJ Standard 0108.01 (1985). *Ballistic Resistant Protective Materials.* US National Institute of Justice.

Novotny W.R., Cepus E., Shahkarami A., Vaziri R., Poursartip A., (2007). Numerical investigation of the ballistic efficiency of multi-ply fabric armours during the early stages of impact. *Int J Impact Engng* 34:71-88.

Pettersson A., Magnusson P., Lundberg P., Nygren M. (2005). Titanium–titanium diboride composites as part of a gradient armour material. *Int J Impact Engng* 32:387-399.

Phoenix S.L., Porwal P.K. (2003). A new membrane model for the ballistic impact response and V50 performance of multi-ply fibrous systems. *Int J Solids Struct* 40:6723-6765.

Roberson C.J. (1995). Ceramic materials and their use in lightweight armour systems. In *Proceedings on Lightweight Armour System Symposium*, Cranfield, England.

Rosenberg Z., Surujon Z., Yeshurun Y., Ashuach Y., Dekel E. (2005). Ricochet of 0.3'' AP projectile from inclined polymeric plates. *Int J Impact Engng* 31:221-233.

Rosenberg Z., Tsaliah J. (1990). Applying Tate's model for the interaction of long rod projectiles with ceramic targets. *Int J Impact Engng* 9:247-251.

Rosenberg Z., Yeshurun Y. (1988). The relation between ballistic efficiency and compressive strength of ceramic tiles. *Int J Impact Engng* 7:357-362.

Roylance D., Wilde A., Tocci G. (1973). Ballistic Impact of Textile Structures. *Textile Research Journal* 43:34-41.

Rumelhart D., McCleland J. (1988). Parallel Distributed Processing VI: Foundations.

Sadanandan S., Hetherington J.G. (1997). Characterisation of ceramic/steel and ceramic/aluminium amours subjected to oblique impact. *Int J Impact Engng* 19:811-819.

Sarva S., Nemat-Nasser S., McGee J., Isaacs J. (2007). The effect of thin membrane restraint on the ballistic performance of armor grade ceramic tiles. *Int J Impact Engng* 34:277-302.

Senf H., Strassburger E., Rothenhäusler H., Gooch W.A., Burkins M.S. (1995). Ballistic resistance of AD995 Al2O3 ceramics against short projectiles at impact velocities between 1000 and 2000 m/s. In *Proceedings on the 15th International Symposium on Ballistics*, Jerusalem.

Senf H., Strassburger E., Rothenhäusler H. (1994). Stress wave induced damage and fracture in impacted glasses. *J Phys IV* 3:741-746.

Shahkarami A., Vaziri R. (2007). A continuum shell finite element model for impact simulation of woven fabrics. *Int J Impact Engng* 34:104-119.

Shi J., Grow D. (2007). Effect of double constraints on the optimization of two-component armor systems. *Int J Impact Engng* 79:445-453.

Shokrieh M., Javadpour G.H. (2008). Penetration analysis of a projectile in ceramic composite armor. *Composites Structures* 82:269-276.

Silva M.A.G., Cismaiu C., Chiorean C.G. (2005). Numerical simulation of ballistic impact on composite laminates. *Int J Impact Engng* 31:289-306.

Smith J.C., McCracken F.L., Schiefer H.F. (1965). Stress–strain relationships in Yarns subjected to rapid impact loading. Part V: wave propagation in long textile Yarns impacted transversally. *Textile Research Journal* 28:288–302.

Song J.W., Geshury A., Ko F.K. (1997). Behavior of gradient designed composites under ballistic impact. In *Proceedings of the 11th International Conference on Composite Materials*, Gold Coast, Australia.

Song J.W., Egglestone G.T. (1987). Investigation of the PVB/PF ratios on the crosslinking and ballistic properties in glass and aramid fiber laminate systems. In *Proceedings of the 19th SAMPE International Tech Conference*, 108-119.

Sternberg J. (1989). Material properties determining the resistance of ceramics to high velocity penetration. *J Appl Phys* 65:3417-3424.
Strassburger E., Senf H., Rothenhäusler H. (1994). Fracture propagation during impact in three types of ceramics. *J Phys IV* 3:653-658.
Tan V.B.C., Ching T.W. (2006). Computational simulation of fabric armour subjected to ballistic impacts. *Int J Impact Engng* 32:1737-1751.
Tan V.B.C., Shim V.P.W., Zeng X. (2005). Modelling crimp in woven fabrics subjected to ballistic impact. *Int J Impact Engng*, 32:561-574.
Tan V.B.C., Zeng X.S., Shim V.P.W. (2008). Characterization and constitutive modeling of aramid fibers at high strain rates. *Int J Impact Engng* 35:1303-1313.
Tate A. (1967). A theory for the deceleration of long rods after impact. *J Mech Phys Solids* 15:387-399.
Tham C.Y., Tan V.B.C., Lee H.P. (2008). Ballistic impact of a KEVLAR helmet: Experiment and simulations. *Int J Impact Engng* 35:304-318.
The Hague Convention (1899). Declaration III, Laws of War: Declaration on the Use of Bullets Which Expand or Flatten Easily in the Human Body. The Hague.
Tobin L., Iremonger M. (2006). *Modern body armour and helmets: an introduction.* Argos press, Canberra.
Van Hoof J., Deutekom M.J., Worswick M.J., Bolduc M. (1999). Experimental and numerical analysis of the ballistic response of composite helmet materials. In *Proceedings of 18th International Symposium on Ballistics*, San Antonio, Texas, USA.
Waszczyszyn Z., Ziemianski L. (2001). Neural networks in mechanics of structures and materials - new results and prospects of applications. *Comput Struct* 79:2261–2276.
Wilkins M.L., Cline C.F., Honodel C.A. (1969). *Fourth progress report on light armor program UCRL-50694.* Lawrence Radiation Laboratory, University of California, Livermore, CA, USA,
Wilkins M.L., Cline C.F., Honodel C.A. (1967). *Fourth progress report on light armor program UCRL-50284.* Lawrence Radiation Laboratory, University of California, Livermore, CA, USA,
Wilkins M.L., Cline C.F., Honodel C.A. (1967). *Fourth progress report on light armor program UCRL-50349.* Lawrence Radiation Laboratory, University of California, Livermore, CA, USA,
Wilkins M.L., Cline C.F., Honodel C.A. (1968). *Fourth progress report on light armor program UCRL-50460.* Lawrence Radiation Laboratory, University of California, Livermore, CA, USA.
Wilkins M.L. (1978). Mechanics of penetration and perforation. *Int J Engng Sci* 16:793–807.
Woodward R.L. (1990). A simple one-dimensional approach to modelling ceramic composite armour defeat. *Int J Impact Engng* 9:455-474.
Woodward R.L., Baxter B.J. (1994). Ballistic evaluation of ceramics: influence of test conditions. *Int J Impact Engng* 15:119-124.

Woolsey P. (1992). Ceramic Materials Screening by Residual Penetration Ballistic Testing. In *Proceedings on the 13th International Symposium on Ballistics*, Stockholm.

Yaziv D., Rosenberg Z., Partom Y. (1986). Differential ballistic efficiency of appliqué armor. In *Proceedings on the 9th International Symposium on Ballistics*, Shrivenham, UK.

Zaera R. (1997). Modelo analítico para el diseño de sistemas de protección frente a impacto formados por paneles de cerámica/metal. PhD Thesis, Universidad Politécnica de Madrid.

Zaera R., Sánchez-Gálvez V. (1998-a). Analytical modelling of normal and oblique ballistic impact on ceramic/metal lightweight armours. *Int J Impact Engng* 21:133-148.

Zaera R., Sánchez-Gálvez V. (1998-b). Using an analytical model of simulation in the design of light-weight amours. *Simulation* 70:175-181.

Zaera R., Sánchez-Sáez S., Pérez-Castellanos J.L., Navarro C. (2000). Modelling of the adhesive layer in mixed ceramic/metal armours subjected to impact. *Composites: Part A* 31:823-833.

Zukas J.A., Nicholas T., Swift H.F., Greszczuk L.B., Curran D.R. (1982). *Impact Dynamics*. John Wiley & Sons, New York.